About the Cover

Endothelial nitric oxide synthase (eNOS) staining in cultured bovine coronary venular endothelial cells. Cells were grown on glass coverslips and, once confluent, were fixed, and incubated with mouse monoclonal antibody against eNOS, followed by a secondary antimouse IgG antibody conjugated with fluorescein.

The green positive labeling of eNOS documented a prevalent presence of the protein in the cytoplasm, with the highest intensity in the perinuclear region, associated with vesicular and tubular membrane structures of the endoplasmic reticulum and the Golgi apparatus.

The book cover image was taken through confocal microscopy in Prof. Morbidelli Lab. of Pharmacology and Angiogenesis, Dept of Life Sciences, University of Siena (Italy).

Therapeutic Application of Nitric Oxide in Cancer and Inflammatory Disorders

Edited by

Lucia Morbidelli

Department of Life Sciences, University of Siena, Siena, Italy;
Noxamet Ltd, Milan, Italy

Benjamin Bonavida

Department of Microbiology, Immunology, and Molecular Genetics,
David Geffen School of Medicine, University of California,
Los Angeles (UCLA), Los Angeles, CA, United States

ELSEVIER

ACADEMIC PRESS
An imprint of Elsevier

Academic Press is an imprint of Elsevier
125 London Wall, London EC2Y 5AS, United Kingdom
525 B Street, Suite 1650, San Diego, CA 92101, United States
50 Hampshire Street, 5th Floor, Cambridge, MA 02139, United States
The Boulevard, Langford Lane, Kidlington, Oxford OX5 1GB, United Kingdom

Notices

Knowledge and best practice in this field are constantly changing. As new research and experience broaden our understanding, changes in research methods, professional practices, or medical treatment may become necessary.

Practitioners and researchers must always rely on their own experience and knowledge in evaluating and using any information, methods, compounds, or experiments described herein. In using such information or methods they should be mindful of their own safety and the safety of others, including parties for whom they have a professional responsibility.

To the fullest extent of the law, neither the Publisher nor the authors, contributors, or editors, assume any liability for any injury and/or damage to persons or property as a matter of products liability, negligence or otherwise, or from any use or operation of any methods, products, instructions, or ideas contained in the material herein.

Library of Congress Cataloging-in-Publication Data
A catalog record for this book is available from the Library of Congress

British Library Cataloguing-in-Publication Data
A catalogue record for this book is available from the British Library

ISBN: 978-0-12-816545-4

Publisher: Stacy Masucci
Acquisition Editor: Rafael Teixeira
Editorial Project Manager: Sandra Harron
Production Project Manager: Punithavathy Govindaradjane
Cover Designer: Matthew Limbert

Typeset by SPi Global, India

Contents

Contributors

Enrique Aranda Instituto Maimónides de Investigación Biomédica de Córdoba (IMIBIC), Córdoba; Centro de Investigación Biomédica en Red de Cáncer (CIBERONC), Madrid; Unidad de Gestión Clínica de Oncología Médica, Hospital Universitario Reina Sofía, Córdoba, Spain

Fereidoun Azizi Endocrine Research Center, Research Institute for Endocrine Sciences, Shahid Beheshti University of Medical Sciences, Tehran, Iran

Zahra Bahadoran Nutrition and Endocrine Research Center, Research Institute for Endocrine Sciences, Shahid Beheshti University of Medical Sciences, Tehran, Iran

Stavroula Baritaki Division of Surgical Oncology, School of Medicine, University of Crete, Crete, Greece

Elisabetta Barresi Department of Pharmacy, University of Pisa, Pisa, Italy

Benjamin Bonavida Department of Microbiology, Immunology, and Molecular Genetics, David Geffen School of Medicine, University of California, Los Angeles (UCLA), Los Angeles, CA, United States

Claudio Bucolo Department of Biomedical and Biotechnological Sciences, Section of Pharmacology, School of Medicine; Center for Research in Ocular Pharmacology—CERFO, University of Catania, Catania, Italy

Vincenzo Calderone Department of Pharmacy; Interdepartmental Research Centre "Nutraceuticals and Food for Health (NUTRAFOOD)," University of Pisa, Pisa, Italy

Konstantin Chegaev Department of Science and Drug Technology, University of Turin, Turin, Italy

Stefan Chlopicki Jagiellonian Centre for Experimental Therapeutics (JCET), Jagiellonian University, Krakow, Poland

Valerio Ciccone Department of Life Sciences, University of Siena, Siena, Italy

Valentina Citi Department of Pharmacy, University of Pisa, Pisa, Italy

Emilio Clementi Department of Biomedical and Clinical Sciences, "L. Sacco," Università degli Studi di Milano, Milan, Italy

Federico Da Settimo Department of Pharmacy, University of Pisa, Pisa, Italy

Rafaella C. da Silva Department of Biochemistry, UNIFESP, São Paulo, Brazil

Massimo Donadelli Department of Neurosciences, Biomedicine and Movement Sciences, University of Verona, Verona, Italy

Sandra Donnini Department of Life Sciences, University of Siena, Siena, Italy

Filippo Drago Department of Biomedical and Biotechnological Sciences, Section of Pharmacology, School of Medicine; Center for Research in Ocular Pharmacology—CERFO, University of Catania, Catania, Italy

Józef Dulak Department of Medical Biotechnology, Faculty of Biochemistry, Biophysics and Biotechnology, Jagiellonian University, Krakow, Poland

Nelson Durán Center for Natural and Human Sciences (CCNH), Nanomedicine Research Unit (NANOMED), Universidade Federal do ABC (UFABC), Santo André; NanoBioss; Laboratory of Urogenital Carcinogenesis and Immunotherapy, Department of Structural and Functional Biology, Universidade Estadual de Campinas, Campinas, Brazil
Miriam Durante Department of Life Sciences, University of Siena, Siena, Italy
Suhendan Ekmekcioglu Department of Melanoma Medical Oncology, The University of Texas MD Anderson Cancer Center, Houston, TX, United States
Fiorella Faienza Department of Biology, University of Rome Tor Vergata, Rome, Italy
Wagner J. Fávaro NanoBioss; Laboratory of Urogenital Carcinogenesis and Immunotherapy, Department of Structural and Functional Biology, Universidade Estadual de Campinas, Campinas, Brazil
Letícia Ferraz Center for Natural and Human Sciences (CCNH), Nanomedicine Research Unit (NANOMED), Universidade Federal do ABC (UFABC), Santo André, Brazil
Giuseppe Filomeni Redox Signaling and Oxidative Stress Research Group, Cell Stress and Survival Unit, Center for Autophagy, Recycling and Disease (CARD), Danish Cancer Society Research Center, Copenhagen, Denmark; Department of Biology, University of Rome Tor Vergata, Rome, Italy
Elena Forte Department of Biochemical Sciences, Sapienza University of Rome, Rome, Italy
Maria Frosini Department of Life Sciences, University of Siena, Siena, Italy
Fabio Fusi Department of Life Sciences, University of Siena, Siena, Italy
Elena Gazzano Department of Oncology, University of Torino, Torino, Italy
Asghar Ghasemi Endocrine Physiology Research Center, Research Institute for Endocrine Sciences, Shahid Beheshti University of Medical Sciences, Tehran, Iran
Sevda Gheibi Endocrine Physiology Research Center, Research Institute for Endocrine Sciences, Shahid Beheshti University of Medical Sciences, Tehran, Iran
Hanieh Gholami Endocrine Physiology Research Center, Research Institute for Endocrine Sciences, Shahid Beheshti University of Medical Sciences, Tehran, Iran
Paola Giglio Department of Biology, University of Rome Tor Vergata, Rome, Italy
Alessandro Giuffrè CNR Institute of Molecular Biology and Pathology, Rome, Italy
Elizabeth A. Grimm Department of Melanoma Medical Oncology, The University of Texas MD Anderson Cancer Center, Houston, TX, United States
Sajad Jeddi Endocrine Physiology Research Center, Research Institute for Endocrine Sciences, Shahid Beheshti University of Medical Sciences, Tehran, Iran
Khosrow Kashfi Department of Molecular, Cellular and Biomedical Sciences, Sophie Davis School of Biomedical Education, City University of New York School of Medicine, New York, NY; Avicenna Therapeutics, Inc, Raleigh, NC, United States
Sun-Hee Kim Department of Melanoma Medical Oncology, The University of Texas MD Anderson Cancer Center, Houston, TX, United States
Won Jong Kim Department of Chemistry, Pohang University of Science and Technology (POSTECH), Pohang, Republic of Korea
Joanna Kopecka Department of Oncology, University of Torino, Torino, Italy
Luca Lazzeri Council for Agricultural Research and Economics, Research Centre for Cereal and Industrial Crops (CREA CI), Bologna, Italy

Gian Marco Leggio Department of Biomedical and Biotechnological Sciences, Section of Pharmacology, School of Medicine; Center for Research in Ocular Pharmacology—CERFO, University of Catania, Catania, Italy

Francesca Malagrinò Department of Biochemical Sciences, Sapienza University of Rome; CNR Institute of Molecular Biology and Pathology, Rome, Italy

Alma Martelli Department of Pharmacy; Interdepartmental Research Centre "Nutraceuticals and Food for Health (NUTRAFOOD)," University of Pisa, Pisa, Italy

Ludovica Mascolo Department of Biochemical Sciences, Sapienza University of Rome, Rome, Italy

Daniela Mastronicola Department of Biochemical Sciences, Sapienza University of Rome; CNR Institute of Molecular Biology and Pathology, Rome, Italy

Parvin Mirmiran Nutrition and Endocrine Research Center, Research Institute for Endocrine Sciences, Shahid Beheshti University of Medical Sciences, Tehran, Iran

Hugo P. Monteiro Department of Biochemistry, UNIFESP, São Paulo, Brazil

Lucia Morbidelli Department of Life Sciences, University of Siena, Siena; Noxamet Ltd, Milan, Italy

Kevin Myant The Institute of Genetics and Molecular Medicine (IGMM), University of Edinburgh, Western General Hospital, Edinburgh, Scotland

Dipankar Nandi Department of Biochemistry; Centre for Biosystems Science and Engineering, Indian Institute of Science, Bangalore, India

Dai Ogata Department of Melanoma Medical Oncology, The University of Texas MD Anderson Cancer Center, Houston, TX, United States

Eleonora Pagnotta Council for Agricultural Research and Economics, Research Centre for Cereal and Industrial Crops (CREA CI), Bologna, Italy

Andreas Papapetropoulos National and Kapodistrian University of Athens, Laboratory of Pharmacology, Faculty of Pharmacy, Athens, Greece

Dongsik Park Department of Chemistry, Pohang University of Science and Technology (POSTECH), Pohang, Republic of Korea

Sanmoy Pathak Department of Biochemistry, Indian Institute of Science, Bangalore, India

Milena T. Pelegrino Center for Natural and Human Sciences (CCNH), Nanomedicine Research Unit (NANOMED), Universidade Federal do ABC (UFABC), Santo André; NanoBioss, Universidade Estadual de Campinas, Campinas, Brazil

Jon Peñarando Instituto Maimónides de Investigación Biomédica de Córdoba (IMIBIC), Córdoba, Spain

Cristiana Perrotta Department of Biomedical and Clinical Sciences, "L. Sacco," Università degli Studi di Milano, Milan, Italy

Moisés Luzia Pinto CERENA, Departamento de Engenharia Química, Instituto Superior Técnico, Universidade de Lisboa, Lisboa, Portugal

Rosana Vieira Pinto CERENA, Departamento de Engenharia Química, Instituto Superior Técnico, Universidade de Lisboa; CQB and CQE, Departamento de Química e Bioquímica, Faculdade de Ciências, Universidade de Lisboa, Ed. C8, Campo Grande, Lisboa, Portugal

Eugenia Piragine Department of Pharmacy, University of Pisa, Pisa, Italy

Valeria Pittalà Department of Drug Sciences, University of Catania, Catania, Italy

Chiara Bianca Maria Platania Department of Biomedical and Biotechnological Sciences, Section of Pharmacology, School of Medicine, University of Catania, Catania, Italy

Adriana K.C.A. Reis Department of Chemistry, UNIFESP, Diadema, Brazil
Chiara Riganti Department of Oncology, University of Torino, Torino, Italy
Salvatore Rizza Redox Signaling and Oxidative Stress Research Group, Cell Stress and Survival Unit, Center for Autophagy, Recycling and Disease (CARD), Danish Cancer Society Research Center, Copenhagen, Denmark
Tiago Rodrigue Center for Natural and Human Sciences (CCNH), Nanomedicine Research Unit (NANOMED), Universidade Federal do ABC (UFABC), Santo André, Brazil
Antonio Rodriguez-Ariza Instituto Maimónides de Investigación Biomédica de Córdoba (IMIBIC), Córdoba; Centro de Investigación Biomédica en Red de Cáncer (CIBERONC), Madrid; Unidad de Gestión Clínica de Oncología Médica, Hospital Universitario Reina Sofía, Córdoba, Spain
Barbara Rolando Department of Science and Drug Technology, University of Turin, Turin, Italy
Jason Roszik Department of Melanoma Medical Oncology, The University of Texas MD Anderson Cancer Center, Houston, TX, United States
Menachem Rubinstein Department of Molecular Genetics, The Weizmann Institute of Science, Rehovot, Israel
Iris C. Salaroglio Department of Oncology, University of Torino, Torino, Italy
Salvatore Salomone Department of Biomedical and Biotechnological Sciences, Section of Pharmacology, School of Medicine; Center for Research in Ocular Pharmacology—CERFO, University of Catania, Catania, Italy
Simona Saponara Department of Life Sciences, University of Siena, Siena, Italy
Gurusamy Saravanakumar Department of Chemistry, Pohang University of Science and Technology (POSTECH), Pohang, Republic of Korea
Jan Scicinski Avicenna Therapeutics, Inc, Raleigh, NC, United States
Amedea B. Seabra Center for Natural and Human Sciences (CCNH), Nanomedicine Research Unit (NANOMED), Universidade Federal do ABC (UFABC), Santo André; NanoBioss, Universidade Estadual de Campinas, Campinas, Brazil
Arnold Stern New York University School of Medicine, New York, NY, United States
Claudiu T. Supuran University of Florence, Neurofarba Department, Section of Pharmaceutical and Nutraceutical Sciences, Firenze, Italy
Sabrina Taliani Department of Pharmacy, University of Pisa, Pisa, Italy
Lara Testai Department of Pharmacy; Interdepartmental Research Centre "Nutraceuticals and Food for Health (NUTRAFOOD)," University of Pisa, Pisa, Italy
Luisa Ugolini Council for Agricultural Research and Economics, Research Centre for Cereal and Industrial Crops (CREA CI), Bologna, Italy
Federica Vannini Institute of Life Sciences, Scuola Superiore Sant'Anna, Pisa, Italy
Greta Varchi Institute of Organic Synthesis and Photoreactivity, Italian National Research Council, Bologna, Italy
Massimo Venturelli Department of Neurosciences, Biomedicine and Movement Sciences, University of Verona, Verona, Italy; Department of Internal Medicine, University of Utah, Salt Lake City, United States
Taru Verma Centre for Biosystems Science and Engineering, Indian Institute of Science, Bangalore, India

João B. Vicente Instituto de Tecnologia Química e Biológica António Xavier, Universidade Nova de Lisboa, Oeiras, Portugal
Shikha Yadav Department of Biochemistry, Indian Institute of Science, Bangalore, India
Marina Ziche Department of Life Sciences, University of Siena, Siena, Italy
Karim Zuhra Department of Biochemical Sciences, Sapienza University of Rome; CNR Institute of Molecular Biology and Pathology, Rome, Italy

About the Editors and Acknowledgements

Lucia Morbidelli, Ph.D

Professor Lucia Morbidelli is an associate professor of pharmacology at the Department of Life Sciences, University of Siena (Italy). Professor Morbidelli's research experience is within the pharmacology of angiogenesis and microcirculation by investigations of a plethora of in vitro and in vivo models. Her background is based on the definition of the role of nitric oxide and prostanoid pathways in the process of angiogenesis induced by growth factors and drugs. In the last decade, she focused her attention to the gaseous transmitters nitric oxide and hydrogen sulfide by charactering their activities in the maintenance of endothelial survival and functionality and in the protection of the cardiovascular system integrity and function. She is an author of over 140 scientific publications in peer-reviewed journals and additional reviews and chapters in books on the molecular, biochemical, and cellular aspects of angiogenesis and endothelial pharmacology.

Professor Morbidelli was a guest editor of the special issue of Pharmacological Research "Drugs and druggable signaling in angiogenesis" with six invited reviews published during 2016. Apart from scientific achievements, the international collaborations are related to the participation at the COST Action BM1005 European Network on Gaseous transmitters (coordinator Prof. Andreas Papapetropoulos) in the period 2012–15 and as a member responsible for pharmacological countermeasures of the ESA topical team "Tissue Healing in Space: Techniques for Promoting and Monitoring Tissue Repair and Regeneration" (coordinators Dr. Carlo Iorio and Monica Monici) since 2017.

Professor Morbidelli's current research topics are related to (1) the cross talks and interplays of NO- and H_2S-producing pathways, within their cardiovascular protective properties; (2) the identification of drugs and medical devices (natural extracts and nutraceuticals) to recover endothelial and tissue functions following injury or disease conditions related to inflammatory and oxidative stress; and (3) the characterization of synthetic biomaterials for biocompatibility and to promote healing or to induce tissue regeneration (i.e., in extreme conditions).

Benjamin Bonavida, Ph.D

Benjamin Bonavida, Ph.D, is currently a Distinguished Research Professor at the University of California, Los Angeles (UCLA). His research career, thus far, has focused on basic immunochemistry and cancer immunobiology. His research investigations have ranged from the mechanisms of cell-mediated killing, sensitization of resistant tumor cells to chemo-/immunotherapy, characterization of resistant factors in cancer cells, cell signaling pathways mediated by therapeutic anticancer antibodies, and characterization of a dysregulated NF-κB/Snail/YY1/RKIP/PTEN loop in many cancers that regulates cell survival, proliferation, invasion, metastasis, and resistance. He has also investigated the role of nitric oxide in cancer and its potential antitumor activity. Many of the above studies are centered on the clinical challenging features of cancer patients' failure to respond to both conventional and targeted therapies. The development and activity of various targeting agents, their modes of action, and their resistance are highlighted in many refereed publications.

Acknowledgment

Editors acknowledge the assistant, Jazelle Bautista, who has diligently worked on both the editing and formatting of the various manuscripts in their volume. Ms. Bautista is a senior undergraduate student at UCLA.

Therapeutic Application of Nitric Oxide in Cancer and Inflammatory Disorders

The main scope of this book is to update the rapidly evolving field of NO donors/inhibitors as potential therapeutics, used alone or in combination, for various diseases. Several internationally renowned scientists/clinicians have contributed significantly in this book. It is clear that the various NO donors/inhibitors and their metabolic pathways have demonstrated important physio-pathologic roles in mammals; these represent novel targets with remarkable therapeutic applications in areas ranging from cardiovascular diseases, neuro-degenerative diseases, inflammation, and cancer.

This book's topic, titled, *Therapeutic Application of Nitric Oxide in Cancer and Inflammatory Disorders*, is developed in conjunction with a conference held in Siena, Italy, October 4–5, 2018. In addition to reviews by various scientists, in this book, several abstracts are published that represented various presentations at the conference.

The editors strongly believe that this book is unique in its contents and should be considered as a reference for scientists, clinicians, students, pharmaceutical companies, and others interested in this new field of therapeutics.

Preface

The main objective of this book's topic, titled "Therapeutic Application of Nitric Oxide in Cancer and Inflammatory Disorders," is in conjunction with a conference held in Siena, Italy, October 4–5, 2018, and attended by established scientists/clinicians with worldwide reputations. These are leaders on the investigations of the mediator nitric oxide (NO) and various NO donors/inhibitor agents and their metabolic pathways that demonstrate important physiopathologic roles in mammals (humans and animals); these NO donors/inhibitors represent novel therapies and remarkable therapeutic applications in areas ranging from cardiovascular diseases, neurodegenerative disorders, inflammatory diseases, and cancer.

The current status on the potential uses and applications of therapeutic NO donors/inhibitors in cancer and inflammatory diseases has not advanced until very recently as a result of extensive preclinical findings and very few clinical findings. The recent developments of novel NO donor/inhibitor compounds and conjugates and their preclinical antitumor/antiinflammatory activities have opened an era of a new class of therapeutics and their potentials in the treatment of highly resistant cancers and metastases and various inflammatory diseases. This book consists of reviews by both invited speakers at the conference and also invited nonparticipants at the conference. Hence, this book provides significant novel contributions in this field of NO with the proposed developments of new anticancer/antiinflammatory therapeutic agents.

This book consists of 15 review chapters by experts in their field and also a list of abstracts of presentations delivered at the conference. The highlights of each chapter contribution are as follows:

Dr. Seabra and colleagues (Chapter 1) review NO-releasing engineering nanoparticles in the context of their use in the reversal of resistance of cancer cells to chemotherapy. The rationale of the use of these complexes derives from their functions in increasing blood perfusion and leaky tumor blood vessels, thus enhancing the penetration of chemotherapeutic drugs to kill cancer cells. In this chapter, they describe the various designed NO-releasing nanoparticles for cancer treatment. The released NO is controllable, and the combination use of NO nanoparticles and chemotherapy is discussed for the treatment of drug-resistant tumors.

Dr. Baritaki (Chapter 2) reviews her collaborative findings and those of others in the pleiotropic role that NO exerts in cancer cell models. The use of high concentrations of NO donors resulted in various antitumor manifestations, including the inhibition of cell growth, the reduction of cell viability, the inhibition of invasiveness, the epithelial-mesenchymal transition (EMT) and metastasis, and the reversal of tumor cell resistance to both chemotherapy and immunotherapy. The underlying mechanisms by which NO mediates these multiple antitumor activities are being discussed along with the identification of several therapeutic targets.

Drs. Ciccone and Morbidelli (Chapter 3) review the antitumor therapeutic potential of a novel NO donor belonging to the metal-nonoate family. In a tumor model, they report the antitumor activity of this compound [Ni(*salpipNONO*)] in an in vivo mouse model that demonstrated a reduced tumor survival, clonogenicity, and inhibition of tumor invasiveness along with apoptosis. This compound was also able to exert anti-angiogenic activity. They attributed the effect observed by mechanisms involving the classic NO/cGMP pathway and ROS production.

Dr. Morbidelli and colleagues (Chapter 4) review the role of endothelium-derived NO in angiogenesis. Endothelial cells are intricate in angiogenesis as derived NO regulates blood vessel tone and blood fluidity and maintains a noninflammatory microenvironment in normal conditions. However, under nonnormal conditions (tumor growth, inflammation, etc.), high levels of produced NO contribute to a nonphysiological neovascularization. They proposed that NO donors may exert antiangiogenic activity and, thus, of therapeutic significance.

Dr. Kim and collaborators (Chapter 5) have investigated the role of nitric oxide in the tumor microenvironment in melanoma. In this review, the authors discuss the role of nitric oxide production in the tumor microenvironment affecting the antitumor cell-mediated immune response. Also, they discussed the role of arginine in the environment and present data on the roles of arginase and NOS enzymes as immunosuppressing agents and suppressing antitumor T-cell-mediated immunity.

Drs. Kashfi and Vannini (Chapter 6) review the contrasting dual roles of iNOS expression and NO levels in human cancers that result in either a protumorigenic or antitumorigenic effect. They discussed both their reported studies and their own findings with respect to iNOS levels in various cancers and whether NO inhibitors or donors are recommended for certain types of cancers.

Dr. Scicinski (Chapter 7) reviews the role of three different drugs with different mechanisms of NO release and with different applications. These three drugs are, namely, NBS-1120 (releases NO and H2S), JS-K (an arylating NO donor), and RRx-001 (releases NO indirectly). He describes the activities and safeties of each of the above drugs as well as their mechanisms of action and their potential therapeutic applications.

Dr. Peñarando and colleagues (Chapter 8) discuss NO scavengers as a means that target NO produced from any source. They illustrate the NO scavenger c-PTIO activity that enhances tumor immunotherapy and reduces tumor growth, as well as impairing the cancer stem cell (CSC) phenotype. Their findings are discussed in colorectal cancer, and their studies were performed both in vitro and in vivo.

Dr. Rizza (Chapter 9) discusses the role of S-nitrosoglutathione reductase (GSNOR) in tumor progression. This enzyme is the main denitrosylating system in the cell and controls the amount of S-nitrosylated proteins. Interestingly, cell transformation in liver cancer is a result, in part, of an apparent S-nitrosylation signaling for GSNOR deficiency. Dr. Rizza and colleagues have found that GSNOR deficiency results in mitochondrial dysfunction, a common feature in cancer cells. Also, they report that GSNOR inhibition in cancer promotes tumor progression, EMT, and metastasis.

Dr. Park and colleagues (Chapter 10) review the limitations of the clinical application of NO donors in cancer. Hence, they review the various NO delivery nanoplatforms that have been designed in which they improved the released NO concentrations at the tumor site, thereby enhancing their therapeutic effects both in chemotherapy and radiotherapy.

Dr. Venturelli (Chapter 11) reviews the roles of NO in vascular dysfunction during aging and Alzheimer's disease. The discussion is centered on the role of NO depletion in reducing extracranial blood flow resulting in the impairment of cortical and peripheral circulation with a consequent detriment of cognitive function in humans with Alzheimer's disease. In the absence of any therapeutic drugs for Alzheimer's disease, Dr. Venturelli discusses a potential new therapeutic approach that considers the role of NO in homeostasis.

Dr. Supuran (Chapter 12) reviewed the role of hypoxia-induced carbonic anhydrase (CA) that is overexpressed in tumors and minimally expressed in normal tissues. Inhibitors of CA coupled with NO donors are discussed as targeting cancer cell agents for several pathological diseases, including neuropathic pain and cerebral ischemia.

Dr. Yadav and colleagues (Chapter 13) reviewed the various roles of NO during sepsis and inflammatory disorders. They discuss the pleiotropic roles in the micro- and macrocirculations and how it relates to various immune pathologies. They also present the potential application of NO therapeutics in an effort to alleviate and treat the various sepsis symptoms.

Drs. Rosana and Moisés Pinto (Chapter 14) reviewed the potential therapeutic applications of NO-releasing porous materials. They discuss a novel approach using zeolites and metal-organic frameworks (MOFs) to both store and deliver NO. These materials provide a safe storage and packing of NO within the solid and the delivery of NO in a controlled rate. They also investigate several microporous titanosilicates for toxicity, bioavailability, stability, etc. They also discuss the functions of these NO donors in the control of mitochondrial respiration and cell migration.

In addition to the above review contributions, there is also a list of several abstracts of presentations at the conference in Siena. They are listed alphabetically and provide additional, current information on NO donors/inhibitors in various pathological conditions. The readers may refer to those abstracts and also may communicate with the authors via e-mails that are also listed.

The editors Drs. Morbidelli and Bonavida wish to acknowledge the excellent assistance of Jazelle Bautista for her excellent editing and formatting of the various contents of this volume.

Benjamin Bonavida
UCLA

Lucia Morbidelli
University of Siena

PART 1

Applications of Nitric Oxide in Cancer

CHAPTER 1

Nitric Oxide-Releasing Engineered Nanoparticles: Tools for Overcoming Drug Resistance in Chemotherapy

Amedea B. Seabra*,†, Milena T. Pelegrino*,†, Letícia Ferraz*, Tiago Rodrigue*, Wagner J. Fávaro†,‡, Nelson Durán*,†,‡

*Center for Natural and Human Sciences (CCNH), Nanomedicine Research Unit (NANOMED), Universidade Federal do ABC (UFABC), Santo André, Brazil

†NanoBioss, Universidade Estadual de Campinas, Campinas, Brazil

‡Laboratory of Urogenital Carcinogenesis and Immunotherapy, Department of Structural and Functional Biology, Universidade Estadual de Campinas, Campinas, Brazil

ABSTRACT

The endogenous free radical nitric oxide (NO) plays pivotal physiological and pathophysiological functions, including in cancer biology. NO donors have been extensively used for anticancer activities, and more recently, the combination of NO donors and nanomaterials has been shown to be a promising approach to generate controllable therapeutic concentrations of NO at the target site (tumor cells). Several important publications showed the promising uses of NO-releasing nanomaterials in increasing tumor perfusion, due to the improvement of the enhanced permeability and retention effect, allowing chemotherapeutic drugs to penetrate leaky tumor blood vessels. In this scenario, the coadministration of NO-releasing nanomaterials and chemotherapeutic drugs has emerged as a promising approach to realize the full clinical potential of NO in cancer therapy. Recently, an increasing number of reports have demonstrated that low concentrations of NO (those that if administrated alone would not be able to promote apoptosis) contribute to the reversal of multidrug resistance. Thus, this observation has motivated and inspired researchers to combine NO donors allied with nanomaterials for the improvement of chemotherapy. Therefore, this chapter describes the recent progress (last 3 years) in the design of different kinds of NO-releasing nanomaterials for cancer treatment. Most of these nanocarriers are able to release controllable amounts of NO on demand, upon light, pH, or wave exposure. In addition, the combination of NO-releasing nanomaterials with classical chemotherapeutic drugs in the treatment of cancer is also presented and discussed. The possible mechanisms of NO to sensitize tumor cells and the impact of nanomaterials in

Therapeutic Application of Nitric Oxide in Cancer and Inflammatory Disorders. https://doi.org/10.1016/B978-0-12-816545-4.00001-3

cancer treatment are also discussed. Finally, the perspectives and challenges in the design of efficient NO-releasing nanomaterials in cancer treatment in real clinical settings are highlighted.

ABBREVIATIONS

ADD	adjudin
Alkynyl-JSK	O^2-(2,4-dinitrophenyl) 1-[4-(propargyloxycarbonyl)piperazin-1-yl]diazen-1-ium-1,2-diolate
AMS	amine-modified mesoporous silica
B16F10	mouse melanoma
Bcl-2	B-cell lymphoma 2
$CaCO_3$	calcium carbonate
CS	chitosan
CT26	mouse colonic cancer cells
DETA	diethylenetriamine
DOX	doxorubicin
EPR	enhanced permeability and retention
Fe^{2+}	ferrous ion
FLIP	FLICE-inhibitory protein
GSH	glutathione
GSNO	*S*-nitrosoglutathione
HDPNs	double-layered polymer nanoparticles
HepG2	human hepatocellular carcinoma
HMTNPs	hollow mesoporous titanium dioxide nanoparticles
HT29	colon carcinoma cells
K562	human chronic myeloid leukemia
Lucena-1	a vincristine-resistant K562 cell line
MCF-7	breast cancer cells
MCF-7/ADR	drug-resistant human breast cancer cells
MDR	multidrug resistance
N_2O_3	peroxynitrite
NAP	*N*-acetyl-D-penicillamine thiolactone
NIR	near-infrared
NO	nitric oxide
NO_3^-	nitrate
NO-Dex	hydrophobic nitrated dextran
NONOate	diazeniumdiolates
NO-NPs	NO nanoparticles
NOS	nitric oxide synthase

NOx	NO-related species
PEG	polyethylene glycol
P-gp	P-glycoprotein
RBS	Roussin's black salt, a photosensitive NO donor
ROS	reactive oxygen species
SDT	sonodynamic therapy
SH	sulfhydryl groups
Si-DETA	(3-trimethoxysilylpropyl)diethylenetriamine
SPION@hMSN	superparamagnetic iron oxide-encapsulated mesoporous silica nanoparticles
TPGS	*D*-α-tocopherol polyethylene glycol 1000 succinate
TPZ	tirapazamine
UCNPs	NaYF4:Yb,Eu upconversion
US	ultrasound

Conflict of Interest

No potential conflicts of interest were disclosed.

INTRODUCTION: OVERVIEW OF NITRIC OXIDE IN CANCER BIOLOGY

Nitric oxide (NO) is a free radical involved in several physiological and pathophysiological processes. It is a small-sized and uncharged molecule with a relatively high lipophilicity, which allows its diffusion through biological membranes without the need of membrane channels or receptors [1]. NO is involved in various physiological processes, such as the control of vascular tone, inhibition of platelet adhesion and aggregation, smooth muscle cell replication, immune responses, neuronal communication, wound healing, cell differentiation, and apoptosis [2]. Also, NO is a key molecule in the immune system, and it is considered important for the natural host defense against pathogens, such as bacteria, fungi, and viruses [3]. NO may bind with high affinity to the ferrous ion (Fe^{2+}) and in the iron-sulfur centers of hemoproteins, and both copper and zinc are targets of NO actions [4]. Considering its properties and reactivity, the study of NO activities and the molecular pathways involving NO signaling would provide useful tools as diagnostics, therapeutics, and prognosis of different pathologies, including cancer [5].

In mammals, NO is enzymatically synthesized by the action of nitric oxide synthase (NOS) from the oxidation of the amino acid L-arginine to L-citrulline [6]. The protein family of NOS comprises inducible NOS, endothelial NOS, and neuronal NOS [7]. The first is calcium-independent and produces relatively high doses of NO for a longer period of time. These three NO isoforms may be involved in the process of proliferation or inhibition of tumor growth. High NO concentrations may be cytostatic or cytotoxic to tumor cells, whereas low NO concentrations may have the opposite effect and promote tumor growth [8, 9]. NO synthesized endogenously or donated by exogenous sources generates modifications in signaling proteins and, depending on its concentration, may present both angiogenic and genotoxic properties. NO can cause DNA damage. Peroxynitrite, which is formed from the reaction between NO and superoxide anion, can oxidize and nitrate DNA, causing breaks in the single strand of DNA by attacking the sugar-phosphate skeleton, and N_2O_3 can nitrosate amines, leading to the formation of *N*-nitrosamines [9, 10]. As a consequence of DNA damage, there is an accumulation of p53 in the cells and activation of poly (DNA ribose) polymerase, which in turn may induce apoptosis [9]. In fact, NO can inhibit tumor cell growth, leading to cell death by the apoptotic pathway [11]. The expression of NOS is associated with metabolic alterations, neoplasticity, angiogenesis, chemoresistance, and immune evasion. NO also regulates posttranslational protein modifications and genome-wide epigenetic alterations. NO can activate or amplify tumor-suppressing effects acting in several antioncogenic pathways, for example, activating p53 or suppressing epigenetic modifications that can suppress metastasis and chemoresistance. NO can act as a protumorigenic agent, for example, via metabolic hypoxia, S-nitrosation of certain cell proteins, irregular epigenetic modifications, increased inflammation processes, and reprogramming the tumor microenvironment metabolism [12].

NO acts as a modulator of the tumor microenvironment and reprogramming stromal nonmalignant cells to support tumor progression. In addition, NO affects mitochondrial physiology, exerting concentration-dependent effects on the mitochondrial respiration, ATP formation, cytochrome *c* release, and the generation of reactive oxygen and nitrogen species. At low concentrations, NO reversibly inhibits complex IV of the respiratory chain (cytochrome oxidase) through binding to its copper center. It also inhibits a segment of complex II, without promoting cytochrome *c* release. In contrast, at high concentrations, NO promotes autoxidation of ubiquinol with the concomitant increased production of superoxide, hydrogen peroxide, and peroxynitrite. These reactive species may damage complexes I and II, impairing ATP synthesis, leading to the release of cytochrome *c* and promotion of apoptosis [13–16].

Different classes of enzymes contain cysteines in their active sites that might undergo S-nitrosation, which can inhibit enzymatic activities. Examples are cathepsin B (lysosomal proteolytic enzyme), aldolase (glycolysis), gamma-glutamylcysteine synthetase glutathione, and glyceraldehyde-3-phosphate dehydrogenase (glycolysis and gluconeogenesis) [17]. Antiapoptotic proteins such as B-cell lymphoma 2 (Bcl-2) and FLICE-inhibitory protein (FLIP) [18, 19] are also targets of *S*-nitrosation. FLIP is a key regulator of the extrinsic pathway of apoptosis; this antiapoptotic protein prevents procaspase 8 from recruiting subsequent apoptosis inducers [20]. Bcl-2 is a key protein of the intrinsic pathway of apoptotic cell death, which forms heterodimers with proapoptotic proteins and inhibits the formation of the transition pore of mitochondrial permeability and cytochrome *c* release [21]. The upregulation of cathepsin B expression has been reported in some types of cancer characterizing a tumor phenotype with increased invasiveness and metastatic potential [22], and its inactivation may contribute to the reduction of metastatic capacity and tumor malignancy.

Depending on the redox environment, superoxide can react with NO to form peroxynitrite anions that can nitrate protein tyrosine residues. The nitration of the amino acid results from the addition reaction of the $-NO_2$ group, generally at the 3-position of the phenolic ring of a tyrosine residue generating 3-nitrotyrosine [23]. Nitrotyrosine formation is widely employed as a biomarker for nitrosactive stress [24]. Examples of nitration target proteins are superoxide dismutase and actin [25]. Nitration of the manganese-dependent superoxide dismutase leads to enzyme inactivation, which may increase superoxide levels in the mitochondria [26].

As NO is involved in several physiological and pathophysiological processes, to increase its lifetime, researchers have been using low-molecular-weight molecules capable of acting as NO donors/generators, such as diazeniumdiolates (NONOates), ruthenium nitrosyl complexes, and *S*-nitrosothiols (RSNOs) [2]. Increased knowledge about the role of NO in cancer biology allowed significant progress to be made related to the possible use of NO donors in cancer chemotherapy. A promising strategy in cancer treatment is the incorporation of NO donors into nanomaterials for the sustained release of therapeutic amounts of NO [27–30].

NANOTECHNOLOGY IN CANCER: THE ENHANCED PERMEABILITY AND RETENTION EFFECT ALLIED TO NO DONORS

According to the European Union Commission, the definition of nanomaterial is as follows: "A natural, incidental or manufactured material containing particles, in an unbound state or as an aggregate or as an agglomerate and where, for 50% or more of the particles in the number size distribution, one or more external dimensions is in the size range 1–100 nm" [31]. In the last decades, several important publications demonstrated that nanotechnology can provide a promising platform in medicine for carrying and delivering chemotherapeutic drugs by reducing side effects, increasing drug accumulation at the tumor site, and improving blood circulation. Since NO is a free radical with a relatively short half-life in biological medium, efficient NO-releasing nanocarriers have been designed to promote on-demand generation of therapeutic amounts of NO, with minimum side effects.

Nanoparticles have been extensively applied in cancer diagnosis and treatment not only to carry NO donors but also as a vehicle for several drugs. Therapeutic nanoparticles are designed to accumulate at the tumor tissue/organ, where the active agent is released, increasing the therapeutic efficacy by reducing the dose and incidence of the treatment. Once administered, nanoparticles have the tendency to accumulate more in the tumor tissue/organ compared with healthy tissues [32]. This tendency is based on the enhanced permeability and retention (EPR) effect. EPR is a consequence of the fact that solid tumors are characterized by leaky blood vessels and hypoxia, which lead to the extravasation of nanoparticles. Angiogenesis is deregulated in tumors, in addition to the superior generation of vascular permeability-enhancing factors, including vascular endothelial growth factor [33]. Moreover, functional lymphatic drainage is seriously compromised in solid tumors. In this sense, the removal of extravasated nanomaterial from the target tumor site is significantly reduced. This effect prolongs the retention of nanomaterials within the pathological organ. In addition to the EPR effect, engineered nanomaterials can be prepared with functional groups on their surface (folic acid and hyaluronic acid) to further target the nanocarrier to tumor organs [34].

In this context, the advantages of nanomaterials over standard low-molecular-weight drugs are the ability to target the drug directly to the desired site of application (tumors), with reduction in their renal excretion, prolonging their circulation times and, therefore, decreasing volume of distribution.

NO is involved in cancer biology, and NO donors have been applied in cancer treatment [28, 29, 35]. NO has a dichotomous action in cancer, as tumor progressor or suppressor, depending on its concentration, flux, duration, and environmental redox chemistry [16]. In general, high NO concentrations (micromolar range) have proapoptotic effects and anticancer effects, whereas low NO concentrations (pico- to nanomolar range) promote tumor growth [11, 36, 37]. Therefore, control of the precise amount and duration of NO released from a donor must be carefully addressed in cancer therapy. In this sense, nanotechnology might improve the efficacy in the generation of precise concentrations of NO, in some cases directly to the tumor site, where NO can have cytotoxic effects [27–30].

Taken altogether, nanoparticles can increase the therapeutic efficacy of NO donors.

COMBINATION OF NO-RELEASING NANOMATERIALS AND CONVENTIONAL ANTICANCER CHEMOTHERAPIES

Currently, NO-mediated anticancer therapy might be achieved by either direct killing or chemosensitization of tumor cells [38]. Recently, the potential of NO to overcome multidrug resistance (MDR) has been demonstrated in several publications. In cancer, MDR is an obstacle in the achievement of an efficient treatment. Cancer cells develop a broad spectrum of defense mechanisms, leading to cell resistance to one or more anticancer drugs. These mechanisms involve decreased drug uptake, increased drug efflux, the partial or total inactivation of the drug, the promotion of DNA repair mechanisms, and the promotion of detoxification pathways. NO has a key function in overcoming MDR.

As stated before, high NO concentrations (micro- to millimolar range) lead to apoptosis and cell death, a direct killing effect of NO. However, low NO concentrations (pico- to nanomolar up to low micromolar range) have shown a potent chemosensitizing effect, which can reverse MDR. Therefore, this relatively new action of NO in cancer biology has recently inspired researchers to develop combined approaches based on (i) NO donors/generators, (ii) nanomaterials, and (iii) traditional chemotherapeutic drugs.

The mechanisms of NO to overcome MDR are still under intense investigation and be summarized as follows:

(i) NO and NO-related species (NOx) can induce the nitration and denaturation of several important proteins involved in DNA repair [38]. NO impairs the cell's ability to repair DNA and, consequently, enhances the cytotoxicity of anticancer drugs leading to DNA damage.
(ii) NO promotes the depletion of glutathione (GSH), which is found at higher levels in drug-resistant cancer cell lines compared with drug-sensitive cancer cells. GSH is a potent inhibitor of platinum (Pt)-contained anticancer drugs, such as cisplatin.
(iii) NO promotes the glutathionylation of histone, reversing MDR mechanisms since these nuclear proteins are important in regulating gene transcription.
(iv) As a potent vasodilator, NO has been shown to overcome hypoxia-induced MDR effects, which are observed in solid tumors.
(v) NO can sensitize NF-κB-associated pathways in resistant cancer cells. NF-κB modulates cell survival, metastatic pathways, and drug resistances.

In this context, the following section describes selected examples based on the recent progress (last 3 years) in either the use of NO-releasing nanomaterials or the combination of NO-releasing nanomaterials with traditional chemotherapeutic drugs against cancer (in vitro and/or in vivo studies). These selected examples highlight the recent tendency in the design of new, efficient NO delivery systems alone or in combination with chemotherapeutic drugs to generate desired amounts of NO in a controlled time and spatial manner.

SELECTED EXAMPLES OF NO-RELEASING NANOMATERIALS FOR CANCER TREATMENT

This section presents selected publications from the last 3 years (2016–18) that describe recent progress in the design of NO-releasing nanomaterials for in vitro and in vivo cancer treatments. Most of the publications showed the successful combination of NO-generating nanomaterials with traditional chemotherapeutic drugs, such as cisplatin or doxorubicin (DOX). Alternatively, some publications report the use of external stimuli, such as light, pH environment, or ultrasound irradiation, to selectively generate therapeutic amounts of NO at the target site of application.

The anticancer effects of NO-releasing engineered nanospheres combined with DOX were evaluated in vitro and in vivo [39]. The nanoparticles were composed of NaYF4:Yb,Eu upconversion (UCNPs) coated with chitosan (CS) through carboxylic acid conjugation with *N*-hydroxysuccinimide. The employed NO donor was Roussin's black salt (RBS, a photosensitive NO donor), combined with DOX leading to the formation of UCNPs(DOX)@CS-RBS. The nanocarrier is able to absorb near-infrared (NIR) photons by applying a safe power density of 0.7 W/cm^2, converting them into visible photos, and leading to NO release from the RBS moiety. In addition, a low pH medium can trigger DOX release from the nanomaterial upon the stretching of CS-oleoyl chains (Fig. 1). Low pH observed in lysosomes and endosomes in cancer might efficiently trigger DOX release from the nanomaterial. The anticancer activity of the nanomaterial was demonstrated in vitro in mouse colonic cancer cells (CT26) and drug-resistant human breast cancer cells (MCF-7/ADR). The cancer cell viability was assayed after 48 h of incubation with UCNPs(DOX)@CS-RBS nanospheres and compared with free DOX. Under NIR irradiation at 980 nm for 5 or 3 min, the IC_{50} values of UCNPs(DOX)@CS-RBS nanospheres were found to be 0.417 ± 0.011 and 0.274 ± 0.008 µg/mL, respectively. In contrast, higher IC_{50} values were observed for nanospheres in the dark or free DOX. In vivo studies were performed by using mice bearing CT26 tumors. These results demonstrated that a combination of DOX and NO release from the engineered nanospheres can have anticancer effects [39].

It should be noted that there is an increasing interest in the design of nanocarriers that efficiently increase tumor permeability. A promising approach in

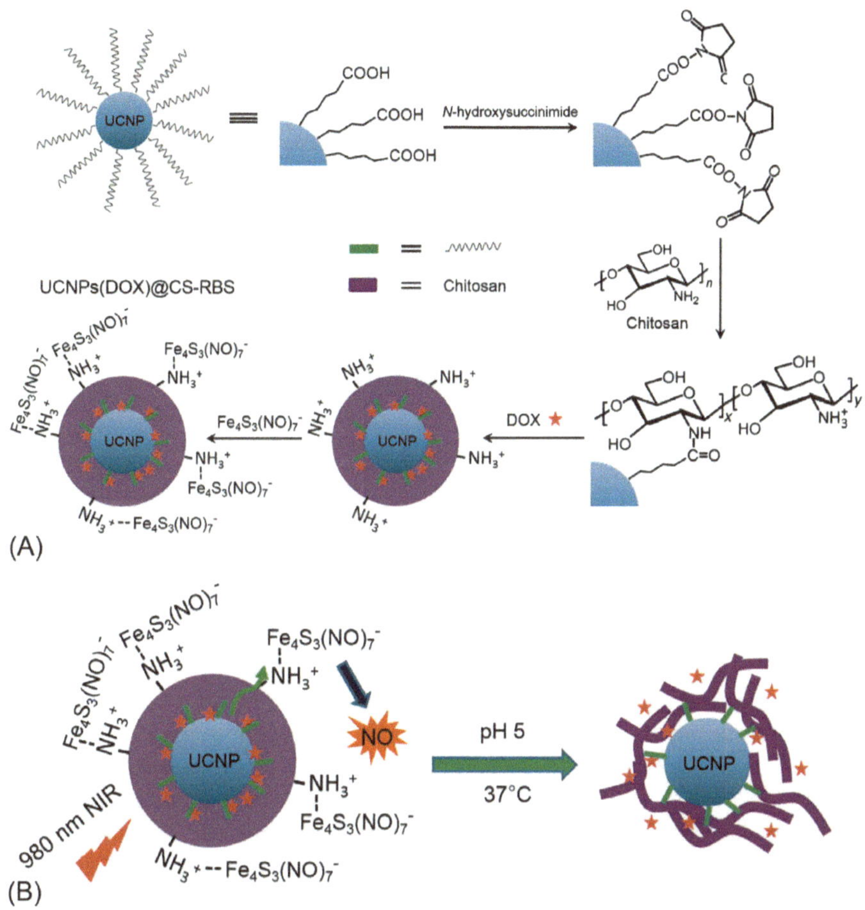

FIG. 1

(A) Schematic representation of the preparation of NaYF4:Yb,Eu upconversion (UCNPs) coated with chitosan (CS) through carboxylic acid conjugation with *N*-hydroxysuccinimide. The employed NO donor was Roussin's black salt (RBS, a photosensitive NO donor), combined with DOX leading to the formation of UCNPs(DOX)@CS-RBS. (B) NIR-triggered NO release and pH-responsive DOX release from UCNPs(DOX)@CS-RBS. *Reproduced from Tan L, Huang R, Li X, Liu S, Shen YM. Controllable release of nitric oxide and doxorubicin from engineered nanospheres for synergistic tumor therapy. Acta Biomater 2017;57:498–510 with permission from Elsevier.*

this sense is the use of NO nanoparticles (NO-NPs) to provide a local vascular dilation at the tumor site. In a similar approach, DOX was combined with NO-NPs for tumor-site-specific delivery of NO, due to the known EPR effect of the nanoparticles [40]. The nanoparticles (hydrodynamic size of 134.0 ± 3.2 nm) were composed of hydrophobic nitrated dextran (NO-Dex) and hydrophilic polyethylene glycol (PEG). The engineered nanocarrier spontaneously releases NO in the presence of glutathione, an abundant molecule found intracellular in tumors. Once in contact with cancerous cells, NO and DOX are simultaneously released by DOX-NO-NPs. Blood vessels were dilated by the released NO, increasing vessel permeability and improving the EPR effect and, thus, the antitumor effect of DOX directly at the tumor site (Fig. 2). In vivo experiments demonstrated the localization of the nanoparticles at the tumor site after systemic administration of the nanomaterial and a significant increase of the local blood flow and vascular permeability due to the local release of NO. Thus, the combination of NO donors and a chemotherapeutic drug (DOX) not only enhances the EPR effect due to the vasodilation promoted by NO via the cyclic GMP signaling pathway but also causes a synergistic anticancer activity by upregulating p53 genes and blocking P-glycoproteins (P-gps).

The encapsulation of DOX into NO-NPs further enhanced the antitumor activity of the nanomaterial, suggesting the high potency of NO-releasing NPs allied to DOX as EPR enhancers to improve clinical outcomes. The authors evaluated the toxic effects of the nanoparticles in human colon carcinoma cells (HT29), and NO internalization by cancerous cells was demonstrated by confocal microscopy. In vivo experiments in HT29 tumor-bearing mice showed the dilation of blood vessels in tumor tissues after the administration of the nanoparticles, as assayed by the ultrasound power Doppler technique. In addition, animals treated with NO-NPs demonstrated a significant increase in nanoparticle accumulation in the tumors in comparison with control groups. The combination of NO-NPs and DOX reduced tumor size. Indeed, after 14 days of nanoparticle administration, the tumor volumes were found to be 544, 394, 309, and 150 cm^3 for the saline, free DOX, DOX-control NP, and DOX-NO-NP groups, respectively (Fig. 3). These results suggest the superior EPR effect of DOX-NO-NPs. Finally, this nanocarrier was found to generate a high NO concentration (46.59 μM of NO equivalent per milligram NPs), which is suitable for its anticancer effects [40].

In an alternative approach, the NO donor *S*-nitrosoglutathione (GSNO) was loaded into calcium carbonate ($CaCO_3$)-mineralized nanoparticles obtained by the anionic block copolymer (PEG-poly(L-aspartic acid))-templated mineralization [41]. The GSNO-containing nanomaterial has a hydrodynamic diameter of 248.8 ± 12.1 nm and a negative zeta potential (-15.2 mV). The acid environment of the endosomes was found to trigger NO release in a pH-responsive nanocarrier (Fig. 4). In vitro studies in breast cancer (MCF-7) cells showed that GSNO-releasing nanoparticles enhanced the therapeutic activity of DOX. Transmission electron microscopy was used to demonstrate the endocytosis of GSNO-containing nanoparticles, and the intracellular levels of NO were assayed by confocal laser scanning electron microscopy. The results showed the generation of NO in the cytosol. Pretreating MCF-7 cells with

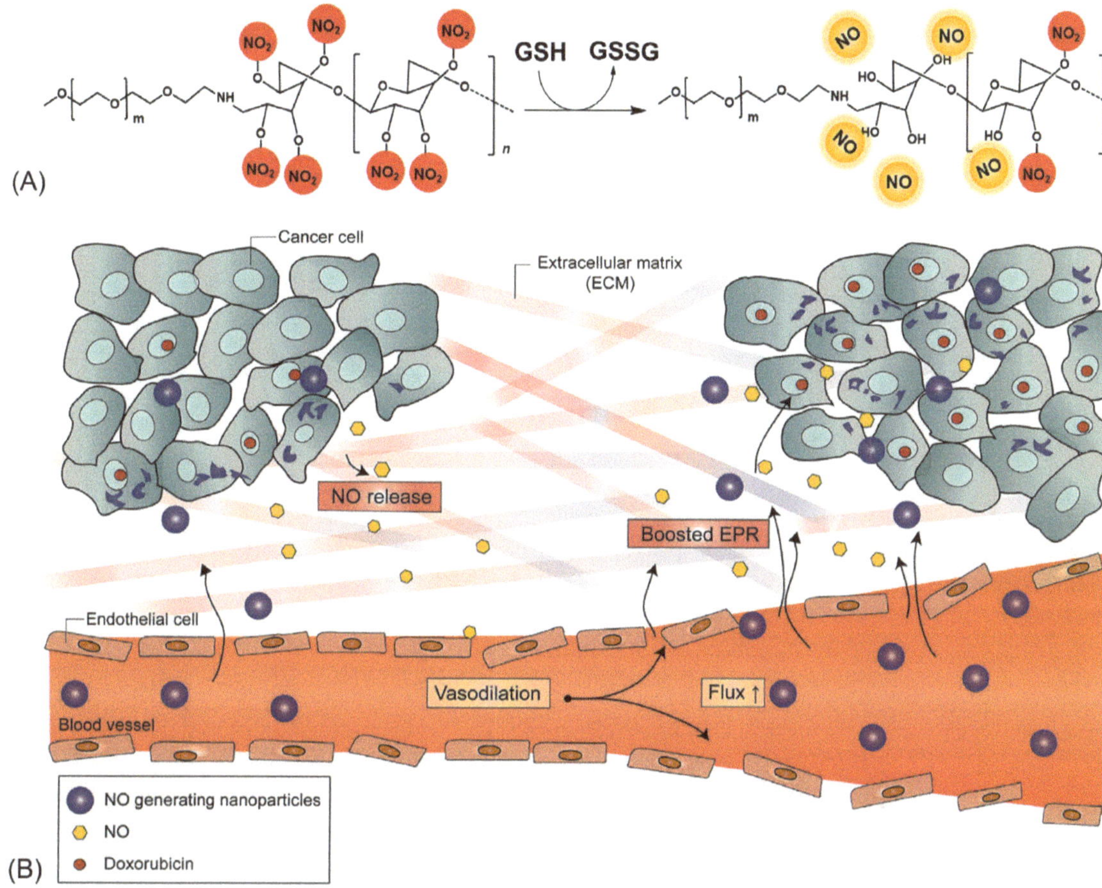

FIG. 2

(A) Endogenous glutathione (GSH), abundant in tumor cells, reduces nitrite to NO from the nanomaterial. (B) Upon the administration of DOX-NO-NPs, due to the EPR effect, the nanoparticles are taken up by tumor cells, where NO is locally generated, acting as a local vasodilator. The nanoparticles locally release DOX. Blood flow is enhanced in dilated blood vessels, boosting the EPR effect and facilitating the nanoparticle accumulation at the tumor site. *Reproduced from Deepagan VG, Ko H, Kwon S, Rao NV, Kim SK, Um W, Lee S, Min J, Lee J, Choi KY, Shin S, Suh M, Park JH. Intracellularly activatable nanovasodilators to enhance passive cancer targeting regime. Nano Lett 2018;18:2637–2644 with permission from the American Chemical Society.*

GSNO nanoparticles for 12 h prior to DOX administration improved the anticancer activity of DOX itself by 11.7%–32.8% [41].

Recently, the codelivery of NO, DOX, and adjudin (ADD) from micelles (average size of 10.0 ± 0.5 nm) to combat MDR and to promote metastasis inhibition was reported [42]. D-α-tocopherol polyethylene glycol 1000 succinate (TPGS)-NO was used as NO donor [43], yielding DOX-ADD@TPGS-NO

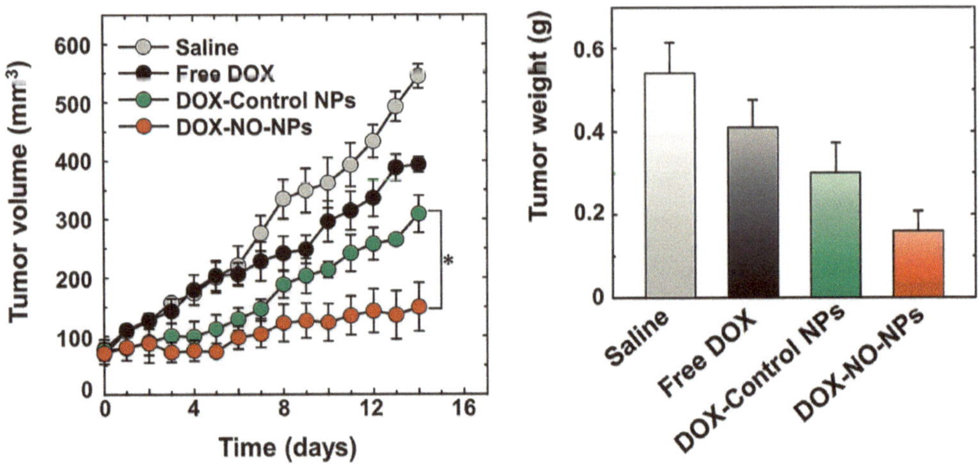

FIG. 3

Changes in tumor volume and in tumor weights in HT29 tumor-bearing mice treated with saline, free DOX, DOX-control NPs, and DOX-NO-NPs, as indicated in the figure. *Reproduced from Deepagan VG, Ko H, Kwon S, Rao NV, Kim SK, Um W, Lee S, Min J, Lee J, Choi KY, Shin S, Suh M, Park JH. Intracellularly activatable nanovasodilators to enhance passive cancer targeting regime. Nano Lett 2018;18:2637–2644 with permission from the American Chemical Society.*

micelles. Micelles have great advantages in the drug delivery system due to their high loading capacity and relatively easy preparation, which save energy and time. DOX is the main toxic agent supplemented with ADD (a male contraceptive agent able to promote mitochondrial inhibition), while TPGS-NO can inhibit metastasis and P-gp leading to mitochondrial dysfunction. Mitochondrial function can be further disrupted by ADD. It should be noted that drug efflux is a key problem in MDR and related with membrane-mediated drug transport (P-gp). Interestingly, NO can modulate P-gp; therefore, cotreatment of NO and traditional anticancer drugs might reverse cell resistance [44]. The toxicity of the nanomaterial was evaluated in drug-resistant MCF-7/ADR cells (breast cancer cells) [42]. DOX-ADD@TPGS-NO micelles were the most cytotoxic on MDR breast cancer cell lines; the authors assigned this effect to the combined effects of the NO donor and ADD on P-gp inhibition and DOX toxicity. In vivo studies were performed in the 4T1 tumor model, with in situ metastasis. The results showed a decreased resistance factor upon treatment with DOX-ADD@TPGS-NO micelles, superior circulation of the nanocarrier, and enhanced micelle accumulation at the tumor site [42].

The combination of the NO donor GSNO-containing liposomes and DOX was designed to overcome extracellular matrix and vascular endothelium barriers [45]. The liposomes were prepared by using conventional protocols,

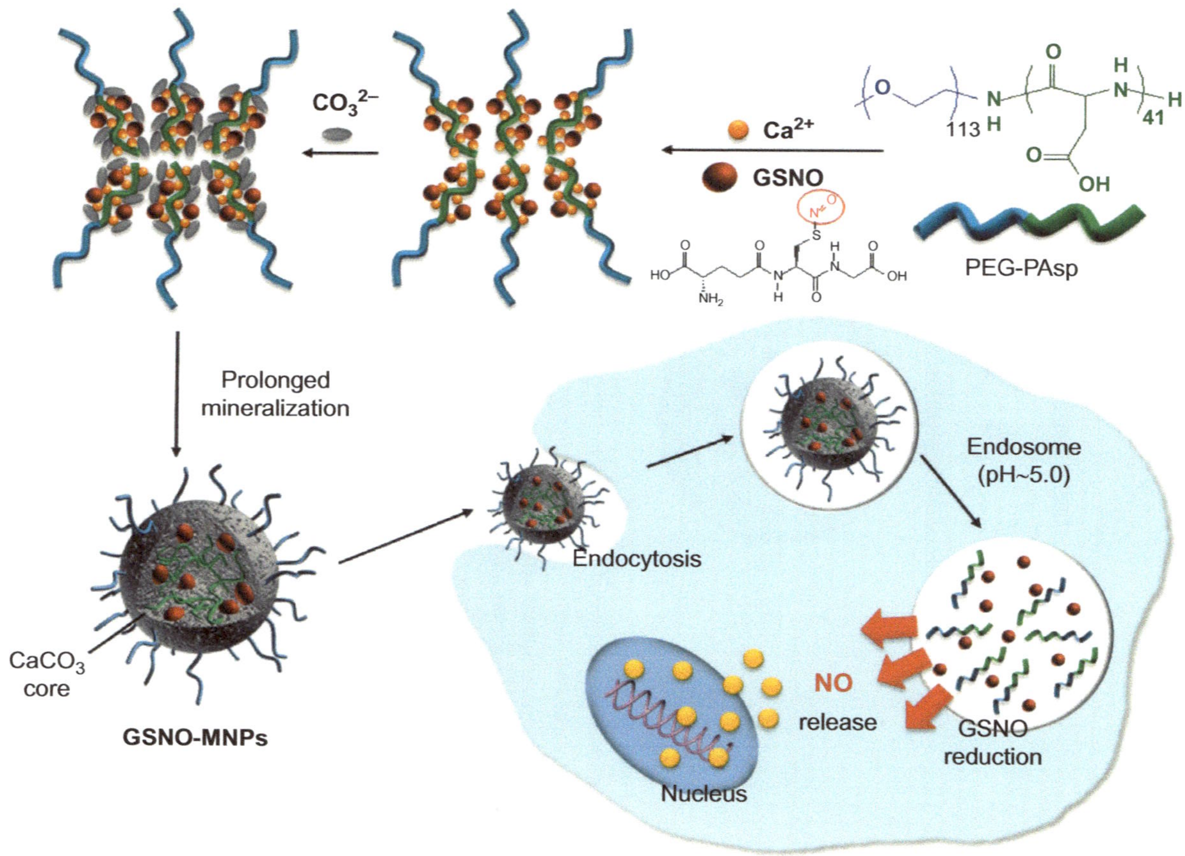

FIG. 4

Schematic representation of the preparation of GSNO loaded into calcium carbonate ($CaCO_3$)-mineralized nanoparticles obtained by anionic block copolymer (PEG-poly(L-aspartic acid))-templated mineralization and intracellular NO release trigged by acid pH. Reproduced from Ref. 41 with permission from Elsevier.

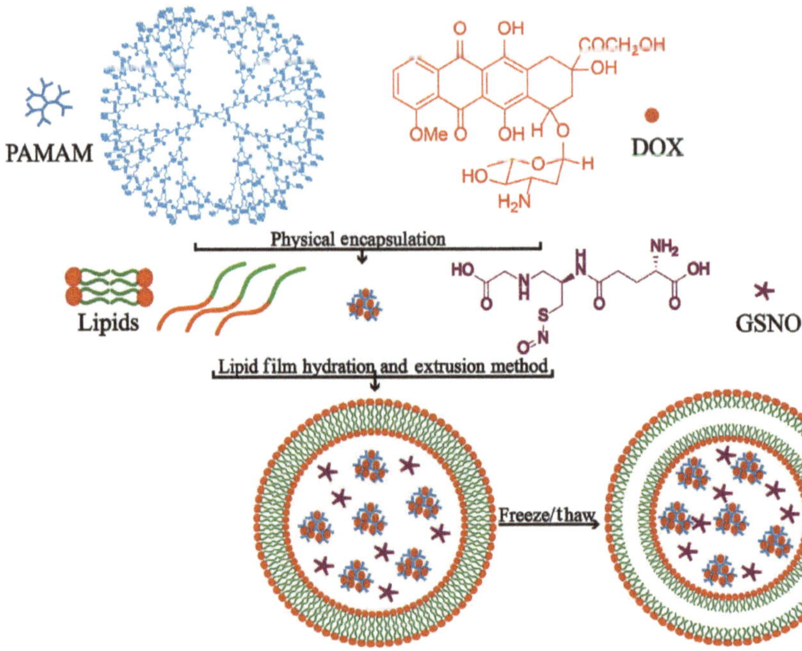

FIG. 5

Schematic representation of the preparation of ultrasound-responsive liposome containing GSNO and DOX. *Reproduced from Wang B, Zhai Y, Shi J, Zhuang L, liu W, Zhang H, Zhang H, Zhang Z. Simultaneously overcome tumor vascular endothelium and extracellular matrix barriers via a non-destructive size-controlled nanomedicine. J Control Release 2017;268:225–236 with permission from Elsevier.*

containing cholesterol, dipalmitoyl phosphatidylcholine, and polyethylene glycol, among other chemicals, in addition to polyamidoamine loaded with DOX (Fig. 5). In a separate protocol, DOX- containing polyamidoamine particles were prepared and, then, mixed with GSNO-loading liposomes; the final nanocarrier had an average size of 200 nm and a nonspherical shape. In vitro studies showed the uptake of the liposomes through endocytoses, whereas an in vivo study demonstrated that the nanocarrier reduced tumor volumes fourfold in pancreatic tumors and threefold in breast tumors, compared with control groups [45].

The codelivery of NO and DOX was also achieved by combining the NO prodrug O^2-(2,4-dinitrophenyl) 1-[4-(propargyloxycarbonyl)piperazin-1-yl]diazen-1-ium-1,2-diolate (alkynyl-JSK), an amphiphilic block copolymer [46], with a nanocarrier having a surface functionalized with poly(galactose) for tumor targeting. As expected, the authors reported a synergist effect of NO and DOX codelivery from nanoparticles to human hepatocellular carcinoma

(HepG2) cells, and the cell death mechanism might involve early and late stages of cell apoptosis [46].

Fan et al. [47] demonstrated the anticancer effect of biocompatible polymeric nanomaterial containing NO and paclitaxel for anti-MDR tumor therapy. The nanomaterial was composed of an amphiphilic copolymer of polyethylenimine bound to NO and poly(L-lactide) (PLLA) as organic aliphatic polyester. The engineered nanomaterial had a size of 137 nm, and the amount of NO release was found to be 13 μM after 20 h at 37°C. The cytotoxicity of the nanomaterial was evaluated against the MDR ovarian cancer cell line OVCAR-8/ADR (IC_{50} of 32 μg of the nanomaterial). In vivo studies showed that the nanoparticles remained at the tumor site for more than 48 h after administration. The nanoparticles might be phagocytized by monocytes/macrophages and eliminated from the body through the spleen and liver metabolisms [47].

Codelivery of NO and paclitaxel from micelles was obtained by combining with D-α-tocopherol polyethylene glycol 1000 succinate (TPGS) (Fig. 6) [48]. Nitrate (NO_3^-) was used as the NO donor upon its reduction. In vitro and in vivo studies showed the efficient internalization of the micelles into MCF-7/ADR cells, and the pharmacokinetic results suggested a low clearance of the nanocarrier assigned to the escape of the nanomaterial from the reticuloendothelial system. NO generated from the nanomaterial not only dilated the blood vessels and enhanced the EPR effect but also caused apoptosis in mice bearing an S180 tumor [48].

An interesting strategy describes the preparation of RSNO-modified hollow mesoporous titanium dioxide nanoparticles (HMTNPs) containing tirapazamine (TPZ) [49]. The nanomaterial is able to generate reactive oxygen species (ROS) upon sonodynamic therapy (SDT) administration, leading to cellular toxicity. NO release from the RSNO moiety is triggered upon the formation of ROS, which further enhanced the cellular toxicity. The nanomaterial releases TPZ that produces cytotoxic ROS. Fig. 7 shows the schematic representation of this approach: the nanomaterial reaches the tumor cells via the EPR effect. Once at the targeted site, SDT generates ROS in hypoxic tumor tissue, and TPZ is activated leading to cell toxicity. In addition, generated ROS would sensitize RSNO decomposition with free NO release. The potential side effects might be significantly reduced in the absence of hypoxic tissues and no SDT stimulation. The nanoparticles have an average size of 100 nm, with a negative zeta potential (-21.3 ± 2.6 mV), with a surface area up to 169.6 m^2/g and pore size of 3.9 nm. Under 80 s of SDT irradiation, the nanomaterial (40 μg/mL) is able to release 18.5 μM of NO. At this concentration, NO is expected to have cytotoxic effects. In vivo studies were performed with BALB/c nude mice bearing MCF-7 tumors. The authors observed a significant decrease of tumor volume. Indeed, in the control group, the average of tumor volume was found to be

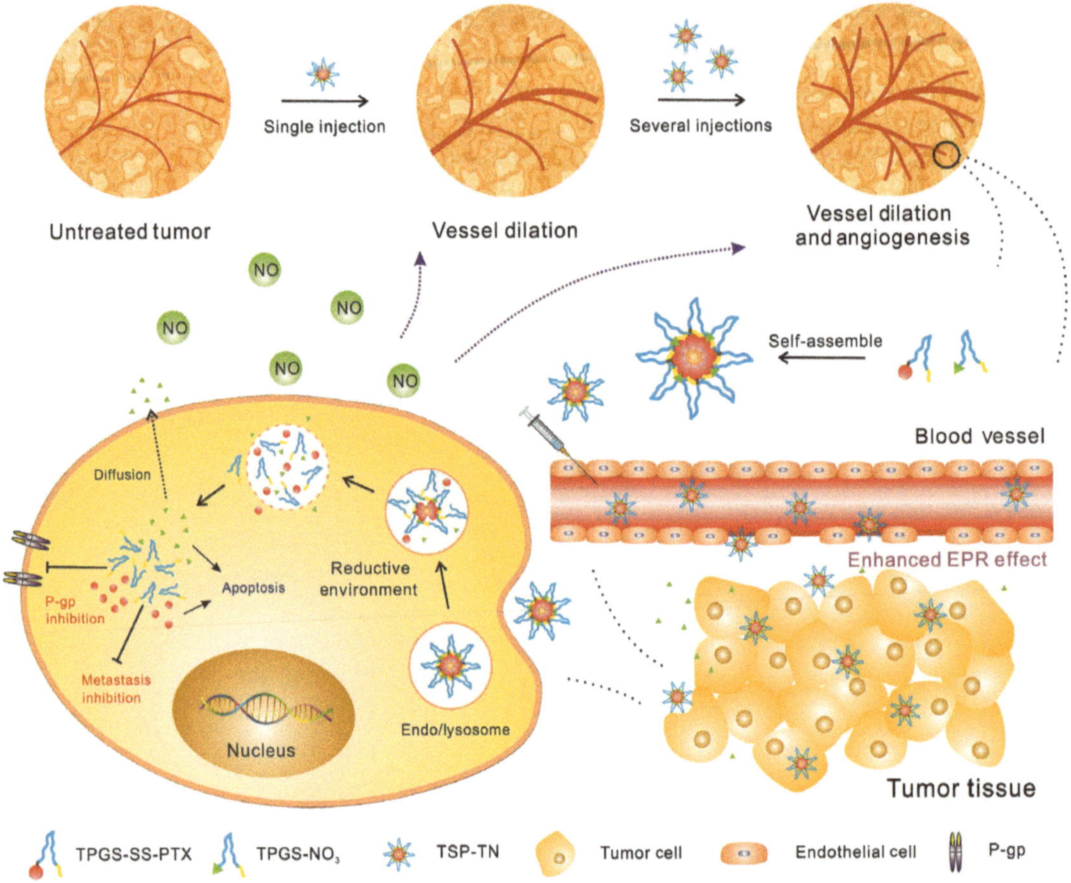

FIG. 6

Schematic representation of the enhanced tumor therapy with reversed MDR and inhibited metastasis by paclitaxel drug codelivery and in situ vascular-promoting strategy via NO release. *Reproduced from Yin M, Tan S, Bao Y, Zhang Z. Enhanced tumor therapy via drug co-delivery and in situ vascular-promoting strategy. J Control Release 2017;258:108–120 with permission from Elsevier.*

743.6 mm^3. This value was reduced to 579.1, 480.3, and 415.2 mm^3 upon tumor treatment with HMTNPs + SDT, HMTNPs-SNO + SDT, and TPZ, respectively. Interestingly, the TPZ/HMTNPs-SNO group combined with SDT irradiation reduced the tumor volume to 79.3 mm^3 at the study end point, indicating the promising application of this approach [49].

Our group described the preparation of an RSNO from *S*-nitrosomercaptossunic acid (*S*-nitroso-MSA) containing CS nanoparticles (size of 108.40 ± 0.96 nm) [50]. Initially, the cytotoxicity of *S*-nitroso-MSA containing CS nanoparticles was demonstrated in human chronic myeloid leukemia cells (K562), Lucena-1, a vincristine-resistant K562 cell line; mouse melanoma (B16F10); and HepG2

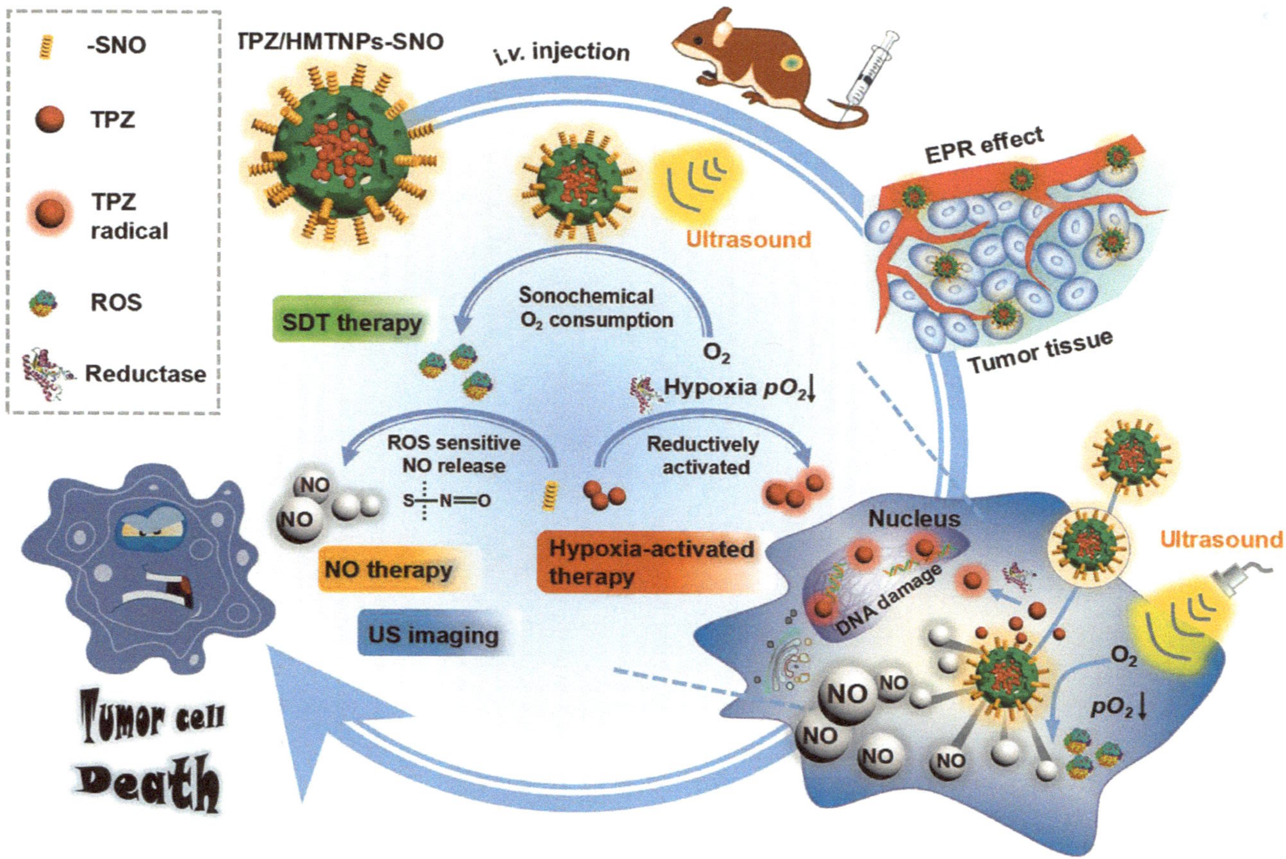

FIG. 7

Schematic representation of the ultrasound-mediated nanoplatform of TPZ/HMTNPs-SNO: the nanomaterial reaches the tumor cells via the EPR effect. Once at the targeted site, SDT generates ROS; under hypoxic tumor tissue, TPZ is activated leading to cell toxicity. In addition, generated ROS would sensitize RSNO decomposition with free NO release. *Reproduced from Feng Q, Li Y, Yang X, Zhang W, Hao Y, Zhang H, Hou L, Zhang Z. Hypoxia-specific therapeutic agents delivery nanotheranostics: a sequential strategy for ultrasound mediated on-demand tritherapies and imaging of cancer. J Control Release 2018; 275:192–200 with permission from Elsevier.*

cell lines. Interestingly, NO-releasing nanoparticles showed slight toxicity to non-tumorigenic melanocytes at toxic concentrations to melanoma B16F10 cells [50]. In a second study, the mechanisms of toxicity of S-nitroso-MSA containing CS nanoparticles to melanoma cells were further investigated [51]. The cytotoxicity was found to be caspase-dependent apoptosis, ROS formation, cysteine residue S-nitrosylation of cellular proteins, and tyrosine nitration [51].

Near-infrared (NIR) laser combined with NO-releasing nanomaterial was used to attenuate the MDR cancer by combining heat-sensitive NO donors and photothermal agents in nanomaterial [52]. The nanomaterial was composed of Fe_3O_4@polydopamine@mesoporous silica, functionalized with sulfhydryl (SH) groups, which were further nitrosated by reacting with -SH groups leading to the formation of S-NO residues (NO donor). The obtained nanoparticles had an average size of 230 nm, and the NO release was trigged upon 5 min of 808 nm laser irradiation (0.3 W/cm^2), due to the conversion of NIR photons into heat. The advantage of NIR (650–900 nm) is its deep tissue penetration. The NO generated could inhibit the expression of P-gp, which is considered a key player in drug resistance due to its ability to pump drugs out of cells. The toxicity of the NO-releasing nanomaterial was demonstrated in MCF-7/ADR cell lines and in MCF-7/ADR tumor-bearing BALB/c nude mice. Overall, the results showed the tumor growth inhibition in MDR cancer with minimum side effects under NIR irradiation [52]. Similarly, NO release from superparamagnetic iron oxide-encapsulated mesoporous silica nanoparticles (SPION@hMSN) under ultrasound (US) was reported by Jin et al. [53] (Fig. 8). Due to the superparamagnetism of the nanoparticles, the NO-releasing nanomaterial can be guided to the target site of application under the application of an external magnetic field. The toxicity of the nanomaterial was demonstrated in HeLa cells (Fig. 9) [53].

Li et al. [54] reported that NO metal coordination with iron (Fe) resulted into photolabile NO-release nanomaterials (size of 12.9 ± 2.6 nm) under light exposure. The authors coupled the NO donor to Roussin's black salt (RBS) with horse spleen apoferritin. Apoferritin is able to specifically bind through the transferrin receptor-overexpressing tumor cells. The NO release under light irradiation and cellular internalization was demonstrated in HeLa and MCF-7 cell lines [54].

RSNO-containing hollow double-layered polymer nanoparticles (HDPNs) were synthesized for tumor therapy [55]. The nanoparticle synthesis involved the preparation of the hollow by reaction with the thiol-containing molecule *N*-acetyl-D-penicillamine (NAP) thiolactone, followed by nitrosation of thiol groups yielding RSNO-containing NPs (Fig. 10, panel A). The amount of NO release from the prepared NPs was in the micromoles per milligram range, and the toxicity of the nanoparticles was evaluated in MCF-7 and HepGe cell lines and compared with normal breast cells and normal hepatocellular cells,

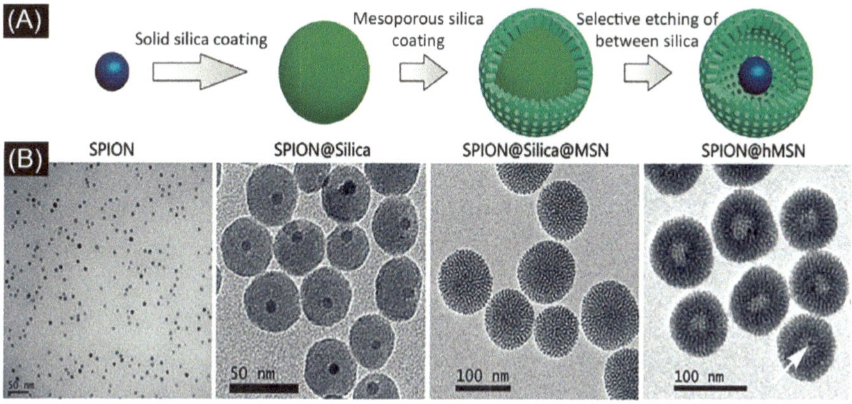

FIG. 8
Schematic representation of the preparation of SPION@hMSN carrier (A) and transmission electron microscopy images. (B) of the corresponding intermediate and final products. *Reproduced from Jin Z, Wen Y, Hu Y, Chen W, Zheng X, Guo W, Wang T, Qian Z, Sub BL, He Q. MRI-guided and ultrasound-triggered release of NO by advanced nanomedicine. Nanoscale 2017;9:3637–3645 with permission from The Royal Society of Chemistry.*

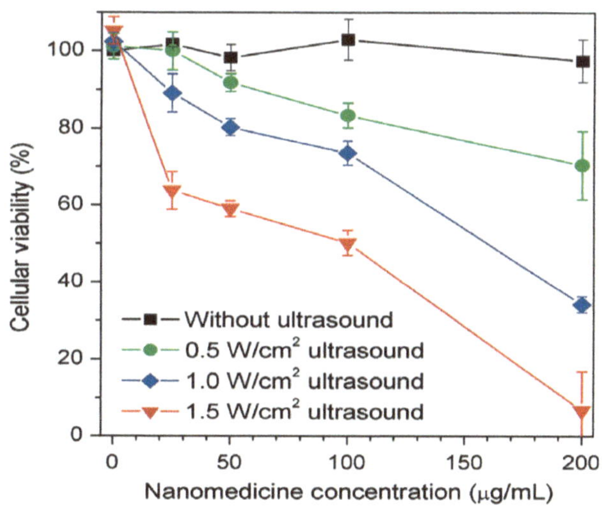

FIG. 9
Cell viability of HeLa cells treated with SPION@hMSN (at different concentrations) under ultrasound (US) for 5 min, under different powers, as indicated in the figure. *Reproduced from Jin Z, Wen Y, Hu Y, Chen W, Zheng X, Guo W, Wang T, Qian Z, Sub BL, He Q. MRI-guided and ultrasound-triggered release of NO by advanced nanomedicine. Nanoscale 2017;9:3637–3645 with permission from The Royal Society of Chemistry.*

respectively. Interestingly, the results showed a significant toxicity toward the two cancer cell lines, whereas a discrete cell killing activity was observed for normal breast and hepatocellular cells. These results suggested the tumor targeting of the NPs, which were further internalized by tumor cells, decreasing GSH levels and promoting NO release and apoptosis inside the tested cells. The NO uptake was followed by confocal microscopy, and the results demonstrated that once internalized into cancer cells, the level of NOx species (NO_2, NO_2^-, $ONOO^-$, and N_2O_3) was increased, affecting the expression and activity

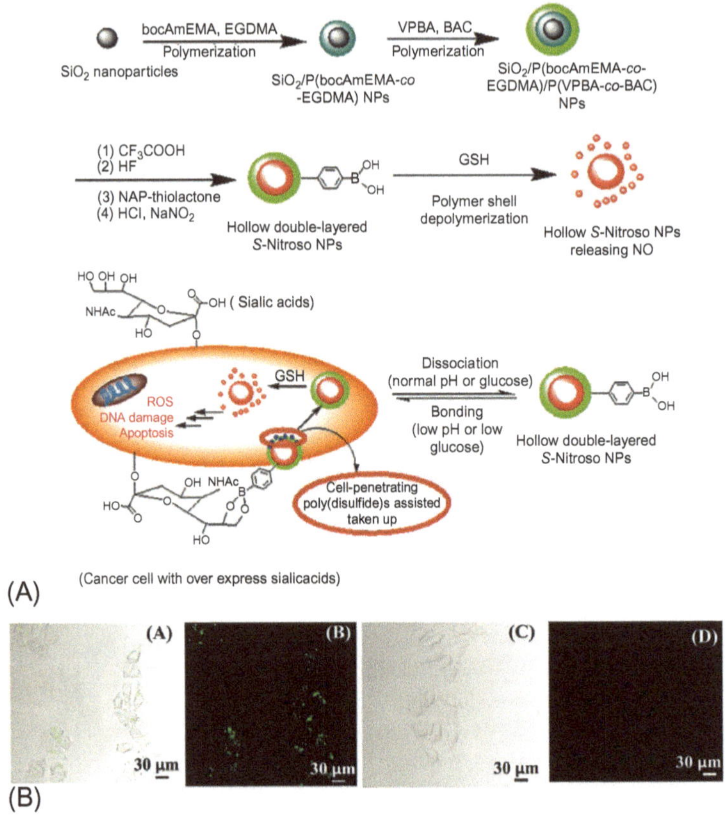

FIG. 10

Panel A: schematic representation of the preparation of RSNO-containing hollow double-layered polymer nanoparticles (HDPNs) for killing tumor cells. Panel B: NOx uptake by HepG2 cells assayed by the NO chromophore 4,5-diaminofluorescein diacetate (DAF-2DA): (A and B) Treated with 100 mg mL^{-1} S-nitroso NPs for 30 min and then washed with PBS buffer. Next, treated with 10 mM DAF-2DA and incubated with medium for 1 h. (C and D) Treated with 10 mM DAF-2DA and incubated with medium for 1 h only (control). *Reproduced from Liu T, Hu J, Ma X, Kong B, Wang J, Zhang Z, Guo DS, Yang X. Hollow double-layered polymer nanoparticles with S-nitrosothiols for tumor targeted therapy. J Mater Chem B 2017;5:7519–7528 with permission from The Royal Society of Chemistry.*

of important proteins involved in the cell cycle and apoptosis (Fig. 10, panel B) [55].

The combination of NO, NPs, and HIV protease inhibitors (HIV-PI) has been demonstrated to have potent anticancer effects [56]. HIV-PI, which was classically used to treat HIV-related Kaposi's sarcoma, has recently emerged as a new therapeutic approach for the treatment of cancer. The HIV-PI saquinavir was loaded in folic acid-conjugated NPs for cancer treatment. Maksimovic-Ivanic et al. [56] discussed the recent progress on the design and application of NO-loaded nanoparticles in combination with HIV-PIs.

NO- and cisplatin-releasing amine-modified mesoporous silica (AMS) NPs (particle size of 56 ± 7 nm) were used against non-small cell lung cancer [57]. To this end, (3-trimethoxysilylpropyl)diethylenetriamine (Si-DETA) and tetraethyl orthosilicate were prepared by a condensation reaction. Diethylenetriamine/NO (DETA) adduct was used as a NO donor in the NPs via exposition of amine groups of AMS NPs to NO gas (60 psi). Cisplatin was also absorbed, leading to cisplatin-AMS. The advantages of using mesoporous silica NPs in drug delivery are their high surface area, tunable mesoporosity, and biocompatibility. The NP released 50 μmol/g of NO after 6 days. The authors demonstrated that NO released from the NPs led to tumor cell sensitization to cisplatin, resulting in cell death. Indeed, the toxicity of NO- and cisplatin-releasing AMS NPs was significantly superior compared with control groups in A549 and H592 cell lines. In addition, platinum internalization by treated cells was measured, and the results showed that the combination of NO and cisplatin into a single particle increased the metal uptake by 32.4% in A549 and 46.6% in H596. These results clearly demonstrated the synergistic effects of NO donors and chemotherapeutic drugs in cancer cells [57].

Recently, Singh et al. [58] reported the NO released from iron oxide magnetic nanoparticles (Fe_3O_4)@gold (Au) core-shell nanoparticles for anticancer activity (average size of 50 nm). Fe_3O_4 and Au NPs were chemically synthesized in separated protocols and then mixed, followed by the functionalization of the Fe_3O_4-Au NP surface with thiazolidine-4-carboxylic acid or thioproline, which was further nitrosated leading to the formation of the NO donor *N*-nitrosothioproline. NO release from the nanomaterial can be triggered by light irradiation at different wavelengths. The authors did not comment on the formation of possible toxic products after the release of NO. The cytotoxicity of the NO-releasing Fe_3O_4-Au NPs was demonstrated in HeLa cells [58].

An interesting approach was designed by combining amine-modified carbon dots with ruthenium nitrosyl (NO donor) as a multifunction nanocarrier for lysosome targeting, coupled with photothermal therapy (NIR irradiation at 808 nm) [59]. The nanomaterial had a spherical shape and size of 7–9 nm. In vitro study in HeLa cells in the dark and under irradiation showed a significant decrease in cell viability (up to value of 27% of cell survival). It should be noted that under dark conditions, the authors reported 85% cell survival, indicating that NIR irradiation can be used to modulate NO release and, consequently, the toxicity of the nanocarrier [59].

CONCLUSIONS: PERSPECTIVES AND CHALLENGES

Along the years, considerable progress has been made regarding NO and cancer. As NO can have dual roles in cancer biology (as tumor suppressor or promoter), several papers describe the administration of high doses of NO donors (micro- to millimolar range) to tumor cells. A sustained and controlled NO release can be achieved by the combination of NO donors and nanomaterials [27–30]. This strategy is based on the killing effect of high doses of NO.

Recently, an increasing number of reports have demonstrated that low concentrations of NO (those that if administrated alone would cause tumor progression) contribute to the reversal of MDR [15, 35]. Thus, this observation has motivated and inspired researchers to combine NO donors allied with nanomaterials for the improvement of chemotherapy. As demonstrated in this chapter, NO-releasing nanomaterials have a great potential to overcome MDR, improving the response to chemotherapy. To this end, a great variety of different kinds of nanomaterials have been prepared, including polymeric, metallic, and silica-based nanocarriers, with different classes of NO donors (such as RSNOs, diazeniumdiolates, and Ru complexes). Light, waves, or pH can be used to trigger and modulate the local NO generation or the release from engineered nanomaterials, increasing the efficacy of the therapy and decreasing side effects.

Ideally, NO-releasing nanomaterials for cancer treatment should be stable during their blood circulation and generate therapeutic amounts of NO directly at the target site (tumor tissue), with minimum side effects to health tissues for the desired amount of time [38]. The end products of the nanomaterials should not cause adverse effects. In this context, there are still some challenges to be fully overcome in order to propose the clinical use of NO-releasing nanomaterials in cancer therapy. First, there is a need for the technological development that will be able to produce at room temperature a stable NO-releasing nanomaterial for commercial purposes. The most employed NO donors in biomedical applications are RSNOs and diazeniumdiolates, which can be unstable in storage and in the blood circulation. The incorporation of NO donors into nanocarriers significantly improves the stability of the NO donors; however, further effort should be performed to improve this stability for clinical settings. The full characterization of the extension of the effects of NO-releasing nanomaterial in reversing MDR effects should be further characterized. The mechanisms and pathways by which NO acts to reverse the MDR effect need to be investigated deeper, in different tumors, with different NO donors and nanomaterials, coupled with different chemotherapeutic agents. The proper range of NO concentrations required to reverse the MDR effect should be investigated. This information is crucial to develop new NO-releasing nanomaterials for cancer treatment in combination with traditional chemotherapeutic drugs. The possible interactions of the nanomaterials with the chemotherapeutic drug should be demonstrated.

Therefore, from the recent literature, it can be observed that the uses of NO-releasing nanomaterials alone or in combination with traditional chemotherapeutic drugs (such as DOX and cisplatin) have a great impact in cancer treatment and, particularly, in the reversal of MDR effects. We hope that this chapter inspires new avenues of consideration in this promising and exciting fields of research.

Acknowledgments

Support from NanoBioss (MCTI), INOMAT (CNPq/MCTI), CNPq, and FAPESP.

We would like to thank Proof-Reading-Service.com Ltd for revising the text.

References

[1] Dioguardi FS. To give or not to give? Lessons from the arginine paradox. J Nutrigenet Nutrigenom 2011;4:90–8.

[2] Seabra AB, editor. Nitric oxide donors: Novel biomedical applications and perspectives. 1st ed. London: Elsevier; 2017.

[3] Bogdan C. Nitric oxide and the immune response. Nat Immunol 2001;2:907–16.

[4] Chang CF, Diers AR, Hoog N. Cancer cell metabolism and the modulating effects of nitric oxide. Free Radic Biol Med 2015;79:324–36.

[5] Scatena R, Bottoni P, Pontoglio A, Giardina B. Pharmacological modulation of nitric oxide release: new pharmacological perspectives, potential benefits and risks. Curr Med Chem 2010;17:61–73.

[6] Stuehr DJ. Mammalian nitric oxide synthases. Biochim Biophys Acta 1999;1411:217–30.

[7] Wink DA, Vodovotz Y, Laval J, Laval F, Dewhirst MW, Mitchell JB. The multifaceted roles of nitric oxide in cancer. Carcinogenesis 1998;19:711–21.

[8] Xie K, Huang S, Dong Z, Juang SH, Gutman M, Xie QM, Nathan C, Fidleer IJ. Transfection with the inducible nitric oxide synthase gene suppresses tumorigenicity and abrogates metastasis. J Exp Med 1995;181:1333–43.

[9] Xu W, liu LZ, Loizidou M, Ahmed M, Charles IG. The role of nitric oxide in cancer. Cell Res 2002;12:311–20.

[10] Wink DA, Kasprsak KS, Maragos CM, Elespuru RK, Mirsra M, Dunams TM, Cebula TA, Koch WH, Andrews AW, Aleen JS. DNA deaminating ability and genotoxicity of nitric oxide and its progenitors. Science 1991;254:1001–3.

[11] Vannini F, Kashfia K, Nath N. The dual role of iNOS in cancer. Redox Biol 2015;6:334–43.

[12] Salimian RB, Achreja A, Nagrath D. Nitric oxide: the forgotten child of tumor metabolism. Trends Cancer 2017;3:659–72.

[13] Chang CF, Diers AR, Hogg N. Cancer cell metabolism and the modulating effects of nitric oxide. Free Rad Biol Med 2015;79:324–36.

[14] Boyd CS, Cadenas E. Nitric oxide and cell signaling pathways in mitochondrial-dependent apoptosis. J Biol Chem 2002;383:411–23.

[15] Garban HJ, Bonavida B. Nitric oxide sensitizes ovarian tumor cells to Fas-induced apoptosis. Gynecol Oncol 1999;13:257–64.

[16] Burke AJ, Sullivan FJ, Giles FJ, Glynn SA. The yin and yang of nitric oxide in cancer progression. Carcinogenesis 2013;34:503–12.

[17] Stamler JS. Redox signaling: nitrosylation and related target interactions of nitric oxide. Cell 1994;78:931–6.

[18] Iyer AK, Azad N, Wang L, Rojanasakul Y. Role of S-nitrosylation in apoptosis resistance and carcinogenesis. Nitric Oxide 2008;19:146–51.

[19] Iyer AK, Rojanasakul Y, Azad N. Nitrosothiol signaling and protein nitrosation in cell death. Nitric Oxide 2014;42:9–18.

[20] Irmler M, Thome M, Hahne M, Schneider P, Hofmann K, Steiner V, Bodmer JL, Schoroter M, Burns K, Mattamann C, Rimoldi D, French LE, Tschopp J. Inhibition of death receptor signals by cellular FLIP. Lett Nat 1997;388:190–5.

[21] Green DR, Reed JC. Mitochondria and apoptosis. Science 1998;281:1309–12.

[22] Stamler JS, Lamas S, Fang FC. Nitrosylation: the prototypic meeting review redox-based signaling mechanism. Cell 2001;106:675–83.

[23] Meji JT, Haselton CL, Hillman KL, Muralikrishnan D, Ebadi M, Yu L. Differential mechanisms of nitric oxide- and peroxynitrite-induced cell death. Mol Pharmacol 2004;66: 1043–53.

[24] Murray J, Taylor SW, Zhang B, Ghosh SS, Capaldi RA. Oxidative damage to mitochondrial complex I due to peroxynitrite: identification of reactive tyrosines by mass spectrometry. J Biol Chem 2003;278:37223–30.

[25] Ischiropoulos H. Biological tyrosine nitration: a pathophysiological function of nitric oxide and reactive oxygen species. Arch Biochem Biophys 1998;356:1–11.

[26] Macmillian-Crow LA, Crow JP, Kerby JD, Beckman JS, Thompson JA. Nitration and inactivation of manganese superoxide dismutase in chronic rejection of human renal allograft. Proc Natl Acad Sci U S A 1996;93:11853–8.

[27] Seabra AB, Durán N. Nitric oxide-releasing vehicles for biomedical applications. J Mater Chem 2010;20:1624–37.

[28] Seabra AB, Durán N. Nanoparticulated nitric oxide donors and their biomedical applications. Mini Rev Med Chem 2017;17:216–23.

[29] Seabra AB, de Lima R, Calderón M. Nitric oxide releasing nanomaterials for cancer treatment: current status and perspectives. Curr Top Med Chem 2015;15:298–308.

[30] Seabra AB, Justo GZ, Haddad PS. State of the art, challenges and perspectives in the design of nitric oxide-releasing polymeric nanomaterials for biomedical applications. Biotechnol Adv 2015;33:1370–9.

[31] http://ec.europa.eu/environment/chemicals/nanotech/faq/definition_en.htm [Accessed 1st July 2018].

[32] Baetke C, Lammers T, Kiessling F. Nanoparticles for diagnostic imaging and radiotherapy special feature: review article. Br J Radiol 2015;88:20150207.

[33] Greish K. Enhanced permeability and retention (EPR) effect for anticancer nanomedicine drug targeting. Methods Mol Biol 2010;624:25–37.

[34] Nakamura Y, Mochida A, Choyke PL, Kobayashi H. Nanodrug delivery: is the enhanced permeability and retention effect sufficient for curing cancer? Bioconjug Chem 2016;27:225–2238.

[35] Bonavida B, Baritaki S, Huerta-Yepez S, Vega MI, Chatterjee D, Yeung K. Novel therapeutic applications of nitric oxide donors in cancer: roles in chemo- and immunosensitization to apoptosis and inhibition of metastases. Nitric Oxide 2008;19:152–7.

[36] Carpenter AW, Schoenfisch MH. Nitric oxide release: part II. Therapeutic applications. Chem Soc Rev 2012;41:3742–52.

[37] Seabra AB, Durán N. Nitric oxide donors for prostate and bladder cancers: current state and challenges. Eur J Pharmacol 2018;826:158–68.

[38] Kim J, Yung BC, Kim WJ, Kim XCJ, Yung BC, Kim WJ, Chen X. Combination of nitric oxide and drug delivery systems: tools for overcoming drug resistance in chemotherapy. J Control Release 2017;263:223–30.

[39] Tan L, Huang R, Li X, Liu S, Shen YM. Controllable release of nitric oxide and doxorubicin from engineered nanospheres for synergistic tumor therapy. Acta Biomater 2017;57:498–510.

[40] Deepagan VG, Ko H, Kwon S, Rao NV, Kim SK, Um W, Lee S, Min J, Lee J, Choi KY, Shin S, Suh M, Park JH. Intracellularly activatable nanovasodilators to enhance passive cancer targeting regime. Nano Lett 2018;18:2637–44.

[41] Lee HJ, Kim DA, Park DJ, Choi GH, DNm Y, Heo JS, Le SC. pH-responsive mineralized nanoparticles as stable nanocarriers for intracellular nitric oxide delivery. Colloids Surf B Biointerfaces 2016;146:1–8.

[42] Qi Y, Qin X, Yang C, Wu T, Qiao Q, Song Q, Zhang Z. Micelle system based on molecular economy principle for overcoming multidrug resistance and inhibiting metastasis. Mol Pharm 2018;15:1005–16.

[43] Li X, Gao C, Wu Y, Cheng CY, Xia W, Zhang Z. Combination delivery of adjudin and doxorubicin via integrating drug conjugation and nanocarrier approaches for the treatment of drug-resistant cancer cells. J Mater Chem B 2015;3:1556–64.

[44] Zhang X, Guo Z, Liu J, Tian G, Chen K, Yu S, Gu Z. Near infrared light triggered nitric oxide releasing platform based on upconversion nanoparticles for synergistic therapy of cancer stem-like cells. Sci Bull 2017;62:985–96.

[45] Wang B, Zhai Y, Shi J, Zhuang L, liu W, Zhang H, Zhang H, Zhang Z. Simultaneously overcome tumor vascular endothelium and extracellular matrix barriers via a non-destructive size-controlled nanomedicine. J Control Release 2017;268:225–36.

[46] Zhang J, Song H, Ji S, Wang X, Huang P, Zhang C, Wang W, Kong D. NO prodrug-conjugated, self-assembled, pH-responsive and galactose receptor targeted nanoparticles for co-delivery of nitric oxide and doxorubicin. Nanoscale 2018;10:4179–88.

[47] Fan J, Song J, liu Y, Yu G, Ma Y, Deng Y, He N, Zhang F. Synthesis of biocompatible polymeric nanomaterial dually loaded with paclitaxel and nitric oxide for anti-MDR cancer therapy. RSC Adv 2016;6:105871–7.

[48] Yin M, Tan S, Bao Y, Zhang Z. Enhanced tumor therapy via drug co-delivery and *in situ* vascular-promoting strategy. J Control Release 2017;258:108–20.

[49] Feng Q, Li Y, Yang X, Zhang W, Hao Y, Zhang H, Hou L, Zhang Z. Hypoxia-specific therapeutic agents delivery nanotheranostics: a sequential strategy for ultrasound mediated on-demand tritherapies and imaging of cancer. J Control Release 2018;275:192–200.

[50] Pelegrino MT, Silva LC, Watashi CM, Haddad PS, Rodrigues T, Seabra AB. Nitric oxide-releasing nanoparticles: synthesis, characterization, and cytotoxicity to tumorigenic cells. J Nanopart Res 2017;19:57.

[51] Ferraz LS, Watashi CM, Colturato-Kido C, Pelegrino MT, Paredes-Gamero EJ, Weller RB, Seabra AB, Rodrigues T. Antitumor potential of S-nitrosothiol-containing polymeric nanoparticles against melanoma. Mol Pharm 2018;15:1160–8.

[52] Guo R, Tian Y, Wang Y, Yang W. Near-infrared laser-triggered nitric oxide nanogenerators for the reversal of multidrug resistance in cancer. Adv Funct Mater 2017;27:1606398.

[53] Jin Z, Wen Y, Hu Y, Chen W, Zheng X, Guo W, Wang T, Qian Z, Sub BL, He Q. MRI-guided and ultrasound-triggered release of NO by advanced nanomedicine. Nanoscale 2017;9:3637–45.

[54] Li X, Zhang Y, Sun J, Chen W, Wang X, Shao F, Zhu Y, Feng F, Sun Y. Protein nanocage-based photo-controlled nitric oxide releasing platform. ACS Appl Mater Interfaces 2017;9:19519–24.

[55] Liu T, Hu J, Ma X, Kong B, Wang J, Zhang Z, Guo DS, Yang X. Hollow double-layered polymer nanoparticles with S-nitrosothiols for tumor targeted therapy. J Mater Chem B 2017;5:7519–28.

[56] Maksimovic-Ivanic D, Fagone P, McCubrey J, Bendtzen K, Mijatovic S, Nicoletti F. HIV-protease inhibitors for the treatment of cancer: repositioning HIV protease inhibitors while developing more potent NO-hybridized derivatives? Int J Cancer 2017;140:1713–26.

[57] Munaweera I, Shi Y, Koneru B, Patel A, Dang MH, Di Pasqua AJ, Balkus Jr. KJ. Nitric oxide- and cisplatin-releasing silica nanoparticles for use against non-small cell lung cancer. J Inorg Biochem 2015;153:23–31.

[58] Sing N, Patel K, Sahoo SK, Kumar R. Human nitric oxide biomarker as potential NO donor in conjunction with superparamagnetic iron oxide @ gold core shell nanoparticles for cancer therapeutics. Colloids Surf B Biointerfaces 2018;163:246–56.

[59] Xiang HJ, Guo M, An L, Yang SP, Zhang QL, Liu JG. A multifunctional nanoplatform for lysosome targeted delivery of nitric oxide and photothermal therapy under 808 nm near-infrared light. J Mater Chem B 2016;4:4667–74.

CHAPTER 2

Reversal of Multiple Cancer Oncogenic Pleiotropic Properties by NO-Modulating Therapies

Stavroula Baritaki
Division of Surgical Oncology, School of Medicine, University of Crete, Crete, Greece

ABSTRACT

Over the last few decades, nitric oxide (NO) has emerged as a molecule of interest in cancer cell biology. There are several lines of experimental evidence showing that NO, produced by cancer cells or released by surrounding cells in the tumor milieu, contributes critically to various functions in cancer cells and the tumor microenvironment. These functions, among others, include carcinogenesis, tumor cell viability, growth and invasiveness, maintenance of cancer stem cell properties, angiogenesis, immunoediting of immune signals, and tumor resistance to conventional chemo-/immunotherapeutics. NO may affect, either positively or negatively, the above processes in various cancer types, thus creating a considerable controversy and confusion in the understanding of the role of NO in cancer biology. In the context of the contrasting nature of NO in cancer, recent findings have clarified this controversy, namely, that NO exerts pro- or antineoplastic properties in the cancer milieu depending on the NO concentration and the location and timing of its release.

Several studies have shown that NO, at high doses, acts as an antineoplastic agent, when used either as a single agent or in combination with other antineoplastic compounds. These findings demonstrate the importance of NO modulation in the response to cancer therapeutics. We and others have reported that compounds able to release high concentrations of NO, such as NO donors, are able, among others, to overcome tumor cell resistance to immune-mediated cytotoxicity and conventional chemotherapy and to inhibit the initial metastatic steps of tumor cells, known as the "oncogenic epithelial-to-mesenchymal transition." NO mediates the above antineoplastic properties via the modulation of signaling pathways and feedback loops comprised of multiple interactions among gene products involved in the regulation of cellular apoptosis and tumor invasive and migratory properties. We will present findings on (1) the range of antineoplastic activities of NO in cancer cells and the tumor

Therapeutic Application of Nitric Oxide in Cancer and Inflammatory Disorders. https://doi.org/10.1016/B978-0-12-816545-4.00002-5

microenvironment, (2) the various molecular mechanisms by which NO mediates its antineoplastic activities, and (3) novel applications of NO in cancer treatment and prevention.

ABBREVIATIONS

4-HNE	4-hydroxy-2-nonenal
5-FU	fluorouracil
ATP	adenosine triphosphate
Bcl2	B-cell lymphoma 2
Bim	BCL-2-interacting mediator of cell death
BRCA1	breast cancer susceptibility gene 1
CDDP	cisplatin
cGMP	3′, 5′-cyclic monophosphate
CTL	cytotoxic T lymphocytes
DETA/NO	diethylenetriamine/nitric oxide
EPR	enhanced permeability and retention
ERK	extracellular signal-regulated kinase
GAPDH	glyceraldehyde 3-phosphate dehydrogenase
GIT-27NO	NO-donating compound (S,R)-3-phenyl-4,5-dihydro-5-isoxazole acetic acid
GluR6	ionotropic glutamate receptor 6
GSH	glutathione
GSNO	*S*-nitrosoglutathione
GST	glutathione *S*-transferase
H4K12	acetylation of histone 4 at lysine 12
HIF-1	hypoxia-inducing factor-1
HUVEC	human umbilical vein endothelial cell
IFN-γ	interferon-γ
IL-1β	interleukin-1β
JS-K	4-[2-(2,4-dinitrofinoxy)-1-oxidodyazinyl]-1-piperazine carboxylate acid, ethyl ester
L-NAME	L-*N*-nitro-L-arginine methyl ester
L-NMMA	nitric oxide synthase inhibitor NG-monomethyl-L-arginine
LPS	lipopolysaccharides
MEK	mitogen-activated protein kinase
MKP-1	mitogen-activated protein kinase (MAPK) phosphatase 1
MMP-9	metaloproteinase-9
NADPH	nicotinamide adenine dinucleotide phosphate hydrogen
NF-κB	nuclear factor kappa-light-chain-enhancer of activated B cells
NK	natural killer

NO-NSAID	nitric oxide nonsteroidal antiinflammatory drug
PKC	protein kinase C
PTEN	phosphatase and tensin homolog
Sag-NO	sodium stibogluconate-nitric oxide
SMAC	second mitochondria-derived activator of caspase protein
SNP	sodium nitroprusside
SOD	superoxide dismutase
TGF-β	transforming growth factor-β
TNF-α	tumor necrosis factor-α
Oct4	octamer-binding transcription factor 4

Conflict of Interest

No potential conflicts of interest were disclosed.

INTRODUCTION

Nitric oxide (NO) is a free radical gas with multiple effects on biological systems. In the human body, NO mainly participates in the regulation of vasodilatation, neurotransmission, smooth muscle relaxation, and the immunologic functions against pathogens [1, 2]. Over the last few decades, the impact of NO in human diseases, including metabolic syndromes and cancer, has gained increasing interest, and for many, NO is being considered a potential target for therapeutic implications.

NO is endogenously produced by oxidation of L-arginine that is mediated by the action of NO synthases (NOs) in the presence of the cosubstrates NADPH and O_2 [3]. There have been reported three isoforms of NOs with catalytic activities on NO synthesis, namely, endothelial NOs (eNOs), neuronal NOs (nNOs), and inducible NOs (iNOs or NOS2). Both eNOs and nNOs, known as constitutive NOs, produce low and short-lasting NO concentrations, while iNOs generate greater and long-lasting NO amounts [4]. Although most of the main biological functions of NO are cGMP-dependent, NO may also react with O_2 or superoxide (O_2^-), via a cGMP-independent pathway, leading to generation of reactive nitrogen species (RNS), reactive nitrosative and oxidative oxygen species, peroxynitrite, and dinitrogen trioxide (N_2O_3) [5, 6]. NO is further known to induce nonenzymatic protein modifications via nitration or *S*-nitrosylation [5, 7, 8]. These end products may affect multiple biophysiological functions in the body, causing disturbances in DNA integrity, cell signaling, and normal tissue homeostasis, thus leading to initiation or worsening of various diseases, including cancer [9–11]. Mechanisms by which NO and its derivatives may induce carcinogenesis and cancer progression include the induction of DNA damage, suppression of DNA repair enzymes, posttranslational modification of proteins, enhancement of cell proliferation, angiogenesis, metastasis, and inhibition of apoptosis and antitumor immunity [12–14].

Contrarily to these proneoplastic properties of NO, almost three decades ago, the first *in vitro* evidence of LPS and IFN-γ-induced macrophage-derived nitrite and nitrate cytotoxic activities against bacteria and tumor cells was reported [15]. The dichotomous effects of NO on cancer arise from its ability to regulate "conditionally" various cancer-related events including tumor growth, migration, invasion, survival, angiogenesis, and metastasis [12]. In the era of this controversy, recent studies highlight the role of NO as a pro- or antineoplastic mediator to be largely depended by (1) the NO concentration, (2) the duration of cell exposure to NO and the NO flux, (3) the cellular redox status (oxidant/antioxidant equilibrium), (4) the tissue microenvironment, and (5) the target cell type (normal or malignant) [12, 16, 17]. Ridnour et al. first reported a threshold of NO concentration (200 nM) above which NO had anticancer properties *in vitro*, while lower NO concentrations (<200 nM) favored mainly proneoplastic functions [3, 16, 18–20].

The cell NO-producing capacity is mainly regulated by the expression levels of NOs, particularly iNOs, given its potential of generating high NO amounts. Variations in iNOS expression and tissue distribution have been monitored between normal and cancer cells [21, 22]. Many cancer types show elevated iNOs transcriptional and translational activities, which are often considered indicators of tumor aggressiveness and poor prognosis, through the activation of genotoxic events [20, 23–25]. An additional important factor to consider when assessing the rate of NO synthesis by NOs and its biological implications is the tissue O_2 availability [22]. O_2 may be used by NOs as a substrate for NO production (O_2-dependent enzymatic synthesis) and as a stabilizer of NO steady-state concentration (NOss), thus influencing NO availability and NO-inducing cell signaling [22, 26]. Although the local O_2 concentration appears to influence NOss levels and metabolism in an unpredictable manner, it is generally observed that the local O_2 levels affect positively iNOs expression, which is paradoxically increased under hypoxia [22, 26, 27]. However, the O_2-triggered iNOs induction is not necessarily translated into increases in NO production, since hypoxic tumors most likely do not have abundance of substrates and cofactors needed for the enzymatic NO synthesis [26, 27]. This notion suggests that hypoxic tumors may be unable to synthesize substantial amounts of NO to exert an antitumor effect [22].

The "conditional" and distinct biological effects of NO in cancer biology may be explained by *in vitro* data demonstrating that most of the protein modifications that take place during malignant cell transformation, such as stabilization and phosphorylation of survival proteins, are mediated at specific NO concentration range (above the "safe zone") [12, 22, 28]. In addition, it has been clearly shown that the completion and the preservation of these NO-induced protein modifications are dramatically affected by the duration of cell exposure to NO and the type of the targeted protein [6, 28]. Although it remains elusive whether externally administrated NO in *in vivo* settings maintains this concentration-dependent functional plasticity shown *in vitro*, it is now well accepted that the controllable and predictable application of NO in cancer biology in the form of NO-releasing agents may be of paramount therapeutic significance.

On this context, this review summarizes the current knowledge on the range of antineoplastic activities of NO in the cancer cells and the tumor microenvironment and the underlying molecular mechanisms. Special emphasis is given on NO-mediated antitumorigenic effects on tumor initiation, growth and apoptosis, tumor cell metastasis and stemness, tumor cell resistance to conventional chemotherapy and immune-mediated cytotoxicity, and tumor angiogenesis and immunoediting of immune signals. Furthermore, we present an overview of the status of current and potential NO-based applications in cancer therapeutics.

ANTINEOPLASTIC ACTIVITIES OF NO IN CANCER CELLS AND THE UNDERLYING MOLECULAR MECHANISMS

NO and Tumor Initiation, Growth, and Apoptosis

The bimodal effects of NO on tumor initiation, proliferation, and apoptosis appear to be largely dictated by the concentration of NO involved and a number of other important factors including the cellular redox status and the tumor microenvironment. In contrast to numerous examples of NO-stimulating effects on carcinogenesis and tumor growth, experimental strategies leading to high intratumoral NO generation have been shown to exert antitumor properties, as it relates to the inhibition of tumor initiation and/or promotion of cellular senescence and apoptosis [12].

Tumor-mediated NO-specific cytostatic and cytotoxic effects have been reported for many NO-releasing agents *in vitro* and *in vivo* in a concentration-dependent manner [12, 29–33]. The NO donor diethylenetriamine nitric oxide adduct (DETA/NO) inhibited the growth of leukemia and pheochromocytoma cell lines only in concentrations ranging from 250 to 1000 μM and from 150 to 250 μM, respectively [34, 35]. Similarly, furoxan-based NO-releasing derivatives of glycyrrhetinic acid demonstrating a cytotoxic effect on human hepatocellular carcinoma (HCC) cells, when used at concentrations of 0.25–1.10 μM [36]. *In vivo*, the NO donor GIT-27NO inhibited tumor growth in a human prostate cancer xenograft mouse model in a concentration-dependent manner, while the Sag-NO donor inhibited prostate cancer xenotransplants [37, 38]. The underlying molecular mechanisms of the antiproliferative and cytostatic actions of NO in many tumor types include suppression of DNA synthesis and cell cycle arrest through the salvage pathway, suppression of cellular respiration, and shift in the iron metabolism [3, 8, 11, 12, 22, 39]. Another mechanism that has been described for the NO-mediated inhibition of human neuroblastoma cell proliferation is the upregulation of the Cip1/Waf1 pathway, known to be implicated in the negative control of the cell cycle [40].

Most of the physiological and pathophysiological functions of NO are mainly mediated through the cGMP signaling pathway. Soluble guanylyl cyclase (sGC) is the major receptor for NO that catalyzes the cGMP synthesis. NO binding to β1 subunit of the predominant α1β1 heterodimer isoform of sGC enzyme leads to sGC activity and cGMP production [41]. Although the role of NO/cGMP route in cancer biology has not been clearly elucidated, recent studies examining the NO/sGC/cGMP signaling molecules in human glioma tissues and cell lines found significant reduction of both α1 and β1 subunits of sGC, at mRNA and protein levels, compared with normal controls [21, 42, 43]. While this reduction was associated with an aggressive clinical course of glioma, a genetic restoration of sGC expression was able to inhibit glioma growth both *in vitro* and *in vivo* [41]. These studies further showed that the sGCβ1 subunit by itself may exert significant inhibitory effects on proliferation and colony formation of human glioma cells [41]. Nevertheless, the above findings reveal that disturbances not only in NO availability but also in molecules participating in NO signaling cascades may have a critical impact on cancer cell growth.

Long-lasting endogenous NO overproduction or extracellular administration of high NO doses has been further associated with proapoptotic effects on a plethora of cancer types via different mechanisms [12, 44]. The reported mechanisms include the induction of MKP-1 and survivin downregulation, activation of protein kinase G, direct membrane damage, activation of JNK signaling, activation of caspases, release of cytochrome C and SMAC, and alterations in the expression of apoptosis-associated proteins, such as the Bcl2 family member proapoptotic protein Bim [3, 8, 10, 12, 45–52]. Contrarily, survival signals have been reported to be promoted by low NO fluxes (30–200 nM) via the phosphorylation of Akt and stabilization of HIF-1a in cancer cells [28]. Interestingly, we have shown that high concentrations (>500 μM) of DETA-NONOate sensitize prostate carcinoma and non-Hodgkin's lymphoma (NHL) cell lines to apoptosis by eliminating the DNA binding activities of the transcription factor NF-κB and its downstream target Yin Yang 1 (YY1) [53–58]. Both NF-κB and YY1 play pivotal roles in preventing tumor apoptosis and promoting cancer cell survival and growth in various malignancies [59]. Endogenous or exogenous NO may further mediate its proapoptotic effects via *S*-nitrosylation of Fas, Bcl2, GAPDH, and GLuR6 in several cancer types [7, 12, 47, 48].

Apoptosis may also be induced by NO-mediated activation of p53 phosphorylation [16, 50]. Tumor cell exposure to exogenously administrated NO has been correlated with increased p53-dependent tumor cell cytotoxicity in a dose- and time-dependent manner [8, 60, 61]. However, the outcome of NO-p53 interactions is not always predictable and appears to be environment-specific [62]. While NO-induced DNA damage may result in p53 phosphorylation and cell growth arrest [3, 18], p53 activation may in turn govern NO production by transcriptionally transrepressing iNOs in mouse and human cells [63, 64], thus protecting cells from apoptosis [65]. Concomitantly, high iNOs levels have been positively correlated with p53 gene point mutations (mainly GC to AT), which may contribute to the loss of its repressor activity [8, 22, 66].

NO may affect the local O_2 gradient through a negative feedback mechanism. High NO concentrations have been shown to inhibit mitochondrial respiration, leading to cell apoptosis [67–69]. The underlying mechanism involves the inhibition of ribonucleotide reductase and cellular ATP generation by direct elimination of the mitochondrial electron transport enzymes aconitase and GAPDH [12, 44]. In other cases, the rapid reaction of NO with O_2^- has been reported to have antioxidant properties by preventing the oxidation of micromolecules and reduction of metals that can lead to DNA damage and potentially to carcinogenesis [20, 22].

Noteworthy, NO has also been implicated in the regulation of inflammation-induced carcinogenesis. To evaluate whether NO and p53 cooperatively regulate tumorigenesis, under inflammatory conditions, which are known to influence positive cancer development, Hussain et al. used a p53/iNOs single or double knockout mouse model [70]. Various infectious agents result in iNOs (NOS2) induction by promoting the secretion of proinflammatory cytokines such as TNF-α or IL-1B, leading to NO production, which in turn participates in the inflammatory response [71]. Mice carrying the intact iNOs locus and knocked out of the p53 locus (p53 −/−NOS2 +/+) developed significantly less lymphomas and sarcomas than mice deficient to both loci or the NOS2 locus. Comparison of the spleens and thymuses of the above groups before tumor development showed that the of p53 −/−NOS2 +/+ mice had decreased cell proliferation and a higher apoptotic index, which were mainly attributed to increased expression of death ligands FasL (CD95L) and TRAIL and an elevated cell cycle checkpoint protein $p21^{WAF1}$ [70]. Concomitantly, in an earlier study, genetic ablation of iNOs in Apc(Min/+) mice led to a higher number of small intestinal tumors compared with wild-type counterparts [72]. *In vivo* models that overexpress iNOs have further corroborated the antitumorigenic properties of NO in high concentrations [5, 8, 61]. Overall, the above studies provide substantial evidence that iNOs-mediated NO production may suppress tumorigenesis possibly through the activation of cytotoxic and cytostatic mechanisms.

Inflammation-induced carcinogenesis may also be influenced by NO-mediated epigenetic regulation. In a rat model, induction of colonic inflammation by 2,4,6-trinitrobenzene sulfonic led to intercellular adhesion molecule-1 expression by the translocation of NF-κB to the nucleus. The NO donor GSNO was able to block NF-κB DNA binding activity via transcriptional downregulation of global histone deacetylase 3 and decreased DNA interaction at the intercellular adhesion molecule-1 promoter containing the binding motifs of NF-κB and suppression of H4K12 acetylation, thus suppressing inflammation [73]. More recently, additional effects of NO at epigenetic level were proposed. According to a recent report, NO oxidation of critical cysteine residues can inhibit histone deacetylases (HDACs) and DNA methyltransferases (DNMTs), which results in multiple biological effects, such as the promotion of p53 expression [74, 75].

Interestingly, other studies from the 1990s found that high concentrations of NO could prevent tumor cell survival by obstructing DNA repair through direct chemical alterations of DNA by reactive nitrogen oxide species (RNOs) (an inducer of nitrosative and oxidative stress) or indirect inhibition of DNA repair enzymes [76–80]. The last notion has come back to the spot by recent studies showing conflicting roles for NO in contextually activating pathways involving the DNA repair protein BRCA1 or downregulating BRCA1 expression [40, 76, 81, 82].

Overall, from the available literature, one can conclude that the effects of NO and its derivatives on cancer cell signaling are critically altered by factors such as modulating tumor establishment, survival and progression such as oxygen content, metabolism, and inflammatory response. As such, the efficacy of NO to inhibit the above processes largely depends on the total NO levels achieved in the tumor bed and the way they have been used in terms of the cellular redox potential (oxidant/antioxidant equilibrium), rather than the individual cell NO-producing ability [8]. In fact, once malignant cell transformation has been initiated, it is more likely that the hypoxic tumor cells cannot synthesize adequate amounts of NO to cause autologous cell cytotoxicity and therefore other sources of NO provided by the tumor microenvironment or externally are needed for potentiating any cytotoxic and/or cytostatic NO effects.

NO and Tumor Metastasis

Oncogenic epithelial-to-mesenchymal transition (EMT) is considered to be a hallmark for the initiation of cancer cell metastasis. During EMT, primary tumor cells lose their epithelial cell-cell adhesive features and acquire migratory and invasive properties characteristic of a mesenchymal cell phenotype. As such, EMT enables intravasation of tumor cells into the bloodstream and lymph nodes (micrometastasis) through breaks of the basement membrane. These circulating cancer cells may further migrate to organs far from the primary tumor site (distant metastases), where they undergo mesenchymal-to-epithelial transition (MET) for clonal outgrowth and expansion. At the molecular level, tumor cells undergoing EMT are characterized by the inhibition of epithelial-like gene expression such as E-cadherin and cytokeratin 18 and the elevation of mesenchymal biomarkers including Snail, Twist, vimentin, and N-cadherin [83].

Both exogenous (NO donor and NOS transfection) and endogenous NO has been reported to influence oncogenic EMT in a concentration-dependent context, thus making the role of NO in tumor metastasis contradictory. Its inhibitory effects are mainly derived from the induction of DNA damage and apoptosis in tumor cells, thus suppressing metastasis [13]. We and others have shown that externally administrated high doses of NO by NO donors are able to inhibit oncogenic EMT and suppress metastasis *in vitro* and *in vivo* via the modulation of the epithelial/mesenchymal gene expression ratio in cancer cells [12, 56, 84]. We have reported that treatment of the metastatic prostate cancer cell lines PC3 and DU145 with 1000 μM DETA/NO inhibited cell migration and invasion through the modulation of the NF-κB/Snail/RKIP feedback loop [85]. NO, at the concentration used, was able to decrease directly the DNA binding activities of the transcription factors NF-κB and Snail through *S*-nitrosylation, a covalent modification of a protein cysteine thiol by a NO moiety. Snail, a downstream transcriptional target of NF-κB, promotes the expression of mesenchymal markers while directly repressing the transcription of the metastasis suppressor genes Raf-1 kinase inhibitory protein (RKIP) and E-cadherin [86–91]. NO-mediated RKIP induction in turn is able to cause further inhibition of NF-κB activation and, therefore, Snail expression through physical association with the upstream kinases NIK and TAK, which prevents IkBa phosphorylation and degradation [92]. Therefore, the modulation of NF-κB/Snail/RKIP circuitry by subtoxic NO concentrations resulted in the upregulation of epithelial cell markers (e.g., E-cadherin and cytokeratin-18) and downregulation of their mesenchymal counterparts (e.g., fibronectin, vimentin, and N-cadherin), thus preventing cell growth, migration, and invasion *in vitro* and *in vivo* [59, 84, 85]. Similarly, in a murine model of UV-induced skin tumors, high doses of NO-exisulind reduced EMT via the inhibition of the mesenchymal markers fibronectin, N-cadherin, SNAIL, Slug, and Twist and increase of E-cadherin [93].

In addition, our preliminary findings have shown that NO-mediated inhibition of the transcription factor YY1, derived either directly by *S*-nitrosylation [53] or indirectly by NO-driven S-nitrosylation of its upstream transcriptional activator NF-κB [85, 94, 95], may further repress the expression of mesenchymal markers in an *in vitro* TGF-β-induced EMT tumor model, while it potentiated the induction of E-cadherin and RKIP (unpublished data). Given that YY1 acts as a transcriptional activator of Snail and it is a downstream target of NF-κB [94, 96], we hypothesized that YY1 may be part of an NF-κB/YY1/Snail/RKIP loop that can be directly targeted by NO, in the context of its anti-EMT action [56, 84]. The importance of endogenous NO-mediated modulation of the NF-κB/YY1/RKIP loop in EMT regulation has been further highlighted in studies where treatment of metastatic cell lines with nonoptimal photodynamic therapy (PDT) or a low-dose pheophorbide a, which are known to induce low levels of NO by NOS, led to cell growth and EMT via the regulation of the above circuitry [97]. Accordingly, exogenous or endogenous NO provided by NO donors or the activation of NOs, respectively, inhibited TGF-β-induced EMT in mouse hepatocytes and alveolar cells, in a concentration-dependent manner [98, 99]. Furthermore, kallistatin, an endogenous protein, inhibited TGF-β-induced EMT in fibrosis and cancer animal models by antagonizing TGF-β-induced miR-21 and Snail1 synthesis while stimulating eNOS expression [100].

The NO antimetastatic effects have been further corroborated by additional *in vivo* studies. Cells from iNOS null tumor cell line showed lower proliferation rates and a number of lung metastases when they were transplanted into iNOs +/+ nude mice than into iNOs −/− mice, while NOS2-transfected pancreatic cancer cells did not form tumors or metastases when injected into ectopic or orthotopic xenograft nude mouse models, due to accelerated NO-mediated apoptosis [12, 25, 101]. Accordingly, iNOs overexpression markedly inhibited cell migration *in vitro* and suppressed tumor metastasis *in vivo*, in an oral cancer model [61]. NO-dependent inhibition of MMP-9 expression and MMP-9 mRNA destabilization have been also described in metastatic breast cancer cells [102, 103] and in an orthotopic RCC animal model [104]. Other NO targets include the N-Myc downstream gene product NDRG1 and the protein kinase G, whose upregulation and activation, respectively, suppress tumor cell migration [45, 105].

Summarizing, there is substantial evidence in the literature for NO-mediated anti-EMT actions; however, these effects should be considered in a concentration-dependent context of endogenous or exogenous NO supply.

NO and Tumor Cell Stemness

Cancer stem cells (CSCs) exist in low numbers in the majority of hematologic and solid tumor types and are thought to be responsible for cancer initiation, progression, spread, and recurrence after therapy. CSCs possess properties similar to those of embryonic stem cells (ESCs) and adult stem cells, including pluripotency and self-renewal capacities, thus maintaining tumor survival and growth [106].

NO levels appear to play a critical role in stem cell functions during embryogenesis and later developmental stages, as well as in adult life. During conception, NO participates in egg activation at fertilization, while *in vitro* differentiation of ESCs from cardiomyocytes has been reported upon exposure to high concentrations of chemical NO donors [107, 108]. Exposure to high NO concentrations triggered the differentiation of mouse ESCs in endoderm, through the downregulation of the expression of the transcription factors Nanog and Oct4, which are known to be critically involved in maintaining the pluripotency and self-renewal characteristics of stem cells [109, 110]. Accordingly, treatment of neural stem cells of Niemann-Pick type C disease mice with the NO inhibitor L-NAME promoted stem cell self-renewal capabilities through AKT/GSK3β signaling [111, 112]. The above findings suggest that maintenance of stem cell self-renewal properties is rather achieved in the presence of low NO concentrations in stem cell niche, while high NO concentration leads to the loss of stem cell phenotype [108].

Stem cell behavior at phenotypic and functional level seems to be also affected under conditions of oxidative stress [108]. It is considered that nitrative (RNS) and oxidative (ROS) stress-mediated DNA damage in stem cells may play a key role in infection-related carcinogenesis via chronic inflammation [71]. Although very little is known about the redox homeostasis in CSCs, it is believed that redox balance sustained by low ROS levels and hypoxic conditions are favorable for the preservation of the stemlike properties of CSCs [76, 108]. This notion is supported by the fact that Oct4, one of the main CSC markers, is a redox-sensitive transcription factor [113]. CSCs inherently show elevated antioxidant status, even after administration of ROS-inducing agents [114–119]. ROS levels in gastric CSCs have been found to be regulated by CD44, a cell surface glycoprotein and hyaluronic acid receptor via reduced glutathione (GSH) [76, 120, 121]. It was further showed that reduction of endogenous ROS in CD44+/24− breast CSCs due to higher expression of free radical scavenging systems is associated with higher tumorigenicity and enhanced radioresistance of the tumors [76, 114]. Accordingly, Yang et al. had reported earlier a schematic model of the sequential events that may distinguish the oxidative stress-dependent self-renewal and differentiation capabilities of normal and cancer stem cells [108]. In this model, the levels of NO are dictated as prognostic indicators of the process that is taking place. Under oxidative stress, ROS levels are increased and cellular redox status alters in favor of stem cell differentiation. In this process, together with the activation of differentiation-related transcriptional factors, NO and other oxidative stress-related markers, such as 4-HNE, are also elevated, while antioxidant biomarkers (GSH and SOD) are decreased. Contrarily, normal cellular redox status sustained by low ROX levels maintains stem cell self-renewal properties accompanied by decreased NO and 4 HNE levels and increased GSH and SOD production rates [108]. Consequently, NO levels within the tumor milieu might be a reliable biomarker to predict the cellular oxidative status and the frequency rates of undifferentiated CSC populations inside the tumor bulk that might sustain tumor survival.

Alterations in NO/sGS/cGMP signaling pathway have also been reported in stem cells [41]. Studies in mouse and human embryonic stem cells revealed reduced expression of the b1 subunit of sGC and little or no expression of sGCα1 subunit, as well as translocation of the sGCβ1 subunit into the nucleus where it may play regulatory roles in gene expression and chromatin remodeling [41]. Although relevant studies are absent in CSCs, given the fact that CSC-initiated tumorigenesis and organogenesis share many aspects, similar defects in cGMP-dependent NO signaling in CSCs might critically impact CSC properties [41].

Considering the role of YY1 in oncogenic EMT and the fact that CSCs are characterized by increased migratory and therefore metastatic potential, recent findings demonstrating the direct stimulating activity of YY1 on prostate stem cell antigen (PSCA) promoter in prostatic intraepithelial neoplasia (PIN) and advanced carcinomas highlight the significance of NO-mediated YY1 inhibition in tumor cell stemness [122]. This notion is further supported by our preliminary findings showing overexpression of YY1 and EMT markers, including Snail, in CSCs (CD44+, CD133+, CD24+) isolated from the aggressive PC3 prostate cancer cell line [123].

Summarizing the reported findings, NO concentration levels and activities may affect both normal and cancer stem cell behavior as it relates to maintenance of a stemlike or differentiated cell phenotype. Nevertheless, NO impact on CSC functions and properties should mainly be evaluated in the context of the cellular redox status.

NO and Tumor Resistance to Conventional Chemotherapy and Immune-Mediated Cytotoxicity

Although the treatment of primary tumors with approved conventional chemotherapeutics results in an initial response, the frequency of tumor recurrences is often high as a result of tumors' acquisition of antiapoptotic mechanisms of resistance [59]. Similarly, tumors may acquire further resistance to immune-mediated cytotoxity, as part of the tumor immunoescape plan, thus making the immune system incompetent to provide efficient tumor immunosurveillance. Therefore, the identification of sensitizing agents that can render resistant tumor cells sensitive to respond to various cytotoxic therapies and immune-mediated stimuli is more than needed.

Dose-dependent effects of NO on the tumor cell response to drug- and immune-mediated apoptosis have been reported in the literature. Over the last 15 years, our group has extensively studied the modulating activity of high NO concentrations on cancer cell resistance to apoptotic stimuli and has provided substantial evidence of a negative regulation. Briefly, we have shown that nitric oxide donors, such as DETA/NO, when administrated in concentrations $>500\,\mu M$, inhibit cell survival and antiapoptotic pathways *in vitro* and *in vivo* and sensitize drug-resistant tumor cells to apoptosis by both chemotherapy and immunotherapy [55–59, 124–126]. The underlying molecular mechanisms of DETA/NO-mediated tumor sensitization to apoptosis include, among others, the inhibition of the constitutively activated survival pathway NF-κB and its downstream antiapoptotic gene products such as Bcl-xl and YY1 [55, 57, 95]. YY1 has been previously shown to confer tumor cell drug resistance and to regulate resistance to both Fas ligand and TRAIL-mediated cytotoxicity through direct transcriptional repression of the death receptors Fas and DR5 [53, 127, 128]. Indeed, we have reported that DETA/NO sensitizes tumor cells resistant to TRAIL- and FasL-induced apoptosis, via the inhibition of the death receptor 5 (DR5) and Fas transcriptional repressor, YY1 [58, 126–129]. In addition to NO-mediated YY1 downregulation, Huerta S et al. identified apoptosis-inducing factor (AIF) upregulation to be responsible for the *in vitro* and *in vivo* sensitization of metastatic colon cancer cells to CDPP-induced apoptosis by DETA/NO [124].

DETA/NO may induce direct NF-κB and YY1 *S*-nitrosylation in tumor cells resulting in the loss of their DNA binding activities and activation of intrinsic and extrinsic apoptotic pathways as response to both chemotherapy and immune-mediated cytotoxicity [53–55]. Other groups have demonstrated that S-nitrosylation emerges as an important posttranslational regulation for the TNF-family death receptor signaling pathways. As such, *S*-nitrosylation of the death receptor 4 (DR4) by a specific NO donor promotes apoptosis [130], while *S*-nitrosylation of Fas, following glyceryl trinitrate (GTN) exposure, promotes the redistribution of the receptor to lipid rafts, formation of the death-inducing signal complex (DISC), and induction of cell death [131].

Along with the direct effects of NO on NF-κB/YY1 signaling, we have shown that the induction of RKIP by DETA/NO further contributes to reversal of tumor resistance to chemo-/immune-mediated apoptosis, via the inhibition of the survival signals provided by the constitutive activation of NF-κB and Raf-1/MEK/ERK pathways [55, 85]. Overexpression of RKIP mimics NO in reversing tumor resistance to apoptotic stimuli [56, 132]. There is also evidence that the NO-modulated NF-κB/YY1/RKIP loop can be completed with the addition of two more players, Snail and PTEN, which can be directly or indirectly targeted by NO. Snail, as mentioned in EMT section, is a transcriptional repressor of RKIP, while it is under the transcriptional regulation of NF-κB and its downstream target YY1 [88, 94, 96]. Similarly, PTEN, a phosphatase inhibitor of the PI3K/Akt survival pathway, is also transcriptionally repressed by Snail [133], while it can inhibit YY1 expression [134]. We have hypothesized that tumor cell resistance to both chemotherapeutic drugs and cytotoxic immune lymphocytes/ligands may arise from the dysregulated expressions of the several interrelated gene products that form the NF-κB/Snail/YY1/RKIP/PTEN loop. Briefly, the expressions and activities of NF-κB, Snail, and YY1 are upregulated, whereas the expressions and activities of RKIP and PTEN are downregulated. Interestingly, the modulation of each of the gene products in the loop is sufficient to reverse resistance [55]. These findings establish the presence of the dysregulated NF-κB/Snail/YY1/RKIP/PTEN circuitry in tumor resistance and suggest that tumor cell treatment with NO moieties is able to modify the cross talk among the members of the loop in favor of the inhibition of tumor cell survival and the response to cytotoxic drugs and immune-mediated cytotoxicity. Beyond the role of NO in reversing tumor drug- and immune resistance, NO has further showed efficacy on strengthening the proapoptotic effects of suboptimal PTD and radiotherapy in cancer cells [135]. Altogether, these findings underscore the potential therapeutic application of high NO doses as sensitizing agents for overcoming acquired tumor resistance to conventional chemotherapy and immune-mediated cytotoxicity and to suboptimal radio and phototherapy.

ANTINEOPLASTIC ACTIVITIES OF NO IN TUMOR MILIEU AND THE UNDERLYING MOLECULAR MECHANISMS

The tumor milieu is a complex system of many cells whose contribution in tumor properties may vary [136, 137]. Most solid tumors contain two compartments: the parenchyma (neoplastic cells) and the stroma (nonmalignant supporting tissues including connective tissue, blood vessels, and immune cells). These cells might exert different NO biology, including variations in NO-producing, NO-releasing, and NO-distributing abilities, as well as diverse cGMP-dependent (NO/sGC/cGMP pathway) and cGMP-independent (NO oxidative pathway) signaling, that can influence the overall tumor behavior [41]. Consequently, the impact of NO as an anti- or proneoplastic agent in cancer biology should be considered on the context of tumor microenvironment and not on a single cancer cell basis.

Below, we discuss the role of NO in the negative regulation of the major tumor-supporting functions of tumor microenvironment that include angiogenesis and immunoediting of immune signals.

NO and Tumor Angiogenesis

Angiogenesis is pivotal for tumor growth and expansion to locations other than the primary site. The tumor microenvironment comprises numerous signaling molecules and pathways that influence the angiogenic response [138]. Thus, controlling tumor-associated angiogenesis is a promising approach in limiting cancer progression.

Just like the divergent impact of NO on other tumor-associated functions, NO can either stimulate or inhibit angiogenesis determined by the concentration of NO and the duration of exposure [13]. Various NO-producing moieties have shown dose-dependent antiangiogenic effects in *in vitro* and *in vivo* system models, through multiple mechanisms. Treatment of colon cancer cell lines with NO-NSAID resulted in the downregulation of VEGF and its receptors and induced apoptosis [12, 139]. Inhibition of angiogenesis in microvascular endothelial cells *in vitro* was further shown to be mediated by concentration-dependent SNAP-induced suppression of PKC activity and ERK and c-Jun phosphorylation [140]. Interestingly, SNAP at concentrations >2 mM also abolished activator protein 1 (AP-1) binding activity to the VEGF promoter resulting in its downregulation [22, 140, 141]. In a chick aortic ring assay, JS-K at IC50 values of <0.7 μM, among other effects, inhibited the proliferation and migration of human umbilical vein endothelial cells (HUVEC) and blocked vessel growth, while it suppressed tumor angiogenesis *in vivo* [142]. Similarly, the NO donors sodium nitroprusside (SNP; NaN) and isosorbide mononitrate caused a dose-dependent inhibition of angiogenesis in the chorioallantoic membrane of chick embryos, whereas NOS inhibitors such as L-NAME and L-NMMA stimulated angiogenesis [22, 143–145].

These above findings highlight NO as a downstream regulator of multiple angiogenic effectors through various mechanisms [146]. Therefore, modulation of its concentration within the tumor microenvironment appears to be a promising target for consideration in antiangiogenic therapeutic strategies.

NO and Immunoediting of Immune Signals by the Tumor Microenvironment

Tumor mass is infiltrated by different kinds of immune cells including neutrophils, eosinophils, basophils, NK cells, CD4 and CD8 (CTL) T lymphocytes, regulatory T cells (Tregs), dendritic cells, tumor-associated macrophages (TAMs), and myeloid-derived suppressor cells (MDSCs) [147, 148]. Although the immune surveillance system aims to destroy cancer cells or suppress their outgrowth, the tumor cells are capable of inhibiting the development and efficiency of antitumor immune responses in the cancer milieu [137]. This suppression of effector immune responses is facilitated by cancer cell-promoted immunoediting of immune signals via different mechanisms, including inflammation and the utilization of reactive radicals such as NO [137, 149–151].

Besides tumor cells, production and release of NO into the tumor microenvironment has been reported for neutrophils, eosinophils, basophils, NK cells, TAMs, MDSCs, and dendritic cells [13, 137, 152, 153]. The immunoediting property of reactive NO may alter the functions of T-cell receptors (TCRs), antigen-specific tumor infiltrating lymphocytes (TILs), CTLs, MDSCs, and intratumoral chemokines and the expression of tumor suppressor genes and interferon responses [137].

Although most of the reports found in the literature agree for a positive role of NO in countersignaling tumor immunity, there is substantial evidence showing that NO may exert differential actions in cancer immunoediting depending on its concentration present in the tumor microenvironment. As such, NO concentrations above 1000 nM have been associated with increased tumor phagocytosis, apoptosis, and cell death [18, 137]. Tumor cells can be targeted by increasing NO toxicity, mediated by overloading of nitrogen species in the tumor milieu [137]. Interestingly, in colon and ovarian cancer models, intratumoral injection of iNOs-expressing microencapsulated cells amplified the expression of death receptor Fas and its ligand FasL on the tumor cell surface, thus making the tumor cells susceptible to CTL killing [13]. Similarly, the NO donor AT38 promoted a massive invasion of effector T cells into the tumor microenvironment in prostate cancer to augment tumor regression by immunotherapy [154]. In addition, NK cell's cytotoxic activity is dependent on NO synthesis, and inhibitors of NO synthesis have been shown to impair NK-cell-mediated target cell killing [155].

Nitric Oxide Novel Applications in Cancer Treatment and Prevention

Over the past decade, the utilization of NO-releasing moieties able to shift NO availability in the tumor microenvironment above the threshold for achieving antitumor responses has revolutionized the era of cancer therapeutics. The current advances in NO-based cancer therapy have been recently comprehensively reviewed by Huang et al. and others [3, 5, 74, 156]. Briefly, the experimental and clinical NO-based applications in cancer treatment include (1) NO donors used either as single agents or in combination with conventional chemo- or immunotherapy or radiation, (2) conjugates of NO donors with other drugs, (3) targeted delivery of high amounts of NO to cancer mass through nanoparticles, and (4) the induction of NOS activity by genetic manipulation [74].

NO Donors

Compounds able to release variable amounts of NO under physiological conditions are known as NO donors. Most NO donors facilitate NO liberation *in vitro* and *in vivo* through various enzymatic or nonenzymatic mechanisms

[74]. However, the spontaneous and nonselective release of NO cannot be excluded for many NO donors, thus reducing their *in vivo* efficacy. The main classes of NO-donating compounds include diazeniumdiolates, organic nitrates, S-nitrosothiols, and metal-NO complexes [3, 5, 74, 156].

We have shown that the diazeniumdiolate DETA/NO, as a single agent, could inhibit EMT in metastatic prostate cancer cells *in vitro* and *in vivo*, by altering the ratio of EMT-promoting and EMT-suppressing gene products [85]. In addition, DETA/NO when used in concentrations above 500 μM was able to reverse resistance of prostate cancer cell lines to TRAIL-mediated apoptosis *in vitro* and *in vivo* [58]. The molecular mechanism underlying the NO-immunosensitizing and EMT-suppressing activities in the reported models includes modifications in the expression and activities of the members of the NF-κB/YY1/Snail/RKIP circuitry, as well as upregulation of the DR5 death receptor and activation of the intrinsic apoptotic pathway [55, 56]. Concomitantly, Rapozzi et al. showed that DETA/NO may further sensitize prostate cancer cells to suboptimal phototherapy-mediated apoptosis through the modulation of the same gene circuitry [97]. Nevertheless, the combination of DETA/NO with various chemotherapeutic drugs such as cisplatin, 5FU, doxorubicin, and fludarabine was able to reverse hypoxia-mediated resistance in various types of solid and hematologic malignancies [5, 9]. Furthermore, breast cancer cell pretreatment with DETA/NO resulted in significant and selective tumor response to farnesyltransferase inhibitor, H56582 [157].

Sodium nitroprusside (SNP) is a well-characterized representative of metal-NO complexes that has shown, at least *in vitro*, a large panel of antitumor activities in different cancer types, including tumor sensitization to suboptimal radiotherapy and immune-mediated cytotoxicity and reduction in tumor cell invasion and hypoxic levels [5, 158]. Preclinical and clinical data are also available for the antitumor properties of organic nitrates, such as glyceryl trinitrate (GTN) and glycidyl nitrate (GLYN), when used either as a single treatment or in combination with conventional anticancer agents [159, 160]. Organic nitrates exert antineoplastic actions mainly through improving tumor oxygenation and perfusion [74]. In clinical trials, concurrent administration of GTN and chemotherapeutic drugs such as vinorelbine, cisplatin, or 5-FU or radiation has shown promising and selective antineoplastic effects in non-small-cell lung, prostate, colon, and rectal cancer models [161–164]. Other NO donors, such as S-nitrosothiols, are also studied for their chemosensitizing activity against various cancer types. S-nitrosothiols, including SNAP, are thought to mediate a sensitization effect to ROS-inducing drugs, such as cisplatin, by potentiating iNOs expression and NO toxicity in cancer cells, through AP1 and NF-κB upregulation [76, 165, 166]. Finally, a recently identified NO donor, namely, compound 2, has shown a wide range of tumor-selective antiproliferative and sensitizing effects in various tumor models *in vitro* and *in vivo* and currently is under clinical trials [74, 167, 168].

NO Donor Conjugates

The logistics behind NO donor hybridization with other moieties is to increase the selective, effective, and durable delivery of NO to tumors. The hybridization strategy has several variations depending on the nature of the compound/fragment conjugated to NO donor. Briefly, hybrids of NO donors can be generated after conjugation—via a cancer targeted linker—with (1) cytotoxic drugs with anticancer activities, (2) nonsteroidal anti-inflammatory drugs (NO-NSAIDs), (3) enzyme inhibitors, (4) active natural products (NO-ANPs), and (5) inhibitors of multidrug resistance (MDR). Representative NO-NSAIDs are the NO-aspirin conjugates, such as NCX-4016 and NCX-4040, which exert enhanced antiproliferative and proapoptotic activities in tumor cell models, through NF-κB inhibition and the induction of redox-dependent signaling as well as G2/M phase cell cycle arrest [169, 170]. Examples of NO donor conjugates with cytotoxic drugs are the diazeniumdiolate/5-FU [171], NO/platinum-based compounds [172, 173], NO/thalidomide [174], and NO/tamibarotene [175]. All the above hybrids have shown synergistic anticancer activities against various cancer models, which were greater than those observed after single-agent treatments [74]. An active natural product that has been used for the construction of NO-ANP hybrids is oleanolic acid (OA). NO-OA conjugates, in various forms, possess strong and selective tumor suppressing activities *in vitro* and *in vivo*, through several mechanisms, including the inhibition of the Wnt/β-catenin and COX-2 signaling pathways [74, 176].

In addition, several structural modifications have frequently been conducted to NO donors, especially diazeniumdiolates, in order to improve their stability and selectivity against cancer cells. One mechanism by which this is achieved is by anchoring the O_2^- anion of diazeniumdiolates to a fragment which can directly act against cancer cells, a process known as O_2^- protection. The protective fragment can be removed following activation by specific enzymes that are usually overexpressed in cancer cells, resulting in selective release of free cytotoxic NO within tumor cells. Among the enzymes used for protective fragment activation and removal from diazeniumdiolates are esterases, GSH/GST, DT-diaphorase, and β-galactosidase [74, 177, 178].

A complete list of currently available NO donor conjugates is thoroughly described by Huang et al. [74]

NO-Releasing Nanoparticles

Another recent advancement in enhancing NO donor stability, selectivity, durability, and effectiveness in terms of NO amounts released is the development of NO-donating nanoparticle platforms. These systems can load large amounts of NO in a sustained manner and undergo passive targeting to tumor site via several mechanisms, including EPR effect and regulation of microenvironmental factors such as temperature and pH. The platforms ideally consist of biocompatible and biodegradable materials including liposomes; quantum dots; and polymeric, silica, or metallic nanoparticles [179, 180]. An additional unique advantage of the high surface area of the NO-donating nanoparticle systems is their ability to carry a cytotoxic drug, in addition to NO donor, thus exerting a synergistic and site-specific antitumor effect [181].

NOS Gene Targeting

Last but not least, modulation of iNOS expression in cancer cells by gene therapy has recently attracted significant attention. The use of various iNOS gene delivery systems for specific and localized iNOs expression in tumor sites has resulted in intratumoral production of elevated NO concentrations able to exert cytotoxic and antimetastatic effects in a plethora of *in vitro* and *in vivo* cancer models [74, 182, 183].

CONCLUSIONS

NO concentration, duration of cell exposure to NO, and the cellular redox status are thought to be the critical determinants of the dichotomous role of NO in cancer biology. NO supply is maintained not only by tumor cells but also from additional cells present in the tumor microenvironment, including immune cell populations and tumor-supporting stroma cells. In general, NO concentrations above the "safe zone," seem to exert antineoplastic effects including suppressed tumorigenesis, cell cycle arrest and inhibition of tumor growth and metastatic potential, reversal of tumor resistance to conventional chemotherapy and immune-mediated cytotoxicity, and inhibition of angiogenesis and cancer cell stemness. The underlying molecular mechanisms of NO-mediated antitumorigenic activities include posttranslational modifications in critical oncoproteins and metastasis inducers, activation of apoptosis-related proteins such as p53, actions at epigenetic level, and reactions with other free radicals such as O2, in order to increase cellular redox status. Overall, NO potency to suppress tumor-promoting and tumor-supporting events largely depends on two main factors: (1) the total NO levels produced in the tumor bed by all the cells present within the tumor milieu and (2) the way these levels have been used for determining the cellular redox potential (oxidant/antioxidant equilibrium).

The antineoplastic nature of NO has been exploited therapeutically via various strategies with remarkable results in preclinical models. Treatment with NO-donating agents such as NO donors, NO nanoparticles, or iNOS induction *in vivo* may result in significant increase of intratumoral NO concentration above the "safe zone." NO at these concentrations maybe be effective alone or in combination with conventional anticancer therapies as a chemo-/immune-/radiosensitizing agent. Although the clinical trials of NO application as an anticancer agent are still limited, the future of NO delivery approaches in cancer therapeutics holds a great deal of promise.

Acknowledgment

The author would like to acknowledge Professor Benjamin Bonavida for providing significant scientific input in the preparation of this manuscript.

References

[1] Basudhar D, Ridnour LA, Cheng R, Kesarwala AH, Heinecke J, Wink DA. Biological signaling by small inorganic molecules. Coord Chem Rev 2016;306:708–23.

[2] Ignarro LJ. Physiology and pathophysiology of nitric oxide. Kidney Int Suppl 1996;55:S2–5.

[3] Huerta S. Nitric oxide for cancer therapy. Future Sci OA 2015;1:FSO44.

[4] Michel T, Feron O. Nitric oxide synthases: which, where, how, and why? J Clin Invest 1997;100:2146–52.

[5] Huerta S, Chilka S, Bonavida B. Nitric oxide donors: novel cancer therapeutics (review). Int J Oncol 2008;33:909–27.

[6] Thomas DD, Ridnour LA, Isenberg JS, Flores-Santana W, Switzer CH, Donzelli S, Hussain P, Vecoli C, Paolocci N, Ambs S, Colton CA, Harris CC, Roberts DD, Wink DA. The chemical biology of nitric oxide: implications in cellular signaling. Free Radic Biol Med 2008;45:18–31.

[7] Aranda E, Lopez-Pedrera C, De La Haba-Rodriguez JR, Rodriguez-Ariza A. Nitric oxide and cancer: the emerging role of S-nitrosylation. Curr Mol Med 2012;12:50–67.

[8] Korde Choudhari S, Sridharan G, Gadbail A, Poornima V. Nitric oxide and oral cancer: a review. Oral Oncol 2012;48:475–83.

[9] Cheng H, Wang L, Mollica M, Re AT, Wu S, Zuo L. Nitric oxide in cancer metastasis. Cancer Lett 2014;353:1–7.

[10] Choi BM, Pae HO, Jang SI, Kim YM, Chung HT. Nitric oxide as a pro-apoptotic as well as anti-apoptotic modulator. J Biochem Mol Biol 2002;35:116–26.

[11] Choudhari SK, Chaudhary M, Bagde S, Gadbail AR, Joshi V. Nitric oxide and cancer: a review. World J Surg Oncol 2013;11:118.

[12] Burke AJ, Sullivan FJ, Giles FJ, Glynn SA. The yin and yang of nitric oxide in cancer progression. Carcinogenesis 2013;34:503–12.

[13] Fukumura D, Kashiwagi S, Jain RK. The role of nitric oxide in tumour progression. Nat Rev Cancer 2006;6:521–34.

[14] Hussain SP, He P, Subleski J, Hofseth LJ, Trivers GE, Mechanic L, Hofseth AB, Bernard M, Schwank J, Nguyen G, Mathe E, Djurickovic D, Haines D, Weiss J, Back T, Gruys E, Laubach VE, Wiltrout RH, Harris CC. Nitric oxide is a key component in inflammation-accelerated tumorigenesis. Cancer Res 2008;68:7130–6.

[15] Stuehr DJ, Marletta MA. Mammalian nitrate biosynthesis: mouse macrophages produce nitrite and nitrate in response to Escherichia coli lipopolysaccharide. Proc Natl Acad Sci U S A 1985;82:7738–42.

[16] Ridnour LA, Thomas DD, Donzelli S, Espey MG, Roberts DD, Wink DA, Isenberg JS. The biphasic nature of nitric oxide responses in tumor biology. Antioxid Redox Signal 2006;8:1329–37.

[17] Villalobo A. Nitric oxide and cell proliferation. FEBS J 2006;273:2329–44.

[18] Ridnour LA, Thomas DD, Switzer C, Flores-Santana W, Isenberg JS, Ambs S, Roberts DD, Wink DA. Molecular mechanisms for discrete nitric oxide levels in cancer. Nitric Oxide 2008;19:73–6.

[19] Wink DA, Ridnour LA, Hussain SP, Harris CC. The reemergence of nitric oxide and cancer. Nitric Oxide 2008;19:65–7.

[20] Wink DA, Vodovotz Y, Laval J, Laval F, Dewhirst MW, Mitchell JB. The multifaceted roles of nitric oxide in cancer. Carcinogenesis 1998;19:711–21.

[21] Bian K, Ghassemi F, Sotolongo A, Siu A, Shauger L, Kots A, Murad F. NOS-2 signaling and cancer therapy. IUBMB Life 2012;64:676–83.

[22] Vasudevan D, Thomas DD. Insights into the diverse effects of nitric oxide on tumor biology. Vitam Horm 2014;96:265–98.

[23] Jenkins DC, Charles IG, Thomsen LL, Moss DW, Holmes LS, Baylis SA, Rhodes P, Westmore K, Emson PC, Moncada S. Roles of nitric oxide in tumor growth. Proc Natl Acad Sci U S A 1995;92:4392–6.

[24] Vakkala M, Kahlos K, Lakari E, Paakko P, Kinnula V, Soini Y. Inducible nitric oxide synthase expression, apoptosis, and angiogenesis in in situ and invasive breast carcinomas. Clin Cancer Res 2000;6:2408–16.

[25] Wei D, Richardson EL, Zhu K, Wang L, Le X, He Y, Huang S, Xie K. Direct demonstration of negative regulation of tumor growth and metastasis by host-inducible nitric oxide synthase. Cancer Res 2003;63:3855–9.

[26] Hickok JR, Vasudevan D, Jablonski K, Thomas DD. Oxygen dependence of nitric oxide-mediated signaling. Redox Biol 2013;1:203–9.

[27] Hickok JR, Sahni S, Shen H, Arvind A, Antoniou C, Fung LW, Thomas DD. Dinitrosyliron complexes are the most abundant nitric oxide-derived cellular adduct: biological parameters of assembly and disappearance. Free Radic Biol Med 2011;51:1558–66.

[28] Thomas DD, Espey MG, Ridnour LA, Hofseth LJ, Mancardi D, Harris CC, Wink DA. Hypoxic inducible factor 1alpha, extracellular signal-regulated kinase, and p53 are regulated by distinct threshold concentrations of nitric oxide. Proc Natl Acad Sci U S A 2004;101:8894–9.

[29] Heller R, Polack T, Grabner R. Till U. Nitric oxide inhibits proliferation of human endothelial cells via a mechanism independent of cGMP. Atherosclerosis 1999;144:49–57.

[30] Pervin S, Singh R, Chaudhuri G. Nitric oxide-induced cytostasis and cell cycle arrest of a human breast cancer cell line (MDA-MB-231): potential role of cyclin D1. Proc Natl Acad Sci U S A 2001;98:3583–8.

[31] Pena-Altamira E, Petazzi P, Contestabile A. Nitric oxide control of proliferation in nerve cells and in tumor cells of nervous origin. Curr Pharm Des 2010;16:440–50.

[32] Huguenin S, Fleury-Feith J, Kheuang L, Jaurand MC, Bolla M, Riffaud JP, Chopin DK, Vacherot F. Nitrosulindac (NCX 1102): a new nitric oxide-donating non-steroidal anti-inflammatory drug (NO-NSAID), inhibits proliferation and induces apoptosis in human prostatic epithelial cell lines. Prostate 2004;61:132–41.

[33] Huguenin S, Vacherot F, Kheuang L, Fleury-Feith J, Jaurand MC, Bolla M, Riffaud JP, Chopin DK. Antiproliferative effect of nitrosulindac (NCX 1102), a new nitric oxide-donating non-steroidal anti-inflammatory drug, on human bladder carcinoma cell lines. Mol Cancer Ther 2004;3:291–8.

[34] Kumar S, Barthwal MK, Dikshit M. Cdk2 nitrosylation and loss of mitochondrial potential mediate NO-dependent biphasic effect on HL-60 cell cycle. Free Radic Biol Med 2010;48:851–61.

[35] Bal-Price A, Gartlon J, Brown GC. Nitric oxide stimulates PC12 cell proliferation via cGMP and inhibits at higher concentrations mainly via energy depletion. Nitric Oxide 2006;14:238–46.

[36] Lai Y, Shen L, Zhang Z, Liu W, Zhang Y, Ji H, Tian J. Synthesis and biological evaluation of furoxan-based nitric oxide-releasing derivatives of glycyrrhetinic acid as anti-hepatocellular carcinoma agents. Bioorg Med Chem Lett 2010;20:6416–20.

[37] Donia M, Maksimovic-Ivanic D, Mijatovic S, Mojic M, Miljkovic D, Timotijevic G, Fagone P, Caponnetto S, Al-Abed Y, McCubrey J, Stosic-Grujicic S, Nicoletti F. In vitro and in vivo anti-cancer action of Saquinavir-NO, a novel nitric oxide-derivative of the protease inhibitor saquinavir, on hormone resistant prostate cancer cells. Cell Cycle 2011;10:492–9.

[38] Donia M, Mijatovic S, Maksimovic-Ivanic D, Miljkovic D, Mangano K, Tumino S, Biondi A, Basile F, Al-Abed Y, Stosic-Grujicic S, Nicoletti F. The novel NO-donating compound GIT-27NO inhibits in vivo growth of human prostate cancer cells and prevents murine immunoinflammatory hepatitis. Eur J Pharmacol 2009;615:228–33.

[39] Brown GC. Nitric oxide and mitochondrial respiration. Biochim Biophys Acta 1999;1411: 351–69.

[40] Van de Wouwer M, Couzinie C, Serrano-Palero M, Gonzalez-Fernandez O, Galmes-Varela C, Menendez-Antoli P, Grau L, Villalobo A. Activation of the BRCA1/Chk1/p53/p21(Cip1/Waf1) pathway by nitric oxide and cell cycle arrest in human neuroblastoma NB69 cells. Nitric Oxide 2012;26:182–91.

[41] Bian K, Murad F. What is next in nitric oxide research? From cardiovascular system to cancer biology. Nitric Oxide 2014;43:3–7.

[42] Mujoo K, Sharin VG, Martin E, Choi BK, Sloan C, Nikonoff LE, Kots AY, Murad F. Role of soluble guanylyl cyclase-cyclic GMP signaling in tumor cell proliferation. Nitric Oxide 2010;22:43–50.

[43] Zhu H, Li JT, Zheng F, Martin E, Kots AY, Krumenacker JS, Choi BK, McCutcheon IE, Weisbrodt N, Bogler O, Murad F, Bian K. Restoring soluble guanylyl cyclase expression and function blocks the aggressive course of glioma. Mol Pharmacol 2011;80:1076–84.

[44] Brune B. Nitric oxide: NO apoptosis or turning it ON? Cell Death Differ 2003;10:864–9.

[45] Deguchi A, Thompson WJ, Weinstein IB. Activation of protein kinase G is sufficient to induce apoptosis and inhibit cell migration in colon cancer cells. Cancer Res 2004;64: 3966–73.

[46] Mocellin S, Bronte V, Nitti D. Nitric oxide, a double edged sword in cancer biology: searching for therapeutic opportunities. Med Res Rev 2007;27:317–52.

[47] Marshall HE, Stamler JS. Nitrosative stress-induced apoptosis through inhibition of NF-kappa B. J Biol Chem 2002;277:34223–8.

[48] Di JH, Li C, Yu HM, Zheng JN, Zhang GY. nNOS downregulation attenuates neuronal apoptosis by inhibiting nNOS-GluR6 interaction and GluR6 nitrosylation in cerebral ischemic reperfusion. Biochem Biophys Res Commun 2012;420:594–9.

[49] Kim JE, Tannenbaum SR. S-Nitrosation regulates the activation of endogenous procaspase-9 in HT-29 human colon carcinoma cells. J Biol Chem 2004;279:9758–64.

[50] Kim MY. Nitric oxide triggers apoptosis in A375 human melanoma cells treated with capsaicin and resveratrol. Mol Med Rep 2012;5:585–91.

[51] Wang K, Brems JJ, Gamelli RL, Holterman AX. iNOS/NO signaling regulates apoptosis induced by glycochenodeoxycholate in hepatocytes. Cell Signal 2011;23:1677–85.

[52] Olson SY, Garban HJ. Regulation of apoptosis-related genes by nitric oxide in cancer. Nitric Oxide 2008;19:170–6.

[53] Hongo F, Garban H, Huerta-Yepez S, Vega M, Jazirehi AR, Mizutani Y, Miki T, Bonavida B. Inhibition of the transcription factor Yin Yang 1 activity by S-nitrosation. Biochem Biophys Res Commun 2005;336:692–701.

[54] Bonavida B. Regulation of cell death apoptotic pathways by nitric oxide in cancer: reversal of drug/immune resistance. Redox Biol 2015;5:415.

[55] Bonavida B, Garban H. Nitric oxide-mediated sensitization of resistant tumor cells to apoptosis by chemo-immunotherapeutics. Redox Biol 2015;6:486–94.

[56] Bonavida B, Baritaki S. Dual role of NO donors in the reversal of tumor cell resistance and EMT: downregulation of the NF-kappaB/Snail/YY1/RKIP circuitry. Nitric Oxide 2011;24:1–7.

[57] Huerta-Yepez S, Baritaki S, Baay-Guzman G, Hernandez-Luna MA, Hernandez-Cueto A, Vega MI, Bonavida B. Contribution of either YY1 or BclXL-induced inhibition by the NO-donor DETANONOate in the reversal of drug resistance, both in vitro and in vivo. YY1 and BclXL are overexpressed in prostate cancer. Nitric Oxide 2013;29:17–24.

[58] Huerta-Yepez S, Vega M, Escoto-Chavez SE, Murdock B, Sakai T, Baritaki S, Bonavida B. Nitric oxide sensitizes tumor cells to TRAIL-induced apoptosis via inhibition of the DR5 transcription repressor Yin Yang 1. Nitric Oxide 2009;20:39–52.

[59] Bonavida B, Baritaki S, Huerta-Yepez S, Vega MI, Chatterjee D, Yeung K. Novel therapeutic applications of nitric oxide donors in cancer: roles in chemo- and immunosensitization to apoptosis and inhibition of metastases. Nitric Oxide 2008;19:152–7.

[60] Zhao SF, Tong XY, Zhu FD. Nitric oxide induces oral squamous cell carcinoma cells apoptosis with p53 accumulation. Oral Oncol 2005;41:785–90.

[61] Harada K, Supriatno SM, Kawaguchi S, Tomitaro O, Yoshida H, Sato M. Overexpression of iNOS gene suppresses the tumorigenicity and metastasis of oral cancer cells. In Vivo 2004;18: 449–55.

[62] Brennan PA, Conroy B, Spedding AV. Expression of inducible nitric oxide synthase and p53 in oral epithelial dysplasia. Oral Surg Oral Med Oral Pathol Oral Radiol Endod 2000;90:624–9.

[63] Ambs S, Glynn SA. Candidate pathways linking inducible nitric oxide synthase to a basal-like transcription pattern and tumor progression in human breast cancer. Cell Cycle 2011;10: 619–24.

[64] Forrester K, Ambs S, Lupold SE, Kapust RB, Spillare EA, Weinberg WC, Felley-Bosco E, Wang XW, Geller DA, Tzeng E, Billiar TR, Harris CC. Nitric oxide-induced p53 accumulation and regulation of inducible nitric oxide synthase expression by wild-type p53. Proc Natl Acad Sci U S A 1996;93:2442–7.

[65] Ambs S, Ogunfusika MO, Merriam WG, Bennett WP, Billiar TR, Harris CC. Up-regulation of inducible nitric oxide synthase expression in cancer-prone p53 knockout mice. Proc Natl Acad Sci U S A 1998;95:8823–8.

[66] Ambs S, Bennett WP, Merriam WG, Ogunfusika MO, Oser SM, Harrington AM, Shields PG, Felley-Bosco E, Hussain SP, Harris CC. Relationship between p53 mutations and inducible nitric oxide synthase expression in human colorectal cancer. J Natl Cancer Inst 1999;91:86–8.

[67] Buerk DG. Nitric oxide regulation of microvascular oxygen. Antioxid Redox Signal 2007;9: 829–43.

[68] Mason MG, Nicholls P, Wilson MT, Cooper CE. Nitric oxide inhibition of respiration involves both competitive (heme) and noncompetitive (copper) binding to cytochrome c oxidase. Proc Natl Acad Sci U S A 2006;103:708–13.

[69] Thomas DD, Liu X, Kantrow SP, Lancaster Jr. JR. The biological lifetime of nitric oxide: implications for the perivascular dynamics of NO and O2. Proc Natl Acad Sci U S A 2001;98: 355–60.

[70] Hussain SP, Trivers GE, Hofseth LJ, He P, Shaikh I, Mechanic LE, Doja S, Jiang W, Subleski J, Shorts L, Haines D, Laubach VE, Wiltrout RH, Djurickovic D, Harris CC. Nitric oxide, a mediator of inflammation, suppresses tumorigenesis. Cancer Res 2004;64:6849–53.

[71] Kawanishi S, Ohnishi S, Ma N, Hiraku Y, Oikawa S, Murata M. Nitrative and oxidative DNA damage in infection-related carcinogenesis in relation to cancer stem cells. Genes Environ 2016;38:26.

[72] Scott DJ, Hull MA, Cartwright EJ, Lam WK, Tisbury A, Poulsom R, Markham AF, Bonifer C, Coletta PL. Lack of inducible nitric oxide synthase promotes intestinal tumorigenesis in the Apc(Min/+) mouse. Gastroenterology 2001;121:889–99.

[73] Li Q, Sarna SK. Nitric oxide modifies chromatin to suppress ICAM-1 expression during colonic inflammation. Am J Physiol Gastrointest Liver Physiol 2012;303:G103–10.

[74] Huang Z, Fu J, Zhang Y. Nitric oxide donor-based cancer therapy: advances and prospects. J Med Chem 2017;60:7617–35.

[75] Oronsky B, Fanger GR, Oronsky N, Knox S, Scicinski J. The implications of hyponitroxia in cancer. Transl Oncol 2014;7:167–73.

[76] Somasundaram V, Nadhan R, K Hemalatha S, Kumar Sengodan S, Srinivas P. Nitric oxide and reactive oxygen species: Clues to target oxidative damage repair defective breast cancers. Crit Rev Oncol Hematol 2016;101:184–92.

[77] Graziewicz M, Wink DA, Laval F. Nitric oxide inhibits DNA ligase activity: potential mechanisms for NO-mediated DNA damage. Carcinogenesis 1996;17:2501–5.

[78] Laval F, Wink DA. Inhibition by nitric oxide of the repair protein, O6-methylguanine-DNA-methyltransferase. Carcinogenesis 1994;15:443–7.

[79] Laval F, Wink DA, Laval J. A discussion of mechanisms of NO genotoxicity: implication of inhibition of DNA repair proteins. Rev Physiol Biochem Pharmacol 1997;131:175–91.

[80] Wink DA, Nims RW, Darbyshire JF, Christodoulou D, Hanbauer I, Cox GW, Laval F, Laval J, Cook JA, Krishna MC, et al. Reaction kinetics for nitrosation of cysteine and glutathione in aerobic nitric oxide solutions at neutral pH. Insights into the fate and physiological effects of intermediates generated in the NO/O2 reaction. Chem Res Toxicol 1994;7:519–25.

[81] Singh KK, Shukla PC, Quan A, Al-Omran M, Lovren F, Pan Y, Brezden-Masley C, Ingram AJ, Stanford WL, Teoh H, Verma S. BRCA1 is a novel target to improve endothelial dysfunction and retard atherosclerosis. J Thorac Cardiovasc Surg 2013;146:949–60 e4.

[82] Yakovlev VA. Nitric oxide-dependent downregulation of BRCA1 expression promotes genetic instability. Cancer Res 2013;73:706–15.

[83] Puisieux A, Brabletz T, Caramel J. Oncogenic roles of EMT-inducing transcription factors. Nat Cell Biol 2014;16:488–94.

[84] Bonavida B, Baritaki S. Inhibition of epithelial-to-mesenchymal transition (EMT) in cancer by nitric oxide: pivotal roles of nitrosylation of NF-kappaB, YY1 and Snail. Immunopathol Dis Therap 2012;3:125–33.

[85] Baritaki S, Huerta-Yepez S, Sahakyan A, Karagiannides I, Bakirtzi K, Jazirehi A, Bonavida B. Mechanisms of nitric oxide-mediated inhibition of EMT in cancer: inhibition of the metastasis-inducer Snail and induction of the metastasis-suppressor RKIP. Cell Cycle 2010;9:4931–40.

[86] Julien S, Puig I, Caretti E, Bonaventure J, Nelles L, van Roy F, Dargemont C, de Herreros AG, Bellacosa A, Larue L. Activation of NF-kappaB by Akt upregulates Snail expression and induces epithelium mesenchyme transition. Oncogene 2007;26:7445–56.

[87] Zhang K, Zhaos J, Liu X, Yan B, Chen D, Gao Y, Hu X, Liu S, Zhang D, Zhou C. Activation of NF-B upregulates Snail and consequent repression of E-cadherin in cholangiocarcinoma cell invasion. Hepatogastroenterology 2011;58:1–7.

[88] Beach S, Tang H, Park S, Dhillon AS, Keller ET, Kolch W, Yeung KC. Snail is a repressor of RKIP transcription in metastatic prostate cancer cells. Oncogene 2008;27:2243–8.

[89] Datar I, Tegegne H, Qin K, Al-Mulla F, Bitar MS, Trumbly RJ, Yeung KC. Genetic and epigenetic control of RKIP transcription. Crit Rev Oncog 2014;19:417–30.

[90] Lee JY, Kong G. Roles and epigenetic regulation of epithelial-mesenchymal transition and its transcription factors in cancer initiation and progression. Cell Mol Life Sci 2016;73: 4643–60.

[91] Ren G, Baritaki S, Marathe H, Feng J, Park S, Beach S, Bazeley PS, Beshir AB, Fenteany G, Mehra R, Daignault S, Al-Mulla F, Keller E, Bonavida B, de la Serna I, Yeung KC. Polycomb protein EZH2 regulates tumor invasion via the transcriptional repression of the metastasis suppressor RKIP in breast and prostate cancer. Cancer Res 2012;72:3091–104.

[92] Yeung KC, Rose DW, Dhillon AS, Yaros D, Gustafsson M, Chatterjee D, McFerran B, Wyche J, Kolch W, Sedivy JM. Raf kinase inhibitor protein interacts with NF-kappaB-inducing kinase and TAK1 and inhibits NF-kappaB activation. Mol Cell Biol 2001;21:7207–17.

[93] Singh T, Chaudhary SC, Kapur P, Weng Z, Elmets CA, Kopelovich L, Athar M. Nitric oxide donor exisulind is an effective inhibitor of murine photocarcinogenesis. Photochem Photobiol 2012;88:1141–8.

[94] Wang H, Hertlein E, Bakkar N, Sun H, Acharyya S, Wang J, Carathers M, Davuluri R, Guttridge DC. NF-kappaB regulation of YY1 inhibits skeletal myogenesis through transcriptional silencing of myofibrillar genes. Mol Cell Biol 2007;27:4374–87.

[95] Huerta-Yepez S, Vega M, Jazirehi A, Garban H, Hongo F, Cheng G, Bonavida B. Nitric oxide sensitizes prostate carcinoma cell lines to TRAIL-mediated apoptosis via inactivation of NF-kappa B and inhibition of Bcl-xl expression. Oncogene 2004;23:4993–5003.

[96] Palmer MB, Majumder P, Cooper JC, Yoon H, Wade PA, Boss JM. Yin yang 1 regulates the expression of snail through a distal enhancer. Mol Cancer Res 2009;7:221–9.

[97] Rapozzi V, Della Pietra E, Bonavida B. Dual roles of nitric oxide in the regulation of tumor cell response and resistance to photodynamic therapy. Redox Biol 2015;6:311–7.

[98] Pan X, Wang X, Lei W, Min L, Yang Y, Song J. Nitric oxide suppresses transforming growth factor-beta1-induced epithelial-to-mesenchymal transition and apoptosis in mouse hepatocytes. Hepatology 2009;50:1577–87.

[99] Vyas-Read S, Shaul PW, Yuhanna IS, Willis BC. Nitric oxide attenuates epithelial-mesenchymal transition in alveolar epithelial cells. Am J Physiol Lung Cell Mol Physiol 2007;293:L212–21.

[100] Guo Y, Li P, Bledsoe G, Yang ZR, Chao L, Chao J. Kallistatin inhibits TGF-beta-induced endothelial-mesenchymal transition by differential regulation of microRNA-21 and eNOS expression. Exp Cell Res 2015;337:103–10.

[101] Le X, Wei D, Huang S, Lancaster Jr. JR, Xie K. Nitric oxide synthase II suppresses the growth and metastasis of human cancer regardless of its up-regulation of protumor factors. Proc Natl Acad Sci U S A 2005;102:8758–63.

[102] Akool el S, Kleinert H, Hamada FM, Abdelwahab MH, Forstermann U, Pfeilschifter J, Eberhardt W. Nitric oxide increases the decay of matrix metalloproteinase 9 mRNA by inhibiting the expression of mRNA-stabilizing factor HuR. Mol Cell Biol 2003;23:4901–16.

[103] Jespersen C, Doller A, Akool el S, Bachmann M, Muller R, Gutwein P, Muhl H, Pfeilschifter J, Eberhardt W. Molecular mechanisms of nitric oxide-dependent inhibition of TPA-induced matrix metalloproteinase-9 (MMP-9) in MCF-7 cells. J Cell Physiol 2009;219:276–87.

[104] Weiss JM, Ridnour LA, Back T, Hussain SP, He P, Maciag AE, Keefer LK, Murphy WJ, Harris CC, Wink DA, Wiltrout RH. Macrophage-dependent nitric oxide expression regulates tumor cell detachment and metastasis after IL-2/anti-CD40 immunotherapy. J Exp Med 2010;207:2455–67.

[105] Hickok JR, Sahni S, Mikhed Y, Bonini MG, Thomas DD. Nitric oxide suppresses tumor cell migration through N-Myc downstream-regulated gene-1 (NDRG1) expression: role of chelatable iron. J Biol Chem 2011;286:41413–24.

[106] Snyder V, Reed-Newman TC, Arnold L, Thomas SM, Anant S. Cancer stem cell metabolism and potential therapeutic targets. Front Oncol 2018;8:203.

[107] Gouge RC, Marshburn P, Gordon BE, Nunley W, Huet-Hudson YM. Nitric oxide as a regulator of embryonic development. Biol Reprod 1998;58:875–9.

[108] Yang SR, Rahman I, Trosko JE, Kang KS. Oxidative stress-induced biomarkers for stem cell-based chemical screening. Prev Med 2012;54(Suppl):S42–9.

[109] Guo Y, Einhorn L, Kelley M, Hirota K, Yodoi J, Reinbold R, Scholer H, Ramsey H, Hromas R. Redox regulation of the embryonic stem cell transcription factor oct-4 by thioredoxin. Stem Cells 2004;22:259–64.

[110] Mora-Castilla S, Tejedo JR, Hmadcha A, Cahuana GM, Martin F, Soria B, Bedoya FJ. Nitric oxide repression of Nanog promotes mouse embryonic stem cell differentiation. Cell Death Differ 2010;17:1025–33.

[111] Kim SJ, Lim MS, Kang SK, Lee YS, Kang KS. Impaired functions of neural stem cells by abnormal nitric oxide-mediated signaling in an in vitro model of Niemann-Pick type C disease. Cell Res 2008;18:686–94.

[112] Kim SJ, Park JS, Kang KS. Stem cells in Niemann-Pick disease. Dis Markers 2008;24:231–8.

[113] Potts MB, Rola R, Claus CP, Ferriero DM, Fike JR, Noble-Haeusslein LJ. Glutathione peroxidase overexpression does not rescue impaired neurogenesis in the injured immature brain. J Neurosci Res 2009;87:1848–57.

[114] Diehn M, Cho RW, Lobo NA, Kalisky T, Dorie MJ, Kulp AN, Qian D, Lam JS, Ailles LE, Wong M, Joshua B, Kaplan MJ, Wapnir I, Dirbas FM, Somlo G, Garberoglio C, Paz B, Shen J, Lau SK, Quake SR, Brown JM, Weissman IL, Clarke MF. Association of reactive oxygen species levels and radioresistance in cancer stem cells. Nature 2009;458:780–3.

[115] Ghaffari S. Oxidative stress in the regulation of normal and neoplastic hematopoiesis. Antioxid Redox Signal 2008;10:1923–40.

[116] Naka K, Muraguchi T, Hoshii T, Hirao A. Regulation of reactive oxygen species and genomic stability in hematopoietic stem cells. Antioxid Redox Signal 2008;10:1883–94.

[117] Phillips TM, McBride WH, Pajonk F. The response of CD24(-/low)/CD44+ breast cancer-initiating cells to radiation. J Natl Cancer Inst 2006;98:1777–85.

[118] Toyokuni S. Novel aspects of oxidative stress-associated carcinogenesis. Antioxid Redox Signal 2006;8:1373–7.

[119] Trachootham D, Alexandre J, Huang P. Targeting cancer cells by ROS-mediated mechanisms: a radical therapeutic approach? Nat Rev Drug Discov 2009;8:579–91.

[120] Dalerba P, Dylla SJ, Park IK, Liu R, Wang X, Cho RW, Hoey T, Gurney A, Huang EH, Simeone DM, Shelton AA, Parmiani G, Castelli C, Clarke MF. Phenotypic characterization of human colorectal cancer stem cells. Proc Natl Acad Sci U S A 2007;104:10158–63.

[121] Ishimoto T, Nagano O, Yae T, Tamada M, Motohara T, Oshima H, Oshima M, Ikeda T, Asaba R, Yagi H, Masuko T, Shimizu T, Ishikawa T, Kai K, Takahashi E, Imamura Y, Baba Y, Ohmura M, Suematsu M, Baba H, Saya H. CD44 variant regulates redox status in cancer cells by stabilizing the xCT subunit of system xc(-) and thereby promotes tumor growth. Cancer Cell 2011;19:387–400.

[122] Tang S, Mishra M, Frazier DP, Moore ML, Inoue K, Deora R, Sui G, Dubey P. Positive and negative regulation of prostate stem cell antigen expression by Yin Yang 1 in prostate epithelial cell lines. PLoS One 2012;7:e35570.

[123] Baritaki S, Rapozzi V, Pelayo R, Vega M, Huerta-Yepez S, Bonavida B. Yin Yang 1 (YY1) regulates EMT via transcriptional activation of Snail and its overexpression in cancer stem cells (CSCs) in prostate cancer: clinical significance, In Proceedings of the AACR-NCI-EORTC International Conference:Molecular Targets and Cancer Therapeutics, Boston, MA. Philadelphia (PA), Mol Cancer Ther 2013;12(11 Suppl):Abstract nr A53.

[124] Huerta S, Baay-Guzman G, Gonzalez-Bonilla CR, Livingston EH, Huerta-Yepez S, Bonavida B. In vitro and in vivo sensitization of SW620 metastatic colon cancer cells to CDDP-induced apoptosis by the nitric oxide donor DETANONOate: involvement of AIF. Nitric Oxide 2009;20:182–94.

[125] Bonavida B, Khineche S, Huerta-Yepez S, Garban H. Therapeutic potential of nitric oxide in cancer. Drug Resist Updat 2006;9:157–73.

[126] Lee JY, Huerta-Yepez S, Vega M, Baritaki S, Spandidos DA, Bonavida B. The NO TRAIL to YES TRAIL in cancer therapy (review). Int J Oncol 2007;31:685–91.

[127] Baritaki S, Huerta-Yepez S, Sakai T, Spandidos DA, Bonavida B. Chemotherapeutic drugs sensitize cancer cells to TRAIL-mediated apoptosis: up-regulation of DR5 and inhibition of Yin Yang 1. Mol Cancer Ther 2007;6:1387–99.

[128] Garban HJ, Bonavida B. Nitric oxide sensitizes ovarian tumor cells to Fas-induced apoptosis. Gynecol Oncol 1999;73:257–64.

[129] Garban HJ, Bonavida B. Nitric oxide inhibits the transcription repressor Yin-Yang 1 binding activity at the silencer region of the Fas promoter: a pivotal role for nitric oxide in the up-regulation of Fas gene expression in human tumor cells. J Immunol 2001;167:75–81.

[130] Tang Z, Bauer JA, Morrison B, Lindner DJ. Nitrosylcobalamin promotes cell death via S nitrosylation of Apo2L/TRAIL receptor DR4. Mol Cell Biol 2006;26:5588–94.

[131] Plenchette S. Role Of S-nitrosylation in the extrinsic apoptotic signalling pathway in cancer. Redox Biol 2015;5:415.

[132] Baritaki S, Yeung K, Palladino M, Berenson J, Bonavida B. Pivotal roles of snail inhibition and RKIP induction by the proteasome inhibitor NPI-0052 in tumor cell chemoimmunosensitization. Cancer Res 2009;69:8376–85.

[133] Escriva M, Peiro S, Herranz N, Villagrasa P, Dave N, Montserrat-Sentis B, Murray SA, Franci C, Gridley T, Virtanen I, Garcia de Herreros A. Repression of PTEN phosphatase by Snail1 transcriptional factor during gamma radiation-induced apoptosis. Mol Cell Biol 2008;28: 1528–40.

[134] Petrella BL, Brinckerhoff CE. PTEN suppression of YY1 induces HIF-2 activity in von-Hippel-Lindau-null renal-cell carcinoma. Cancer Biol Ther 2009;8:1389–401.

[135] Rapozzi V, Della Pietra E, Zorzet S, Zacchigna M, Bonavida B, Xodo LE. Nitric oxide-mediated activity in anti-cancer photodynamic therapy. Nitric Oxide 2013;30:26–35.

[136] Krummel MF. Illuminating emergent activity in the immune system by real-time imaging. Nat Immunol 2010;11:554–7.

[137] Amin AR. The metabolomics of nitric oxide and reactive nitrogen species in immune editing tumor milieu: influence of nitric oxide-modulating therapies. J Drug Metab Toxicol 2012; https://doi.org/10.4172/2157-7609.S8-002.

[138] Weis SM, Cheresh DA. Tumor angiogenesis: molecular pathways and therapeutic targets. Nat Med 2011;17:1359–70.

[139] Pathi SS, Jutooru I, Chadalapaka G, Sreevalsan S, Anand S, Thatcher GR, Safe S. GT-094, a NO-NSAID, inhibits colon cancer cell growth by activation of a reactive oxygen species-microRNA-27a: ZBTB10-specificity protein pathway. Mol Cancer Res 2011;9:195–202.

[140] Jones MK, Tsugawa K, Tarnawski AS, Baatar D. Dual actions of nitric oxide on angiogenesis: possible roles of PKC, ERK, and AP-1. Biochem Biophys Res Commun 2004;318:520–8.

[141] Tsurumi Y, Murohara T, Krasinski K, Chen D, Witzenbichler B, Kearney M, Couffinhal T, Isner JM. Reciprocal relation between VEGF and NO in the regulation of endothelial integrity. Nat Med 1997;3:879–86.

[142] Kiziltepe T, Anderson KC, Kutok JL, Jia L, Boucher KM, Saavedra JE, Keefer LK, Shami PJ. JS-K has potent anti-angiogenic activity in vitro and inhibits tumour angiogenesis in a multiple myeloma model in vivo. J Pharm Pharmacol 2010;62:145–51.

[143] Pipili-Synetos E, Papageorgiou A, Sakkoula E, Sotiropoulou G, Fotsis T, Karakiulakis G, Maragoudakis ME. Inhibition of angiogenesis, tumour growth and metastasis by the NO-releasing vasodilators, isosorbide mononitrate and dinitrate. Br J Pharmacol 1995;116: 1829–34.

[144] Pipili-Synetos E, Sakkoula E, Haralabopoulos G, Andriopoulou P, Peristeris P, Maragoudakis ME. Evidence that nitric oxide is an endogenous antiangiogenic mediator. Br J Pharmacol 1994;111:894–902.

[145] Pipili-Synetos E, Sakkoula E, Maragoudakis ME. Nitric oxide is involved in the regulation of angiogenesis. Br J Pharmacol 1993;108:855–7.

[146] Cooke JP. NO and angiogenesis. Atheroscler Suppl 2003;4:53–60.

[147] Chouaib S, Umansky V, Kieda C. The role of hypoxia in shaping the recruitment of proangiogenic and immunosuppressive cells in the tumor microenvironment. Contemp Oncol (Pozn) 2018;22:7–13.

[148] Wang JJ, Lei KF, Han F. Tumor microenvironment: recent advances in various cancer treatments. Eur Rev Med Pharmacol Sci 2018;22:3855–64.

[149] Hussain SP, Hofseth LJ, Harris CC. Radical causes of cancer. Nat Rev Cancer 2003;3: 276–85.

[150] Mantovani A, Allavena P, Sica A, Balkwill F. Cancer-related inflammation. Nature 2008;454: 436–44.

[151] Schreiber RD, Old LJ, Smyth MJ. Cancer immunoediting: integrating immunity's roles in cancer suppression and promotion. Science 2011;331:1565–70.

[152] Aiello S, Noris M, Piccinini G, Tomasoni S, Casiraghi F, Bonazzola S, Mister M, Sayegh MH, Remuzzi G. Thymic dendritic cells express inducible nitric oxide synthase and generate nitric oxide in response to self- and alloantigens. J Immunol 2000;164:4649–58.

[153] Allavena P, Sica A, Solinas G, Porta C, Mantovani A. The inflammatory micro-environment in tumor progression: the role of tumor-associated macrophages. Crit Rev Oncol Hematol 2008;66:1–9.

[154] Molon B, Ugel S, Del Pozzo F, Soldani C, Zilio S, Avella D, De Palma A, Mauri P, Monegal A, Rescigno M, Savino B, Colombo P, Jonjic N, Pecanic S, Lazzarato L, Fruttero R, Gasco A, Bronte V, Viola A. Chemokine nitration prevents intratumoral infiltration of antigen-specific T cells. J Exp Med 2011;208:1949–62.

[155] Cifone MG, Festuccia C, Cironi L, Cavallo G, Chessa MA, Pensa V, Tubaro E, Santoni A. Induction of the nitric oxide-synthesizing pathway in fresh and interleukin 2-cultured rat natural killer cells. Cell Immunol 1994;157:181–94.

[156] Miller MR, Megson IL. Recent developments in nitric oxide donor drugs. Br J Pharmacol 2007;151:305–21.

[157] Pervin S, Singh R, Gau CL, Edamatsu H, Tamanoi F, Chaudhuri G. Potentiation of nitric oxide-induced apoptosis of MDA-MB-468 cells by farnesyltransferase inhibitor: implications in breast cancer. Cancer Res 2001;61:4701–6.

[158] Yang L, Lan C, Fang Y, Zhang Y, Wang J, Guo J, Wan S, Yang S, Wang R, Fang D. Sodium nitroprusside (SNP) sensitizes human gastric cancer cells to TRAIL-induced apoptosis. Int Immunopharmacol 2013;17:383–9.

[159] Ning S, Bednarski M, Oronsky B, Scicinski J, Knox SJ. Novel nitric oxide generating compound glycidyl nitrate enhances the therapeutic efficacy of chemotherapy and radiotherapy. Biochem Biophys Res Commun 2014;447:537–42.

[160] Sukhatme V, Bouche G, Meheus L, Sukhatme VP, Pantziarka P. Repurposing drugs in oncology (ReDO)-nitroglycerin as an anti-cancer agent. Ecancermedicalscience 2015;9:568.

[161] Dingemans AM, Groen HJ, Herder GJ, Stigt JA, Smit EF, Bahce I, Burgers JA, van den Borne BE, Biesma B, Vincent A, van der Noort V, Aerts JG. A randomized phase II study comparing paclitaxel-carboplatin-bevacizumab with or without nitroglycerin patches in patients with stage IV nonsquamous nonsmall-cell lung cancer: NVALT12 (NCT01171170)dagger. Ann Oncol 2015;26:2286–93.

[162] Reinmuth N, Meyer A, Hartwigsen D, Schaeper C, Huebner G, Skock-Lober R, Bier A, Gerecke U, Held CP, Reck M. Randomized, double-blind phase II study to compare nitroglycerin plus oral vinorelbine plus cisplatin with oral vinorelbine plus cisplatin alone in patients with stage IIIB/IV non-small cell lung cancer (NSCLC). Lung Cancer 2014;83:363–8.

[163] Yasuda H, Yamaya M, Nakayama K, Sasaki T, Ebihara S, Kanda A, Asada M, Inoue D, Suzuki T, Okazaki T, Takahashi H, Yoshida M, Kaneta T, Ishizawa K, Yamanda S, Tomita N, Yamasaki M, Kikuchi A, Kubo H, Sasaki H. Randomized phase II trial comparing nitroglycerin plus vinorelbine and cisplatin with vinorelbine and cisplatin alone in previously untreated stage IIIB/IV non-small-cell lung cancer. J Clin Oncol 2006;24:688–94.

[164] Illum H, Wang DH, Dowell JE, Hittson WJ, Torrisi JR, Meyer J, Huerta S. Phase I dose escalation trial of nitroglycerin in addition to 5-fluorouracil and radiation therapy for neoadjuvant treatment of operable rectal cancer. Surgery 2015;158:460–5.

[165] Leung FP, Yung LM, Laher I, Yao X, Chen ZY, Huang Y. Exercise, vascular wall and cardiovascular diseases: an update (Part 1). Sports Med 2008;38:1009–24.

[166] Perez-Sala D, Cernuda-Morollon E, Diaz-Cazorla M, Rodriguez-Pascual F, Lamas S. Posttranscriptional regulation of human iNOS by the NO/cGMP pathway. Am J Physiol Renal Physiol 2001;280:F466–73.

[167] Ning S, Bednarski M, Oronsky B, Scicinski J, Saul G, Knox SJ. Dinitroazetidines are a novel class of anticancer agents and hypoxia-activated radiation sensitizers developed from highly energetic materials. Cancer Res 2012;72:2600–8.

[168] Oronsky B, Scicinski J, Ning S, Peehl D, Oronsky A, Cabrales P, Bednarski M, Knox S. RRx-001, a novel dinitroazetidine radiosensitizer. Invest New Drugs 2016;34:371–7.

[169] Thatcher GR, Nicolescu AC, Bennett BM, Toader V. Nitrates and NO release: contemporary aspects in biological and medicinal chemistry. Free Radic Biol Med 2004;37:1122–43.

[170] Chaudhary SC, Singh T, Kapur P, Weng Z, Arumugam A, Elmets CA, Kopelovich L, Athar M. Nitric oxide-releasing sulindac is a novel skin cancer chemopreventive agent for UVB-induced photocarcinogenesis. Toxicol Appl Pharmacol 2013;268:249–55.

[171] Cai TB, Tang X, Nagorski J, Brauschweiger PG, Wang PG. Synthesis and cytotoxicity of 5-fluorouracil/diazeniumdiolate conjugates. Bioorg Med Chem 2003;11:4971–5.

[172] Zhao J, Gou S, Sun Y, Fang L, Wang Z. Antitumor platinum(II) complexes containing platinum-based moieties of present platinum drugs and furoxan groups as nitric oxide donors: synthesis, DNA interaction, and cytotoxicity. Inorg Chem 2012;51:10317–24.

[173] Zhao J, Gou S, Sun Y, Yin R, Wang Z. Nitric oxide donor-based platinum complexes as potential anticancer agents. Chemistry 2012;18:14276–81.

[174] Wang T, Zhang YH, Kong XW, Lai YS, Ji H, Chen YP, Peng SX. Synthesis and biological evaluation of nitric oxide-donating thalidomide analogues as anticancer agents. Chem Biodivers 2009;6:466–74.

[175] Bian H, Feng J, Li M, Xu W. Novel antileukemic agents derived from tamibarotene and nitric oxide donors. Bioorg Med Chem Lett 2011;21:7025–9.

[176] Huang Z, Fu J, Liu L, Sun Y, Lai Y, Ji H, Knaus EE, Tian J, Zhang Y. Glycosylated diazeniumdiolate-based oleanolic acid derivatives: synthesis, in vitro and in vivo biological evaluation as anti-human hepatocellular carcinoma agents. Org Biomol Chem 2012;10:3882–91.

[177] Hrabie JA, Keefer LK. Chemistry of the nitric oxide-releasing diazeniumdiolate ("nitrosohydroxylamine") functional group and its oxygen-substituted derivatives. Chem Rev 2002;102:1135–54.

[178] Keefer LK. Fifty years of diazeniumdiolate research. From laboratory curiosity to broad-spectrum biomedical advances. ACS Chem Biol 2011;6:1147–55.

[179] Maeda H, Nakamura H, Fang J. The EPR effect for macromolecular drug delivery to solid tumors: improvement of tumor uptake, lowering of systemic toxicity, and distinct tumor imaging in vivo. Adv Drug Deliv Rev 2013;65:71–9.

[180] Quinn JF, Whittaker MR, Davis TP. Delivering nitric oxide with nanoparticles. J Control Release 2015;205:190–205.

[181] Chung MF, Liu HY, Lin KJ, Chia WT, Sung HW. A pH-responsive carrier system that generates NO bubbles to trigger drug release and reverse P-glycoprotein-mediated multidrug resistance. Angew Chem Int Ed Engl 2015;54:9890–3.

[182] Tan J, Zeng Q, Jiang XZ, He LY, Wang JR, Yao K, Wang CH. Apoptosis of bladder transitional cell carcinoma T24 cells induced by adenovirus-mediated inducible nitric oxide synthase gene transfection. Chin J Cancer Res 2013;25:593–9.

[183] Ye S, Yang W, Wang Y, Ou W, Ma Q, Yu C, Ren J, Zhong G, Shi H, Yuan Z, Su X, Zhu W. Cationic liposome-mediated nitric oxide synthase gene therapy enhances the antitumor effects of cisplatin in lung cancer. Int J Mol Med 2013;31:33–42.

CHAPTER 3

Antitumor Effect of a Metal-Nonoate Through Angiogenesis Impairment

Valerio Ciccone*, Lucia Morbidelli*,†
**Department of Life Sciences, University of Siena, Siena, Italy*
†Noxamet Ltd, Milan, Italy

ABSTRACT

Nitric oxide (NO) has been reported to induce different effects on tumor biology depending on its concentration and availability. We studied the antitumor activity of Ni(SalPipNONO), a new NO donor belonging to the metal-nonoate family. The antitumor activity of nonoate was characterized on A549 cells (human lung carcinoma). Ni(SalPipNONO) reduced tumor cell survival, clonogenicity, and invasiveness and promoted apoptosis. The effect was due, in part, to the classic NO/cGMP pathway and, in part, to reactive oxygen species (ROS) production. ROS promoted ERK1/2 activation, which in its turn induced p53 upregulation. Finally, we assessed the anti-angiogenic activity of the metal-nonoate. Ni(SalPipNONO) treatment reduced the angiogenic factor expressions of tumor cells (hypoxia-inducible factor-1α (HIF-1α) and vascular endothelial growth factor (VEGF)) that were associated with the impairment of the endothelial cell function that is related to angiogenesis.

Overall, Ni(SalPipNONO) exerts antitumor activity through different mechanisms, involving an anti-angiogenic effect. Further studies should be done to develop this compound alone or in combination with conventional therapy for the treatment of cancer.

ABBREVIATIONS

cGMP	cyclic 3′,5′-guanosine monophosphate
DCFH-DA	2′,7′-dichlorodihydrofluorescein diacetate
HIF-1α	hypoxia-inducible factor-1α
HUVEC	human umbilical vein endothelial cells
MTT	3-(4,5-dimethylthiazol-2-yl)-2,5-diphenyltetrazolium bromide
NAC	*N*-acetyl-L-cysteine
NO	nitric oxide

Therapeutic Application of Nitric Oxide in Cancer and Inflammatory Disorders. https://doi.org/10.1016/B978-0-12-816545-4.00003-7

ODQ	1H-[1,2,4]oxadiazolo[4,3-a]quinoxalin-1-one
PT-α	pifithrin-alpha
ROS	reactive oxygen species
sGC	soluble guanylate cyclase
VEGF	vascular endothelial growth factor

Conflict of Interest

No potential conflicts of interest were disclosed.

INTRODUCTION

From the first studies reported on the effect of nitric oxide (NO) in cancer biology, this mediator emerged as a biphasic modulator, behaving as both an antineoplastic and a proneoplastic stimuli [1]. The bimodal actions of NO can be explained through the duration of NO exposure, the cellular microenvironment, the tumor cell proliferation rate, and the occurrence of oxidizing and reducing processes [1].

The established biochemical/cellular events elicited by NO against tumor development are essentially mediated by the inhibition of cell proliferation and proapoptotic events and vascular effects including anti-angiogenesis [2]. Tumor tissue, characterized by low oxygen tension, presents an activation and stabilization of the hypoxia-inducible factor-1α (HIF-1α) that, in turn, controls the transcription of vascular endothelial growth factor (VEGF), thus promoting angiogenesis, tumor growth, and metastasis. NO has been reported to inhibit the expression of HIF-1α through the activation of HIF-1 prolyl-hydroxylases and its proteasomal degradation [3].

Recently, a new family of metal-nonoates has been developed [4] and characterized for their potential use in cardiovascular diseases and exerted by endothelial dysfunction and achieving a vascular protective effect at nanomolar concentrations [5]. Herein, we have evaluated the antitumor activity of a member of this class, Ni(SalPipNONO), and assessed the antitumor efficacy in two epithelial-derived tumor cell lives, A549 and HT29, representative of lung and colon carcinomas, respectively. Ni(SalPipNONO) was characterized for different underlying mechanisms related to tumor hallmarks and for its anti-angiogenic effects on both tumor and endothelial cells.

RESULTS

To study the effect of nonoate on tumor cell survival, we performed the MTT assay on A549 cells. Ni(SalPipNONO), compared with equimolar concentrations of DETA/NO, was more effective in reducing cell number, in particular in the range of 0.1–1 mM ($p < 0.05$ and $p < 0.001$ vs untreated cells and $p < 0.05$ and $p < 0.001$ Ni(SalPipNONO) vs DETA/NO). Similar results were obtained with the human colon adenocarcinoma cells HT29. Besides cytotoxic/antiproliferative activity, we have studied the capability of the metal-nonoate to interfere with the clonogenicity of A549 cells. The data showed that at 0.5 and 1 mM of Ni(SalPipNONO), the surviving fraction was reduced by >90% with respect to the untreated control ($p < 0.001$ vs untreated cells), while DETA/NO exerted a modest, nonsignificant, effect. To study tumor invasiveness, we performed the chemotaxis assay. After 18 h of exposure to 0.5 mM Ni(SalPipNONO), we detected half of the cells migrating across the filter toward the lower chamber in the presence of a chemoattractant agent (10% serum) ($p < 0.01$ vs untreated cells).

We then evaluated the mechanisms responsible for the tumor inhibitory effects. First, we evaluated the involvement of the classical soluble guanylate cyclase (sGC)/cGMP pathway by the use of 1H-[1,2,4]oxadiazolo[4,3-a]quinoxalin-1-one (ODQ). The preincubation with ODQ (10 μM, 30 min) before Ni(SalPipNONO) treatment only partially reverted the NO donor cytotoxic effect ($p < 0.05$), suggesting that other pathways or chemical derivatives of the molecule could be responsible of the antitumor actions. The oxidative stress and ROS production have been described to contribute to the cytotoxic effect of high doses on NO, due to its radical nature [6]. We measured ROS levels in A549 cells stimulated with the metal-nonoate by means of the fluorophore DCFH-DA, and we found a significant burst in ROS production ($p < 0.001$ vs untreated cells) that was abolished by *N*-acetyl-L-cysteine (NAC) pretreatment.

Concerning the moiety of the molecule responsible for the cytotoxic activity, we observed that the compound with the salicylaldehyde group exerted the greater cytotoxic activity (probably due to its metabolism into salicylic acid), with a synergetic effect with the NONO group. Since the compound devoid of the NONO group did not create an oxidative environment in the cells, a double mechanism of cytotoxic activity could be hypothesized; namely, the NONO group induces an oxidative stress, and the salicylaldehyde exerts an additional mechanism.

From a mechanistic point of view, it is known of the ability of NO to promote apoptosis in cancer cells [7]. Exposure of cells to Ni(SalPipNONO) (0.5 mM, 15 min) doubled the cytochrome *c* levels, through the reactive species generated by NO (like peroxynitrite and ROS), that cause opening of the permeability transition pores of the mitochondria. Moreover, it is known that ROS activate the ERK1/2 [8] pathway and in A549 cells treated with Ni(SalPipNONO), but not with DETA/NO, strongly increased pERK1/2 activity after 30 min of incubation. Activated ERK1/2, in turn, upregulated p53, and in our experiments, the p53 maximum level was obtained after 60 min of incubation with the metal-nonoate. Indeed, the preincubation of the cells with the MEK inhibitor U0126 interfered with p53 upregulation.

The ultimate event downstream of cytochrome *c* release is caspase-3 activation that leads to DNA cleavage and apoptosis. Our data showed a maximum production (triple with respect to basal level) of cleaved caspase-3 after 4 h of exposure with Ni(SalPipNONO), an effect that was completely abolished by preincubation with the p53 inhibitor PT-α ($p < 0.001$ vs Ni(SalPipNONO) alone).

From all the above data, it results that Ni(SalPipNONO) impairs tumor cell migration and survival by activating the apoptotic pathway that is mediated through ROS production and COX-2 activation, besides the partial involvement of the sGC/cGMP conventional signaling.

It is known that the expression of proangiogenic factors is related to any increased risk of mortality in several types of cancer [9]. With A549 cells, we found HIF-1α detectable levels already under normoxic conditions. When A549 cells were treated for 24 h with Ni(SalPipNONO), there was a substantial decrease of HIF-1α expression and a consequent reduction of VEGF levels. These results demonstrate that the metal-nonoate reduces the angiogenetic signals in tumor cells. We then assessed whether the anti-angiogenic property was directly induced in endothelial cells. HUVEC were exposed to VEGF (20 ng/ml) in the presence of Ni(SalPipNONO). At 0.1 mM, the NO donor completely abolished VEGF-induced proliferation. The anti-angiogenic activity of Ni(SalPipNONO) was also evident in an *in vivo* angiogenesis assay using the subcutaneous Matrigel plug implant. After 10 days, the presence of metal-nonoate abolished the vascularization produced by VEGF ($p < 0.001$ vs VEGF alone).

Finally, coculture experiments with tumor cells and endothelial cells were set up to strengthen the above results. The endothelial cell organization on Matrigel layers was evaluated in the presence of tumor cells grown on transwell inserts and treated or not with Ni(SalPipNONO). In the presence of untreated A549, HUVEC formed netlike structures, which were impaired when the tumor cells were treated with the NO donor ($p < 0.001$ vs untreated cells).

CONCLUSIONS

In our *in vitro* study, we demonstrated that the anti-angiogenic activity of Ni (SalPipNONO) is both direct on endothelial cells and indirect on tumor cells, which were not able to upregulate VEGF [10] (Fig. 1).

Our data confirm the potential use of NO donors in the treatment of resistant cancers, and in particular Ni(SalPipNONO), by acting through multiple mechanisms, as a chemotherapeutic agent. The lead compound needs to be further developed in its chemical formula and formulation in order to be safely used alone or in combination with anticancer conventional therapies.

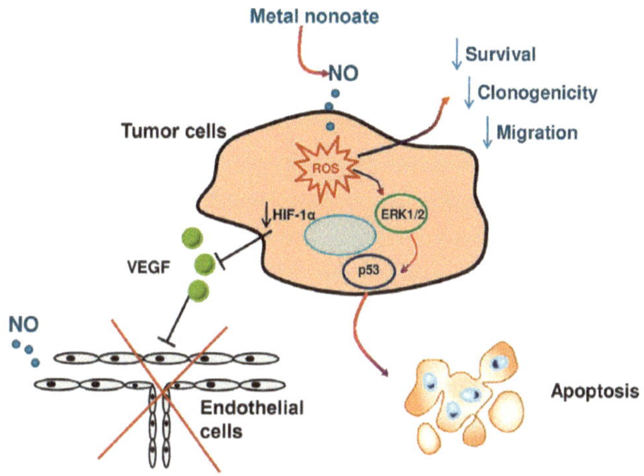

FIG. 1

Schematic representation of the antitumor and anti-angiogenic effects of Ni(SalPipNONO). Nitric oxide released by Ni(SalPipNONO) causes a burst of ROS in tumor cells leading to reduced survival, clonogenicity, and migration of these cells. In addition, ROS activate ERK1/2 that, in turn, induces p53 upregulation and apoptosis. The metal-nonoate reduces the expression of HIF-1α and VEGF in tumor cells that are associated with decreased angiogenic functions in endothelial cells. Additionally, a direct anti-angiogenic effect is observed when endothelial cells are treated with Ni(SalPipNONO).

Acknowledgments

We thank Dr. Martina Monti (University of Siena and Noxamet Ltd) for scientific and technical supervision. We would like to thank Prof. Luigi Casella and Enrico Monzani (University of Pavia and Noxamet Ltd) for providing the metal-nonoates.

References

[1] Burke AJ, Sullivan FJ, Giles FJ, Glynn SA. The yin and yang of nitric oxide in cancer progression. Carcinogenesis 2013;34:503–12.

[2] Morbidelli L. Therapeutic potential of nitric oxide donors in cancer: focus on angiogenesis. Crit Rev Oncogen 2016;21(5–6):447–58.

[3] Sogawa K, Numayama-Tsuruta K, Ema M, Abe M, Abe H, Fujii-Kuriyama Y. Inhibition of hypoxia-inducible factor 1 activity by nitric oxide donors in hypoxia. Proc Natl Acad Sci U S A 1998;95:7368–73.

[4] Ziche M, Donnini S, Morbidelli L, Monzani E, Roncone R, Gabbini R, Casella L. Nitric oxide releasing metal-diazeniumdiolate complexes strongly induce vasorelaxation and endothelial cell proliferation. ChemMedChem 2008;3:1039–47.

[5] Monti M, Ciccone V, Pacini A, Roggeri R, Monzani E, Casella L, Morbidelli L. Anti-hypertensive property of a nickel-piperazine/NO donor in spontaneously hypertensive rats. Pharmacol Res 2016;107:352–9.

[6] Porasuphatana S, Tsai P, Rosen GM. The generation of free radicals by nitric oxide synthase. Comp Biochem Physiol C Toxicol Pharmacol 2003;134:281–9.

[7] Kim PK, Zamora R, Petrosko P, Billiar TR. The regulatory role of nitric oxide in apoptosis. Int Immunopharmacol 2001;1:1421–41.

[8] Lee YJ, Cho HN, Soh JW, Jhon GJ, Cho CK, Chung HY, Bae S, Lee SJ, Lee YS. Oxidative stress-induced apoptosis is mediated by ERK1/2 phosphorylation. Exp Cell Res 2003;291:251–66.

[9] Semenza GL. Targeting HIF-1 for cancer therapy. Nat Rev Cancer 2003;3:721–32.

[10] Ciccone V, Monti M, Monzani E, Casella L, Morbidelli L. The metal-nonoate Ni(SalPipNONO) inhibits in vitro tumor growth, invasiveness and angiogenesis. Oncotarget 2018; 9(17):13353–65.

CHAPTER 4

Therapeutic Implications of the Nitric Oxide Pathway in the Angiogenesis of Tumors and Inflammatory-Related Disorders

Lucia Morbidelli*[,†], Sandra Donnini*, Marina Ziche*

Department of Life Sciences, University of Siena, Siena, Italy

[†]Noxamet Ltd, Milan, Italy

ABSTRACT

Not only is angiogenesis, the development of new vessels from preexisting ones, a prerequisite for the physiological growth, repair, and functioning of our organs, but also it concurs, as examples, with the pathologies of cancer, eye diseases, and chronic degenerative disorders. In most of the pathological conditions, an inflammatory and oxidative state exists, supporting deregulation of angiogenesis. Endothelial cells, the core of angiogenesis, produce nanomolar levels of nitric oxide (NO), guaranteeing proper vessel tone, blood fluidity, and a noninflammatory environment. In contrast, in the presence of a growing tumor or pathological conditions related to inflammation and redox imbalance, higher (micromolar) NO levels are present, contributing to the development of an aberrant neovascularization. Nevertheless, the use of NO donor drugs is proposed in some types of tumors as an antitumor/antiangiogenic strategy. The focus of this review will be on the role and activities elicited by the NOS pathway in pathological angiogenesis, speculating on the likely therapeutic outcomes.

ABBREVIATIONS

ACE	angiotensin-converting enzyme
ADMA	asymmetric dimethylarginine
cGMP	cyclic guanosine 3′,5′-monophosphate
COX-2	cyclooxygenase-2
DDAH	dimethylarginine dimethylaminohydrolase
ECs	endothelial cells
eNOS	endothelial nitric oxide synthase
ERK	extracellular-signal-regulated protein kinase
FGF-2	fibroblast growth factor-2

Therapeutic Application of Nitric Oxide in Cancer and Inflammatory Disorders. https://doi.org/10.1016/B978-0-12-816545-4.00004-9

HIF-1α	hypoxia-inducible factor-1α
HSP90	heat shock protein 90
iNOS	inducible nitric oxide synthase
L-NMMA	*N*-monomethyl-L-arginine
MMP	matrix metalloproteinase
NF-kB	nuclear factor-kB
NO	nitric oxide
PDE	phosphodiesterase
PI-3K	phosphatidylinositol 3-kinase
PKA	protein kinase A
PKC	protein kinase C
PKG	protein kinase G
RNS	reactive nitrogen species
ROS	reactive oxygen species
sGC	soluble guanylate cyclase
TIMP	tissue inhibitor of metalloprotease
TSP-1	thrombospondin-1
VEGF	vascular endothelial growth factor

Conflict of Interest

No potential conflicts of interest were disclosed.

ANGIOGENESIS: DEFINITION AND PATHOPHYSIOLOGICAL ROLES

Angiogenesis, mainly stimulated by hypoxia in growing, remodeling, or ischemic tissues, refers to the growth of new capillaries from the preexisting blood vessels. The neovascular growth involves the proliferation, migration, and differentiation of vascular endothelial cells (ECs) [1,2].

Angiogenesis plays pivotal physiological functions in the formation of the placenta, embryonic development, wound repair, and tissue remodeling and engineering [1]. During embryo growth, angiogenesis (occurring after vasculogenesis) is essential and required for the proper development of a functional circulatory system, providing nutrients and oxygen to every cell of the body and allowing the removal of waste products [3]. In contrast, in the adult, angiogenesis is restricted to few processes, namely, those related to the reproductive cycle, bone repair, and wound healing. In all of these cases, angiogenesis occurs transiently and is finely regulated by the balance between pro- and antiangiogenic factors. However, when dysregulated, the formation of new blood vessels may contribute to numerous oncogenic, ischemic, infectious, and inflammatory diseases [4,5]. The long list of angiogenesis-dependent diseases includes a number of seemingly unrelated diseases, including psoriasis, endometriosis, proliferative retinopathies, rheumatoid arthritis, and many types of cancer. In these angiogenesis-dependent diseases, the so-called angiogenic switch is due to an imbalance of the proangiogenic regulatory factors over antiangiogenic ones [1]. On the other hand, there are pathologies such as ischemic diseases where there is an insufficient blood supply. In principle, compounds able to control the angiogenic balance could be candidates for the pharmacological treatment of angiogenesis-dependent diseases (antiangiogenic compounds) or for those diseases lacking sufficient blood supply (proangiogenic compounds) or requiring endothelialization of compromised vessels. This explains the great interest on angiogenesis that has been maintained for the last decades by pharmaceutical industries and research groups looking for compounds or strategies able to affect this process [6–8].

At present, both anti- and proangiogenic treatments have been developed for human diseases. While the inhibition of angiogenesis has been shown to have success in disorders such as cancer, hemangioma, arthritis, psoriasis, and ophthalmic diseases [9], few treatment protocols have reached the clinic with the aim to stimulate angiogenesis in ischemia-associated diseases [10,11].

Beside the success of antiangiogenic drugs, relevant problems remain to be solved as resistance occurrence and toxic side effects. The research is indeed active in exploring all the factors that affect or promote angiogenesis. In the last decades, the gaseous transmitter nitric oxide (NO) has been identified as one of these factors.

REGULATION OF ENDOGENOUS NO PRODUCTION AND VASCULAR PHYSIOLOGY

NO is a short-lived gaseous-free radical that is produced by the activity of specific NOS isoforms starting from the precursor L-arginine [12]. NO is a highly diffusive hydrophobic molecule, and it is, therefore, a key signaling molecule both in the physiological control of the cardiovascular system and in the inflammation-driven diseases including cancer [13].

For a comprehensive description of NO biosynthesis and NOS isoform regulation and expression, we refer to pertinent literature [14]. In this review, particular attention will be reserved to the endothelial constitutive (eNOS) and inducible (iNOS) isoforms. The reaction performed by the enzyme isoforms is the same: conversion of L-arginine to L-citrulline and NO, in a stoichiometric manner. The first eNOS isoform is mainly present in the vascular endothelium, producing nanomolar levels of NO in calcium- and BH4-dependent manners.

The iNOS isoform is overexpressed in inflammatory, stromal, and vascular cells exposed to inducers and in most of the solid tumors thus far analyzed. iNOS produces micromolar concentrations of NO in a calcium-independent manner. The distinction, however, is not so rigid, since there are evidences that iNOS can be induced in the endothelium by inflammatory mediators, while numerous tumor cells express eNOS [15]. These findings underline the problem of the lack of efficacy and outcome of adverse events by NOS inhibitors designed to be specific for the various enzymes and proposed as therapeutic strategies. Asymmetric dimethylarginine (ADMA) and *N*-monomethyl-L-arginine (L-NMMA) are endogenous, competitive inhibitors of all the NOS isoforms, and NO production is physiologically regulated by methylarginines [16]. Indeed, direct nonselective NOS inhibitors can lead to inherent adverse effects by limiting the NO proinflammatory/defensive properties or by inhibiting the constitutive NOS isoforms [17]. Further, the synthesis of NO can be modulated by the two dimethylarginine dimethylaminohydrolase (DDAH) isoforms acting on the ADMA pathway [18]. DDAH enzymes catalyze the hydrolysis of methylarginines, ADMA and L-NMMA, to citrulline and either dimethylamine or methylamine, respectively [19].

Beside substrate availability, other mechanisms control NO production. First, eNOS dimerization is essential for proper enzymatic activity and NO production. The regulation of eNOS signaling is complex, and although it is mainly regulated by phosphorylation of its amino acid residues at multiple sites on its domains, its interaction with other proteins and intracellular localization provide a further mechanism of modulation.

In the absence of L-arginine or BH4, eNOS is uncoupled, and the monomer form of eNOS synthesizes superoxide instead of NO [20]. Monomeric eNOS transfers the electrons to oxygen, facilitating the formation of superoxide [20]. Proper eNOS dimerization is regulated by the heat shock protein 90 (HSP90) as silencing of HSP90 or its inhibition with geldanamycin destabilizes eNOS dimers, leading to its degradation [20].

NO contributes to cardiovascular functions by multiple mechanisms, such as the regulation of vascular tone (vasodilation), vascular structure (maintenance of a complete endothelial layer and inhibition of smooth muscle cell proliferation), and cell-cell interactions in blood vessels (inhibition of platelet adhesion and aggregation and inhibition of white cell and monocyte adhesion) [21]. NO is involved in the regulation of basal systemic, coronary, and pulmonary vascular tone through the activation of soluble guanylyl cyclase (sGC) and, subsequent by the generation of cyclic guanosine 3′,5′-monophosphate (cGMP) levels and protein kinase G (PKG) activation in smooth muscle cells, downregulation of the vasoconstrictor peptide endothelin-1 and inhibition of norepinephrine release from sympathetic nerve terminals [21–23].

NO production has also been reported to significantly contribute to the endothelium-protective and growth-promoting effects of vasodilating peptides (as substance P and bradykinin) [24,25] and growth/vasopermeabilizing factors [26,27]. Diminished release of NO generated by eNOS contributes to endothelial dysfunction, and nutraceuticals or other natural products with cardiovascular protective properties demonstrated to exert their effects on vascular and lymphatic systems by activating eNOS [28,29]

ROLE OF NO IN ANGIOGENESIS

NO released by vascular endothelium is essential for the maintenance of cardiovascular homeostasis. In addition to its well-established role in regulating vascular tone, NO has also been shown to be a mediator of postnatal angiogenesis. Mice in which eNOS has been genetically ablated develop normally but display a severe form of critical limb ischemia with impaired collateral adaptation and blood flow recovery after femoral artery ligation [30–32], together with impaired wound healing and defective angiogenesis [33]. Mice lacking eNOS show also a reduced severity of oxygen-induced retinopathy [34]. Both angiogenic and arteriogenic responses to hindlimb ischemia were severely blunted in $eNOS^{-/-}$ mice, an effect that could be rescued by genetic delivery of a constitutively active form of eNOS [35].

These seminal data clearly demonstrate the importance of NO in the control of arteriogenesis and angiogenesis. At the same time, prolonged hypoxia and ischemia have been demonstrated to reduce endothelial eNOS expression via post-transcriptional mechanisms that result in *NOS3* transcript destabilization.

HOW NO CONTROLS ANGIOGENESIS

The contribution of NO to angiogenesis modulation is multiple-faced. NO has been shown to mediate angiogenesis by both direct and indirect mechanisms.

Direct Effects of NO on Endothelial Cells

We and others have largely contributed to demonstrate that exposure of ECs to exogenous NO increases DNA synthesis, cell proliferation, and migration of endothelial cells through the sGC-cGMP pathway and through *S*-nitrosylation or nitration [36–40]. Furthermore, NO as an endothelial growth factor itself mediates tumor vascularization [41] and tumor blood flow [42]. A role for NO in the mobilization of stem and endothelial progenitor cells has also been described [43].

During the last decades, various authors have reported antiangiogenic effects of NO [44–46], and this discrepancy might be explained by differences in both the

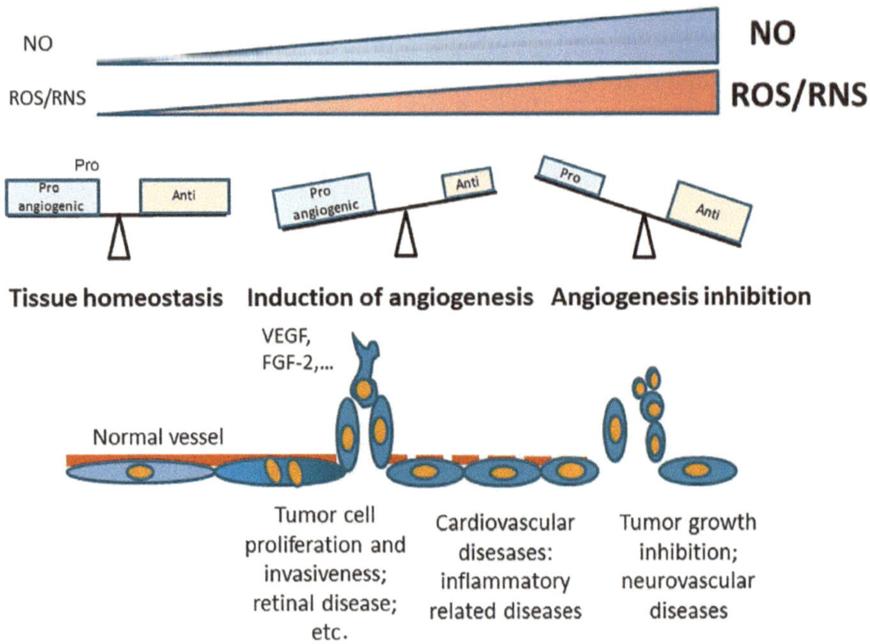

FIG. 1

Summary of the pro- and antiangiogenic activities of NO in relation to its concentration and capacity to produce an ROS/RNS burst. Nanomolar concentrations of NO, being produced by eNOS, are able to sustain endothelial survival and to promote angiogenesis through the upregulation of proangiogenic factors (VEGF and FGF-2) and enzymes as MMPs. Conversely, micromolar levels of NO, deriving from inducible isoform of NOS, are reported to inhibit angiogenesis and to participate in antitumor or degenerative effects through the increased availability and reactivity of ROS and RNS, on their turn responsible of cytotoxic effects and cell apoptosis.

concentration and the duration of NO exposure (Fig. 1). Low concentrations of NO increase but high concentrations inhibit phosphorylation of protein kinase C (PKC), extracellular-signal-regulated protein kinase (ERK) and Jun, and the binding activity of activator protein 1 [47–48].

NO as a Mediator of the Angiogenic Factor Activity

NO has been shown to mediate the function of many angiogenic factors and physiological modulators. VEGF, sphingosine-1-phosphate, angiopoietins, estrogen, shear stress, and metabolic stress activate eNOS through definite signaling events. The capillary-like structure formation by HUVEC stimulated with VEGF was blocked by the NOS inhibitor, L-NAME [27,49]. In vivo, mice with genetic ablation of the eNOS gene (*eNOS*$^{-/-}$) demonstrated reduced capillary growth in implanted collagen plugs and impaired angiogenic responses to hindlimb ischemia compared with control animals, and in both cases, the phenotype could not be rescued by the administration of VEGF [31,50]. In addition, VEGF-induced angiogenesis was shown to be blocked by L-NAME in a rabbit cornea model of angiogenesis [51].

VEGF increases angiogenesis in both $iNOS^{+/+}$ and $iNOS^{-/-}$ mice, but not in $eNOS^{-/-}$ mice, supporting a predominant role of eNOS in VEGF-induced angiogenesis and vascular permeability [52]. Additionally, VEGF can also upregulate eNOS mRNA and protein [53,54].

Mechanistically, the activation of (human) eNOS is associated with phosphorylation at Ser615, Tyr81, Ser633, and Ser1177 [55–58] and has been shown to be mediated by the actions of multiple protein kinases, including Akt/PKB, PKA, c-Src, and AMPK [59–61]. Activation of VEGFR2 in endothelial cells (by agonist binding) results in Akt-dependent phosphorylation of eNOS at Ser1177 [55,62]. In addition, VEGF signaling induces the c-Src-dependent phosphorylation and association of HSP90 with eNOS [52,63–66]. eNOS is additionally activated by other stimuli, including shear stress [67], which have been shown to be dependent on both Akt- and PKA-dependent mechanisms. As downstream signaling, NO mainly promotes angiogenesis through the sGC-cGMP-PKG-dependent pathway(s) [68–71].

Further, data previously reported by us indicate that endogenously produced NO is able to promote metalloprotease (MMP-2) activity in the microvascular endothelium stimulated by VEGF, by promoting MMP-2 gene upregulation and downregulating tissue inhibitor of metalloprotease-1 (TIMP-1) and metalloprotease-2 expression [72]. In the complex, eNOS definitely mediates VEGF- and angiopoietin-1-induced angiogenesis in vivo [31,50,51,73].

Also, inflammatory mediators as prostanoids and free radicals can induce angiogenesis [74]. In cell culture models, eNOS is a central mediator of prostaglandin E_2 (PGE_2) [75]. Moreover, PGE_2 activates the PI-3K/Akt pathway and promotes endothelial cell sprouting (the first step in neoangiogenesis) through the NO/cGMP pathway [75]. The relevance of eNOS-derived NO in PGE_2 activity is also derived from in vivo experiments in the rabbit cornea model [76].

Both exogenously and endogenously generated reactive oxygen species (ROS, in the form of H_2O_2) have been shown in endothelial cells to promote both NO production and NO-dependent signaling in vitro [77] and to stimulate angiogenic responses (proliferation and migration) that are blocked by inhibiting eNOS [78]. Moreover, in vivo, the transgenic overexpression of catalase in mice demonstrated that endogenous, endothelial-derived H_2O_2 mediates the upregulation of eNOS after exercise [79]. In BAECs H_2O_2 (but not O_2^-) was shown to increase both the rate of transcription and the stability of eNOS mRNA in a dose- and time-dependent manner, although the precise mechanisms of the transcriptional regulation remain unknown [80]. These increases were shown to be prevented by both the antioxidant *N*-acetylcysteine (NAC) and the H_2O_2 scavenger catalase.

H_2O_2 was also shown to be able to upregulate eNOS activity through the modulation of VEGF-dependent signaling, as the phosphorylation of eNOS at Ser1177 [81]. Consistently, administration of exogenous ROS to BAECs was sufficient to promote NO production via Akt and $ERK_{1/2}$ activation [82].

NOS Inhibition as Effector of Antiangiogenic Molecules and Antitumor Drugs

In recent years, it has become apparent that ADMA and its associated metabolic enzymes DDAHI and DDAHII can regulate angiogenesis through their effects on NOS substrate bioavailability [83,84]. ADMA is an endogenous methylated form of L-arginine that acts as a competitive inhibitor of all NOS enzymes and is formed from the hydrolysis of proteins that have been methylated on arginine residues. DDAHI and DDAHII are enzymes that metabolize ADMA and, therefore, negate its inhibitory effects on NOS, elevating NO levels [85]. Both in vitro and in vivo studies have demonstrated that DDAH enzymes are involved in angiogenesis. Aortic rings isolated from DDAHI heterozygous mice ($DDAH^{+/-}$) showed reduced sprouting in Matrigel. This effect could be mimicked through the addition of ADMA to ring explants from WT mice, suggesting that endogenous removal of ADMA by DDAHI is required for vessel growth [86]. Moreover, endothelium-specific ablation of DDAHI blunted angiogenesis both in vitro and in vivo [87], while gene overexpression of DDAHI enhanced NO production and angiogenesis in a murine model of hindlimb that resulted in its association with reduced ADMA levels [88].

Furthermore, antiangiogenic factors exert their effects through the inhibition of eNOS/NO signaling.

Endostatin dephosphorylates eNOS at Ser1177 through the protein phosphatase 2A, and a NO donor reverses the inhibitory effect of endostatin on endothelial cell migration [89]. Thrombospondin 1 (TSP1) has been found to limit eNOS activity and NO signaling through directly targeting sGC and, in some cells, through directly inhibiting targets of the second messenger cGMP [90,91]. TSP1- and CD47-null mice show constitutively increased levels of cGMP and phosphorylated eNOS [92]. Animals lacking either the ligand TSP1 or the receptor CD47 have enhanced angiogenesis and increased tissue blood flow in response to a range of stressors including temperature, ischemia, and reperfusion [93].

Somatostatin inhibits tumor angiogenesis through the inhibition of eNOS and MAPK pathway [94]. This is common to other antiangiogenic strategies as 16 kDa prolactin, found to downregulate iNOS expression in endothelial cells [95], and TNP-470, which inhibits nuclear factor-kB (NF-kB) activation and iNOS expression in experimental models of hepatocarcinogenesis [96].

Indirectly, also bevacizumab antiangiogenic effects and side effects are due to eNOS impairment due to reduced VEGF plasma availability [97]. Indeed, the incidence of cardiovascular events (hypertension and stroke) described in patients treated with antiangiogenic agents may be causally related to the derangement of the NOS pathway and induction of endothelial dysfunction.

Recent studies on conventional antitumor agents suggest also for this class of drugs that the clinical efficacy may involve NOS inhibition. Examples are the topoisomerase I inhibitor camptothecin, exerting its antitumor activity, in part, through the inhibition of NO biosynthesis [98], and irinotecan, associated with the downregulation of COX-2, iNOS, HIF-1α, and angiogenesis [99]. Also, the antitumor activity of selected conventional drugs, such as the tubulin-binding agent combretastatin-A4 phosphate (CA4-P), melphalan, tumor necrosis factor, and mitomycin, appears to be considerably potentiated by the simultaneous administration of L-NNA, L-NAME, or the iNOS selective inhibitor aminoguanidine, respectively [100,101].

NO Controls the Angiogenic Modulator Expression or Activity in Endothelial, Tumor, and Stromal Cells

NO is an important modulator of the expression of angiogenic factors. NO can act upstream of VEGF signaling to promote angiogenesis. Thus, increased NO, either administered by exogenous NO donors or due to the induction or overexpression of NOS, acts in vascular smooth muscle cells, in vitro, to increase VEGF expression [102,103]. A NO donor has been demonstrated to induce VEGF expression in tumor cells [104]. In addition, in vivo inhibition of NOS eliminated the increased expression of VEGF (and VEGFR-2) in electrically stimulated capillary growth in rat muscle for up to 4 days [105].

NO and hypoxia are reported to stimulate the VEGF gene expression by enhancing HIF-1α activity [106,107]. Mechanistically, NO has been shown to activate the VEGF promoter via the *cis*-regulatory hypoxia response element (HRE), which mediates the transactivation of VEGF transcription by HIF-1α [108,109].

NO activates the transcription factor HIF-1α, which in turn upregulates VEGF, thereby promoting angiogenesis [109,110]. NO induces HIF-1α synthesis through the MAPK and PI-3K pathways under normoxic conditions [111]. It also impairs normoxic degradation of HIF-1α by inhibiting the action of prolyl hydroxylases [112]. Heme oxygenase 1 also mediates NO-induced VEGF upregulation in human endothelial cells [113]. A further study found the expression of iNOS and/or eNOS in all cases of HIF-1α-positive oral squamous cell carcinomas, which suggests that there is NO-induced HIF-1α accumulation and subsequent tumor-promoting effects in cancer [114].

Recent data establish miR-200b-eNOS regulation as a first hypoxamiR-based mechanism that limits NO bioavailability during hypoxia in endothelial cells and show that hypoxamiRs could become useful therapeutic targets for cardiovascular diseases and other hypoxia-related diseases including various types of cancer [115].

By contrast, NO has also been shown to inhibit hypoxia-induced HIF-1α activation and VEGF expression [116,117], and recently, this mechanism has been demonstrated to occur also in normoxia [118]. NO inhibits the expression of HIF-1α via activating HIF-1-prolyl hydroxylases and its proteasomal degradation [116,119–121]. NO, by reducing HIF-1α-dependent VEGF levels, can thus improve the delivery of antitumor drugs, by vascular normalization and reversion of the oncotic pressure gradient [122].

Moreover, in an ex vivo model of balloon angioplasty in rat thoracic aortas, increased VEGF expression was inhibited by NO donors [123]. All these discrepancies might be explained by the difference in the amount and duration of exposure to NO [124]. At a low level, endogenous or exogenous NO can also serve as an intracellular second messenger to induce the expression of proangiogenic factors in tumor cells, as IL-8, which represents an indirect angiogenesis stimulator [125,126]. In tumor and stromal cells, NO can regulate the expression of inflammatory molecules NF-kB and COX-2 [127,128]. We have contributed to demonstrate that exogenous NO stimulates coronary vein endothelial cell proliferation through the upregulation of fibroblast growth factor-2 (FGF-2) [129]. The altered balance between MMPs and their TIMPs by exogenous and endogenous NO in the microvascular endothelium has also been described [72], and the regulation of TIMP-4 or MMP-13 activation (by nitration) and release by endogenous NO have been reported [37,40,130].

Conversely, NO inhibits the expression of endogenous antiangiogenic factors. A NO donor reduces TSP1 expression [48], and a NO inhibitor increases angiostatin expression in vascular endothelial cells [131]. In particular, downregulation of TSP-1 by NO depends on cGMP and ERK [48]. The expressions of tumor-suppressive interferon-inducible protein-10 and monokyne induced by interferon gamma are suppressed by NO [132]. The resulting altered balance leads to the stimulation of tumor angiogenesis.

CONTADICTORY ROLES OF NO IN TUMOR ANGIOGENESIS

In the last two decades, the role of NO in tumor cell biology and tumor angiogenesis has been firmly established. The elevation of NOS activity in correlation with angiogenesis and tumor growth and aggressiveness has been extensively reported [133]. The general concept regarding cooperation among different inflammatory mediators is indeed universally accepted both in inflammatory-based diseases and in cancer [134].

A solid tumor consists of cancer cells and host-derived cells, including tumor-infiltrating leukocytes and cells of the tumor vasculature, especially endothelial cells. One or more of these cellular constituents may be responsible for the production of NO in the tumor microenvironment [135]. NO produced by endothelial cells and tumor cells acts cooperatively in tumor angiogenesis [136].

The functional roles of tumor-derived NO in cancer progression and spreading represent a complex combination of NO-mediated effects on tumor cell proliferation and invasiveness and the functions of immune/stromal cells infiltrating tumors. It has been proposed that NO promotes tumor growth also by regulating tumor blood flow and maintaining the vasodilatory tone of tumors. In addition to angiogenesis stimulation, NO can promote metastasis by increasing vascular permeability and upregulating MMPs, thus changing the structure of the extracellular matrix. Moreover, it has been reported that NO released by metastatic tumor cells may impair the immune system, which facilitates the escape from immunosurveillance and metastasis of tumor cells [137,138], and that the iNOS/NO pathway in metastatic cancer cells mediates cancer stem cell-like properties induced by oncogenic signals [139].

Early experimental studies have shown that the induction of iNOS in tumor cells promotes angiogenesis (by upregulating VEGF expression), which increases microvascular density and tumor progression [104,140–143]. Inhibition of NOS, by genetic manipulation or pharmacologically, has been shown to reduce VEGF levels and inhibit tumor angiogenesis [41,42,144,145].

The strongest data supporting a fundamental role of NO in tumor angiogenesis derived, however, from the histological examination of clinical samples revealing a significant relationship between high angiogenic activity (i.e., microvessel density or VEGF expression) and iNOS expression in human brain, head and neck, lung, breast, stomach, colon tumors, etc. [146–155 among the others]. Altogether, these findings definitely indicate that tumor cell-derived NO mediates tumor angiogenesis, invasion, and growth.

Not only NOS isoforms but also the enzymes involved in the NOS inhibitor processing control tumor angiogenesis. The activity of DDAH has been detected in various human tumors including those arising in the brain; astrocytomas in particular exhibit high levels of DDAH activity [156]. High expression of DDAHI has also been observed in aggressive triple-negative breast cancer cell lines, where DDAHI was described as a key regulator of cell migration and vasculogenic mimicry [157]. Overexpression of DDAHI has been shown to increase NO production and enhance tumor growth and angiogenesis in C6 gliomas in vivo [158,159]. Pharmacological manipulation of DDAH may be, thus, an attractive therapeutic target for regulating NO in cancer. DDAH inhibitors can help overcome one of the major considerations for NO-focused therapies, which is targeting pathological NO production while sparing the beneficial NO produced constitutively.

Previous studies have looked into DDAH-mediated NO-dependent effects on tumor vascularization. However, other studies have also suggested that DDAH and/or ADMA can exert direct, NO-independent effects on angiogenesis. For example, DDAHII overexpression led to increased VEGF expression in endothelial cells by binding to PKA, increasing phosphorylation and activation of the transcription factor Sp1 [160,161]. These studies imply that the DDAH/ADMA pathway may directly regulate tumor angiogenesis, representing a promising therapeutic target [17].

Beside this plethora of literature reports in favor of the proangiogenic effect of tumor-derived NO, many articles have been published reporting antiangiogenic/antitumor activities by NO [118,138,162–166].

Recently, alteration of the expression of angiogenic genes has been confirmed in hepatoma cells in response to the NO donor JS-K, in particular an increase in TSP-1 and tissue inhibitors of metalloprotease-1 [167], known inhibitors of angiogenesis and tumor cell migration.

A direct antiangiogenic activity by exogenous NO has been reported. An *in vitro* assay with cultured endothelial cells revealed that the NO hybrid compound NCX-4016 significantly inhibited angiogenesis in a concentration-dependent manner, with almost complete inhibition at a 100 μM concentration, suggesting that NCX-4016 can act as an antiangiogenic drug [168].

Based on the above findings, let us hypothesize that the final outcome of the NO effect on tumor angiogenesis may be dependent on the reached local concentration and the availability and reactivity of cytotoxic ROS and reactive nitrogen species (RNS) coming from the biochemical reactivity of high levels of NO (Fig. 1).

INNOVATIVE STRATEGIES FOR CONTROLLED NO AVAILABILITY

Treatments that increase NO levels, either by enhancing endogenous NO or donating exogenous NO, have been studied in the last decades for various disorders [169]. In cardiovascular diseases and in other inflammatory-related diseases, the kinetic and amount of NO necessary to recover tissue loss-of-function are critical to avoid side effects related to the burst NO release, typical of synthetic NO donors.

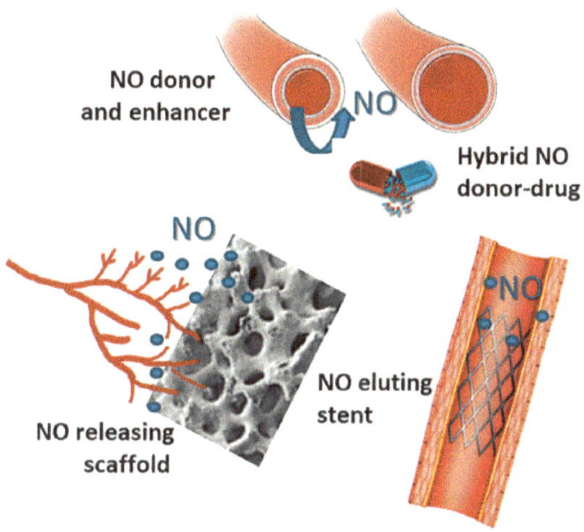

FIG. 2

Examples of innovative approaches for a controlled NO release in the cardiovascular system and during tissue engineering. Innovations regarding novel NO donors, NO enhancers acting in different steps of the eNOS/sGC/cGMP pathway, hybrid molecules of conventional drugs and NO donor moiety in order to have synergistic/complementary actions, synthetic materials to be used as scaffolds or stents to favor tissue biocompatibility and endothelialization, and novel NO donor formulations.

Innovation in the field of NO donors or amplifiers of the NOS pathway passes through different approaches (Fig. 2). First, we have to consider drugs or therapeutic strategies that act through eNOS activation or upregulation. Examples, beside the documented proangiogenic effects of VEGF, are cardiovascular drugs as statins or ACE inhibitors, which exert pleiotropic off-target effects as eNOS activation [170–172]. For ACE inhibitors, the enhanced NO production is likely the result of diminished breakdown of bradykinin. The clinical efficacy of these drugs indeed is accompanied by recovery of endothelial dysfunction [173]. Similarly, nutraceuticals or plant-derived compounds as polyphenols and principles present in the Mediterranean diet demonstrate a protective effect on the cardiovascular system by promoting endothelial function and eNOS upregulation [28,29,174].

An alternative approach is to potentiate the endogenous NO activity by the use of phosphodiesterase V (PDEV) inhibitors as sildenafil, which improves the cGMP/PKG pathway, and it is described for the improvements of endothelial function and antioxidant capacity [175]. Another NO enhancer is cilostazol, a PDEIII inhibitor that has been shown to increase NO production by altering eNOS phosphorylation via PKA and PI-3K/Akt-dependent mechanism [176]. Its efficacy in inducing angiogenesis has been recently reported in different settings [177–179].

Interestingly, in the last decades, an intricate correlation between the NO pathway and another gaseous transmitter, H_2S, has been demonstrated. The two mediators exert similar activities, and they ignite each other's pathway [69]. In the vascular endothelium, H_2S promotes the NOS pathway through the Akt/eNOS phosphorylation [180] or PDEV inhibition [181], while exogenous or endogenous NO has been reported to upregulate the key enzymes responsible for H_2S production [182]. Thus, drugs or natural compounds able to release H_2S could be thought as enhancers of the NOS pathway [183–185].

Additional strategies to increase NO levels include also the reduction of ADMA levels by increasing DDAH enzymatic activity [17]. Drugs as fenofibrate, nebivolol, pioglitazone, and pravastatin have been demonstrated to upregulate DDAH [186].

Beside the development of new NO donors (*see the chapter of Prof. Scicinski*), novel approaches include the synthesis of hybrid molecules derived from the combination of conventional drugs and NO donors, novel NO donor formulations, and biomaterials bearing or uploading NO donors.

By using functional groups present in molecules that already have a potential therapeutic value, hybrid compounds have been developed that have a NO release benefit by means of inclusion of NO donors (e.g., ester nitrates and nitrites, S-nitrosothiols, metal complexes, furoxans, oxadiazoles, diazeniumdiolates, and NO nanoparticles), with a conventional drug. The objective of the design of NO donor hybrid compounds is to achieve a balance between the release of therapeutic amounts of NO, especially at the site of action, and maintaining the native drug activity [187]. Examples are the COX inhibitor-NO, statin-NO, etc. [168,188]. Interestingly, H_2S-NO hybrid molecules have been reported to promote therapeutic synergistic effects against heart failure by multiple mechanisms as the upregulation of endogenous antioxidants, glutathione peroxidase, and heme oxygenase-1 [189].

Among the innovative drug formulations, particular emphasis is given to nanoparticles designed to release NO donors (*see the chapter from Prof. Kim*) or combination of multiple drugs [190] and microporous solids able to uptake and then release NO in a controlled manner [191,192].

A proper NO release is important for tissue engineering, wound repair, and vascular structure recovery. Problems coming from synthetic material are the lack of endothelialization, stimulation of smooth muscle cell proliferation, and platelet or coagulation activation. The combination of biodegradable materials and anticoagulant heparins would provide an anticoagulant and endothelization substrate, supporting the environment for vascular tissue engineering. Interestingly, these materials could be engineered to catalyze the NO donor in blood to release NO, which could enhance endothelial cell growth, while decreasing smooth muscle cell proliferation and platelet adhesion. The final aim could be to provide scaffolds for vascular tissue engineering with an adequate endothelialization without atherogenesis and thrombus formation [193].

The literature is particularly rich in biocompatible polymeric materials able to promote a sustainable NO release in blood or tissues, i.e., for antibacterial activity [194]. Approaches of light activation of vascular scaffold accompanied by high levels of NO release have been described [195].

CONCLUSIONS

After the discovery of NO, several actions in both physiological and pathological conditions have been attributed to this mediator. Most of them are divergent, depending on the NO concentration, the duration of its release, the cell type, and the presence of scavengers or other reactive molecules in the microenvironment that may impair or otherwise amplify NO effects.

From a pharmacological point of view, both if we want to give exogenous NO or to inhibit its endogenous production, problems arise due to the pleiotropic and multiple functions of this important mediator. Innovative strategies are under progress to solve these problems and demonstrate their effectiveness and safety. Further research needs to be done to establish the right formulation to specifically target the cells or tissue of interest or to synergize with other conventional or target therapies to control NO production and NO-dependent angiogenesis disorders.

Acknowledgments

This work has been partially funded by the Italian Ministry of Education, University and Research (MIUR)-PRIN projects.

References

[1] Carmeliet P. Angiogenesis in health and disease. Nat Med 2003;9:653–60.

[2] Carmeliet P. Angiogenesis in life, disease and medicine. Nature 2005;438:932–6.

[3] Fischer C, Schneider M, Carmeliet P. Principles and therapeutic implications of angiogenesis, vasculogenesis and arteriogenesis. Handb Exp Pharmacol 2006;176(Pt 2):157–212.

[4] Carmeliet P, Jain RK. Angiogenesis in cancer and other diseases: from genes to function to therapy. Nature 2000;407:249–57.

[5] Folkman J. What is the evidence that tumors are angiogenesis dependent? J Natl Cancer Inst 1990;82(1):4–6.

[6] Medina MA, Muñoz-Chápuli R, Quesada AR. Challenges of antiangiogenic cancer therapy: trials and errors, and renewed hope. J Cell Mol Med 2007;11:374–82.

[7] Quesada AR, Muñoz-Chápuli R, Medina A. Anti-angiogenic drugs: from bench to clinical trials. Med Res Rev 2006;26:483–530.

[8] Quesada AR, Medina MA, Muñoz-Chápuli R, Ponce ALG. Do not say ever never more: the ins and outs of antiangiogenic therapies. Curr Pharm Des 2010;16:3932–57.

[9] Potente M, Gerhardt H, Carmeliet P. Basic and therapeutic aspects of angiogenesis. Cell 2011;146:873–87.

[10] Shimamura M, Nakagami H, Koriyama H, Morishita R. Gene therapy and cell-based therapies for therapeutic angiogenesis in peripheral artery disease. Biomed Res Int 2013; 2013:186215.

[11] Besnier M, Gasparino S, Vono R, Sangalli E, Facoetti A, Bollati V, Cantone L, Zaccagnini G, Maimone B, Fuschi P, Da Silva D, Schiavulli M, Aday S, Caputo M, Madeddu P, Emanueli C, Martelli F, Spinetti G. miR-210 enhances the therapeutic potential of bone-marrow-derived circulating proangiogenic cells in the setting of limb ischemia. Mol Ther 2018 pii: S1525-0016(18)30262-4.

[12] Schmidt HH, Hofmann H, Schindler U, Shutenko ZS, Cunningham DD, Feelisch M. NO from NO synthase. Proc Natl Acad Sci U S A 1996;93(25):14492–7.

[13] Ignarro LJ. Biosynthesis and metabolism of endothelium-derived nitric oxide. Annu Rev Pharmacol Toxicol 1990;30:535–60.

[14] Förstermann U, Sessa WC. Nitric oxide synthases: regulation and function. Eur Heart J 2012;33(7, 1):829–37.

[15] Ying L, Hofseth LJ. An emerging role for endothelial nitric oxide synthase in chronic inflammation and cancer. Cancer Res 2007;67(4):1407–10.

[16] Leiper JM, Santa Maria J, Chubb A, MacAllister RJ, Charles IG, Whitley GS, Vallance P. Identification of two human dimethylarginine dimethylaminohydrolases with distinct tissue distributions and homology with microbial arginine deiminases. Biochem J 1999;343(Pt 1): 209–14.

[17] Leiper J, Nandi M. The therapeutic potential of targeting endogenous inhibitors of nitric oxide synthesis. Nat Rev Drug Discov 2011;10(4):277–91.

[18] MacAllister RJ, Parry H, Kimoto M, Ogawa T, Russell RJ, Hodson H, Whitley GS, Vallance P. Regulation of nitric oxide synthesis by dimethylarginine dimethylaminohydrolase. Br J Pharmacol 1996;119(8):1533–40.

[19] Ogawa T, Kimoto M, Sasaoka K. Purification and properties of a new enzyme, NG, NG-dimethylarginine dimethylaminohydrolase, from rat kidney. J Biol Chem 1989;264 (17):10205–9.

[20] Chen W, Xiao H, Rizzo AN, Zhang W, Mai Y, Ye M. Endothelial nitric oxide synthase dimerization is regulated by heat shock protein 90 rather than by phosphorylation. PLoS One 2014;9:e105479.

[21] Cannon III RO. Role of nitric oxide in cardiovascular disease: focus on the endothelium. Clin Chem 1998;44:1809–19.

[22] McNamara DB, Bedi B, Aurora H, Tena L, Ignarro LJ, Kadowitz PJ, Akers DL. L-Arginine inhibits balloon catheter-induced intimal hyperplasia. Biochem Biophys Res Commun 1993;193:291–6.

[23] Dubey K, Jackson EK, Luscher TF. Nitric oxide inhibits angiotensin II-induced migration of rat aortic smooth muscle cell. Role of cyclic-nucleotides and angiotensin1 receptors. J Clin Invest 1995;96:141–9.

[24] Ziche M, Morbidelli L, Masini E, Amerini S, Granger HJ, Maggi C, Geppetti P, Ledda F. Nitric oxide mediates angiogenesis in vivo and endothelial cell growth and migration in vitro promoted by substance P. J Clin Invest 1994;94:2036–44.

[25] Parenti A, Morbidelli L, Ledda F, Granger HJ, Ziche M. The bradykinin/B1 receptor promotes angiogenesis by upregulation of endogenous FGF-2 in endothelium via the nitric oxide synthase pathway. FASEB J 2001;15:1487–9.

[26] Ku DD, Zaleski JK, Liu S, Brock TA. Vascular endothelial growth factor induces EDRF-dependent relaxation in coronary arteries. Am J Physiol 1993;265:H586–92.

[27] Morbidelli L, Chang C-H, Douglas JG, Granger HJ, Ledda F, Ziche M. Nitric oxide mediates mitogenic effect of VEGF on coronary venular endothelium. Am J Physiol 1996;270:H411–5.

[28] Kim JH, Auger C, VB1 S-K. Activation of eNOS by polyphenol-rich products and polyphenolic compounds. Curr Pharm Des 2014;20(22):3521–9.

[29] Ciccone V, Monti M, Antonini G, Mattoli L, Burico M, Marini F, Maidecchi A, Morbidelli L. Efficacy of AdipoDren® in reducing interleukin-1 induced lymphatic endothelial hyperpermeability. J Vasc Res 2016;53(5–6):255–68.

[30] Park B, Hoffman A, Yang Y, Yan Y, Tie G, Bagshahi H, Nowicki PT, Messina LM. eNOS affects both early and late collateral arterial adaptation and blood flow recovery after induction of hindlimb ischemia in mice. J Vasc Surg 2010;51(1):165–73.

[31] Murohara T, Asahara T, Silver M, Bauters C, Masuda H, Kalka C, Kearney M, Chen D, Symes JF, Fishman MC, Huang PL, Isner JM. Nitric oxide synthase modulates angiogenesis in response to tissue ischemia. J Clin Invest 1998;101:2567–78.

[32] Aicher A, Heeschen C, Mildner-Rihm C, Urbich C, Ihling C, Technau-Ihling K, Zeiher AM, Dimmeler S. Essential role of endothelial nitric oxide synthase for mobilization of stem and progenitor cells. Nat Med 2003;9:1370–6.

[33] Lee PC, Salyapongse AN, Bragdon GA, Shears 2nd LL, Watkins SC, Edington HD, Billiar TR. Impaired wound healing and angiogenesis in eNOS deficient mice. Am J Physiol 1999;277: H1600–8.

[34] Brooks SE, Gu X, Samuel S, Marcus DM, Bartoli M, Huang PL, Caldwell RB. Reduced severity of oxygen-induced retinopathy in eNOS-deficient mice. Invest Ophthalmol Vis Sci 2001; 42(1):222–8.

[35] Yu J, deMuinck ED, Zhuang Z, Drinane M, Kauser K, Rubanyi GM, Qian HS, Murata T, Escalante B, Sessa WC. Endothelial nitric oxide synthase is critical for ischemic remodeling, mural cell recruitment, and blood flow reserve. Proc Natl Acad Sci U S A 2005;102:10999–1004.

[36] Ziche M, Morbidelli L, Masini E, Geppetti P, Granger HJ, Ledda F. Nitric oxide promotes DNA synthesis and cyclic GMP formation in endothelial cells from postcapillary venules. Biochem Biophys Res Commun 1993;192:1198–203.

[37] Zaragoza C, Soria E, Lopez E, Browning D, Balbin M, Lopez-Otin C, Lamas S. Activation of the mitogen activated protein kinase extracellular-signal-regulated kinase 1 and 2 by the nitric oxide-cGMP-cGMP-dependent protein kinase axis regulates the expression of matrix metalloproteinase 13 in vascular endothelial cells. Mol Pharmacol 2002;62:927–35.

[38] Kawasaki K, Smith Jr. RS, Hsieh CM, Sun J, Chao J, Liao JK. Activation of the phosphatidylinositol 3-kinase/protein kinase Akt pathway mediates nitric oxide-induced endothelial cell migration and angiogenesis. Mol Cell Biol 2003;23:5726–37.

[39] Oliveira CJ, Schindler F, Ventura AM, Morais MS, Arai RJ, Debbas V, Stern A, Monteiro HP. Nitric oxide and cGMP activate the Ras–MAP kinase pathway-stimulating protein tyrosine phosphorylation in rabbit aortic endothelial cells. Free Radic Biol Med 2003;35:381–96.

[40] Donnini S, Monti M, Roncone R, Morbidelli L, Rocchigiani M, Oliviero S, Casella L, Giachetti A, Schulz R, Ziche M. Peroxynitrite inactivates human-tissue inhibitor of metalloproteinase-4. FEBS Lett 2008;582(7):1135–40.

[41] Maeda H, Noguchi Y, Sato K, Akaike T. Enhanced vascular permeability in solid tumor is mediated by nitric oxide and inhibited by both new nitric oxide scavenger and nitric oxide synthase inhibitor. Japan J Cancer Res 1994;85(4):331–4.

[42] Tozer GM, Prise VE, Chaplin DJ. Inhibition of nitric oxide synthase induces a selective reduction in tumor blood flow that is reversible with L-arginine. Cancer Res 1997;57(5):948–55.

[43] Duda DG, Cohen KS, Kozin SV, Perentes JY, Fukumura D, Scadden DT, Jain RK. Evidence for bone marrow-derived endothelial cells incorporation into perfused blood vessels in tumors. Blood 2006;107:2774–6.

[44] Pipili-Synetos E, Sakkoula E, Maragoudakis ME. Nitric oxide is involved in the regulation of angiogenesis. Br J Pharmacol 1993;108:855–7.

[45] Lau Y-T, Ma W-C. Nitric oxide inhibits migration of cultured endothelial cells. Biochem Biophys Res Commun 1996;221:670–4.

[46] Heller R, Polack T, Grabner R, Till U. Nitric oxide inhibits proliferation of human endothelial cells via a mechanism independent of cGMP. Atherosclerosis 1999;144:49–57.

[47] Jones MK, Tsugawa K, Tarnawski AS, Baatar D. Dual actions of nitric oxide on angiogenesis: possible roles of PKC, ERK, and AP-1. Biochem Biophys Res Commun 2004;318:520–8.

[48] Ridnour LA, Isenberg JS, Espey MG, Thomas DD, Roberts DD, Wink DA. Nitric oxide regulates angiogenesis through a functional switch involving thrombospondin-1. Proc Natl Acad Sci U S A 2005;102:13147–52.

[49] Papapetropolous A, Garcia-Cardena G, Madri JA, Sessa WC. Nitric oxide production contributes to the angiogenic properties of vascular endothelial growth factor in human endothelial cells. J Clin Invest 1997;100:3131–9.

[50] Fukumura D, Gohongi T, Kadambi A, Izumi Y, Ang J, Yun CO, Buerk DG, Huang PL, Jain RK. Predominant role of endothelial nitric oxide synthase in vascular endothelial growth factor-induced angiogenesis and vascular permeability. Proc Natl Acad Sci U S A 2001;98:2604–9.

[51] Ziche M, Morbidelli L, Choudhuri R, Zhang H-T, Donnini S, Granger HJ, Bicknell R. Nitric oxide synthase lies downstream from vascular endothelial growth factor-induced but not basic fibroblast growth factor-induced angiogenesis. J Clin Invest 1997;99:2625–64.

[52] Duda DG, Fukumura D, Jain RK. Role of eNOS in neovascularization: NO for endothelial progenitor cells. Trends Mol Med 2004;10:143–5.

[53] Hood JD, Meininger CJ, Ziche M, Granger HJ. VEGF upregulates ecNOS message, protein, and NO production in human endothelial cells. Am J Physiol 1998;43:H1054–8.

[54] Kroll J, Waltenberger J. VEGF-A induces expression of eNOS and iNOS in endothelial cells via VEGF receptor-2 (KDR). Biochem Biophys Res Commun 1998;252:743–6.

[55] Fulton D, Gratton JP, McCabe TJ, Fontana J, Fujio Y, Walsh K, Franke TF, Papapetropoulos A, Sessa WC. Regulation of endothelium-derived nitric oxide production by the protein kinase Akt. Nature 1999;399:597–601.

[56] Fulton D, Ruan L, Sood SG, Li C, Zhang Q, Venema RC. Agonist-stimulated endothelial nitric oxide synthase activation and vascular relaxation. Role of eNOS phosphorylation at Tyr83. Circ Res 2008;102:497–504.

[57] Luo Z, Fujio Y, Kureishi Y, Rudic RD, Daumerie G, Fulton D, Sessa WC, Walsh K. Acute modulation of endothelial Akt/PKB activity alters nitric oxide dependent vasomotor activity in vivo. J Clin Invest 2000;106:493–9.

[58] Michell BJ, Harris MB, Chen ZP, Ju H, Venema VJ, Blackstone MA, Huang W, Venema RC, Kemp BE. Identification of regulatory sites of phosphorylation of the bovine endothelial nitric-oxide synthase at serine 617 and serine 635. J Biol Chem 2002;277:42344–51.

[59] Radisavljevic Z, Avraham H, Avraham S. Vascular endothelial growth factor up-regulates ICAM-1 expression via phosphatidylinositol 3OH-kinase/AKT/nitric oxide pathway and modulates migration of brain microvascular endothelial cells. J Biol Chem 2000;275:20770–4.

[60] Fulton D, Church JE, Ruan L, Li C, Sood SG, Kemp BE, Jennings IG, Venema RC. Src kinase activates endothelial nitric-oxide synthase by phosphorylating Tyr-83. J Biol Chem 2005;280:35943–52.

[61] Michell BJ, Chen Z, Tiganis T, Stapleton D, Katsis F, Power DA, Sim AT, Kemp BE. Coordinated control of endothelial nitric-oxide synthase phosphorylation by protein kinase C and the cAMP-dependent protein kinase. J Biol Chem 2001;276:17625–8.

[62] Dimmeler S, Fleming I, Fisslthaler B, Hermann C, Busse R, Zeiher AM. Activation of nitric oxide synthase in endothelial cells by Akt-dependent phosphorylation. Nature 1999;399:601–5.

[63] Duval M, Le Boeuf F, Huot J, Gratton JP. Src-mediated phosphorylation of Hsp90 in response to vascular endothelial growth factor (VEGF) is required for VEGF receptor-2 signaling to endothelial NO synthase. Mol Biol Cell 2007;18:4659–68.

[64] He H, Venema VJ, Gu X, Venema RC, Marrero MB, Caldwell RB. Vascular endothelial growth factor signals endothelial cell production of nitric oxide and prostacyclin through Flk-1/KDR activation of c-src. J Biol Chem 1999;274:25130–5.

[65] Veikkola T, Karkkainer M, Claesson-Welsh L, Alitalo K. Regulation of angiogenesis via vascular endothelial growth factor receptors. Cancer Res 2000;60:203–12.

[66] Venema RC, Venema VJ, Ju H, Harris MB, Snead C, Jilling T, Dimitropoulou C, Maragoudakis ME, Catravas JD. Novel complexes of guanylate cyclase with heat shock protein 90 and nitric oxide synthase. Am J Physiol 2003;285:H669–78.

[67] Boo YC, Hwang J, Sykes M, Michell BJ, Kemp BE, Lum H, Jo H. Shear stress stimulates phosphorylation of eNOS at Ser(635) by a protein kinase A-dependent mechanism. Am J Physiol Heart Circ Physiol 2002;283:H1819–28.

[68] Parenti A, Morbidelli L, Cui XL, Douglas JG, Hood J, Granger HJ, Ledda F, Ziche M. Nitric oxide is an upstream signal for vascular endothelial growth factor-induced extracellular signal-regulated kinases½ activation in postcapillary endothelium. J Biol Chem 1998;273:4220–6.

[69] Coletta C, Papapetropoulos A, Erdelyi K, Olah G, Modis K, Panopoulos P, Asimakopoulou A, Gero D, Sharina I, Martin E, Szabo C. Hydrogen sulfide and nitric oxide are mutually dependent in the regulation of angiogenesis and endothelium-dependent vasorelaxation. Proc Natl Acad Sci U S A 2012;109:9161–6.

[70] Morbidelli L, Pyriochou A, Filippi S, Vassiliades Y, Roussos C, Zhou Z, Loutrari H, Waltenberger J, Stössel A, Giannis A, Ziche M, Papapetropoulos A. The soluble guanylyl cyclase inhibitor NS-2028 reduces vascular endothelial growth factor-induced angiogenesis and permeability. Am J Physiol Regul Integr Comp Physiol 2010;298(3):R824–32.

[71] Koika V, Zhou Z, Vasileiadis I, Roussos C, Finetti F, Monti M, Morbidelli L, Papapetropoulos A. PKG-I Inhibition attenuates vascular endothelial growth factor-stimulated angiogenesis. Vascul Pharmacol 2010;53(5–6):215–22.

[72] Ziche M, Morbidelli L. Molecular regulation of tumor angiogenesis by nitric oxide. Eur Cytokine Netw 2009;20(4):164–70.

[73] Babaei S, Teichert-Kuliszewska K, Zhang Q, Jones N, Dumont DJ, Stewart DJ. Angiogenic actions of angiopoietin-1 require endothelium-derived nitric oxide. Am J Pathol 2003;162:1927–36.

[74] Finetti F, Solito R, Morbidelli L, Giachetti A, Ziche M, Donnini S. Prostaglandin E2 regulates angiogenesis via activation of fibroblast growth factor receptor-1. J Biol Chem 2008;283 (4):2139–46.

[75] Namkoong S, Lee SJ, Kim CK, Kim YM, Chung HT, Lee H, Han JA, Ha KS, Kwon YG, Kim YM. Prostaglandin E2 stimulates angiogenesis by activating the nitric oxide/cGMP pathway in human umbilical vein endothelial cells. Exp Mol Med 2005;37:588–600.

[76] Ziche M, Morbidelli L, Parenti A, Ledda F. Nitric oxide modulates angiogenesis elicited by prostaglandin E1 in rabbit cornea. In: Samuelsson B, Paoletti R, Ramwell PW, editors. Advances in Prostaglandins, Thromboxane, and Leukotriene Research. 23:New York: Raven Press; 1995. p. 495–7.

[77] Cai H, Griendling KK, Harrison DG. The vascular NAD(P)H oxidases as therapeutic targets in cardiovascular diseases. Trends Pharmacol Sci 2003;24:471–8.

[78] Polytarchou C, Papadimitriou E. Antioxidants inhibit human endothelial cell functions through down-regulation of endothelial nitric oxide synthase activity. Eur J Pharmacol 2005;510:31–8.

[79] Lauer N, Suvorava T, Ruther U, Jacob R, Meyer W, Harrison DG, Kojda G. Critical involvement of hydrogen peroxide in exercise-induced up-regulation of endothelial NO synthase. Cardiovasc Res 2005;65:254–62.

[80] Drummond GR, Cai H, Davis ME, Ramasamy S, Harrison DG. Transcriptional and posttranscriptional regulation of endothelial nitric oxide synthase expression by hydrogen peroxide. Circ Res 2000;86:347–54.

[81] Thomas SR, Chen K, Keaney Jr JF. Hydrogen peroxide activates endothelial nitric-oxide synthase through coordinated phosphorylation and dephosphorylation via a phosphoinositide 3-kinase-dependent signaling pathway. J Biol Chem 2002;277:6017–24.

[82] Cai H, Li Z, Davis ME, Kanner W, Harrison DG, Dudley Jr. SC. Akt-dependent phosphorylation of serine 1179 and mitogen-activated protein kinase kinase/extracellular signal-regulated kinase 1/2 cooperatively mediate activation of the endothelial nitric-oxide synthase by hydrogen peroxide. Mol Pharmacol 2003;63:325–31.

[83] Fiedler LR, Wojciak-Stothard B. The DDAH/ADMA pathway in the control of endothelial cell migration and angiogenesis. Biochem Soc Trans 2009;37:1243–7.

[84] Fiedler LR, Bachetti T, Leiper J, Zachary I, Chen L, Renne T, Wojciak-Stothard B. The ADMA/DDAH pathway regulates VEGF-mediated angiogenesis. Arterioscler Thromb Vasc Biol 2009;29:2117–24.

[85] Palm F, Onozato ML, Luo Z, Wilcox CS. Dimethylarginine dimethylaminohydrolase (DDAH): expression, regulation, and function in the cardiovascular and renal systems. Am J Physiol Heart Circ Physiol 2007;293:H3227–45.

[86] Wojciak-Stothard B, Torondel B, Tsang LY, Fleming I, Fisslthaler B, Leiper JM, Vallance P. The ADMA/DDAH pathway is a critical regulator of endothelial cell motility. J Cell Sci 2007;120:929–42.

[87] Dowsett L, Piper S, Slaviero A, Dufton N, Wang Z, Boruc O, Delahaye M, Colman L, Kalk E, Tomlinson J, Birdsey G, Randi AM, Leiper J. Endothelial dimethylarginine dimethylaminohydrolase 1 is an important regulator of angiogenesis but does not regulate vascular reactivity or hemodynamic homeostasis. Circulation 2015;131:2217–25.

[88] Jacobi J, Sydow K, von Degenfeld G, Zhang Y, Dayoub H, Wang B, Patterson AJ, Kimoto M, Blau HM, Cooke JP. Overexpression of dimethylarginine dimethylaminohydrolase reduces tissue asymmetric dimethylarginine levels and enhances angiogenesis. Circulation 2005;111:431–1438.

[89] Urbich C, Reissner A, Chavakis E, Dernbach E, Haendeler J, Fleming I, Zeiher AM, Kaszkin M, Dimmeler S. Dephosphorylation of endothelial nitric oxide synthase contributes to the anti-angiogenic effects of endostatin. FASEB J 2002;16:706–8.

[90] Isenberg JS, Shiva S, Gladwin M. Thrombospondin-1–CD47 blockade and exogenous nitrite enhance ischemic tissue survival, blood flow and angiogenesis via coupled NO-cGMP pathway activation. Nitric Oxide 2009;21:52–62.

[91] Ramanathan S, Mazzalupo S, Boitano S, Montfort WR. Thrombospondin-1 and angiotensin II inhibit soluble guanylyl cyclase through an increase in intracellular calcium concentration. Biochemistry 2011;50:7787–99.

[92] Bauer EM, Qin Y, Miller TW, Bandle RW, Csanyi G, Pagano PJ, Bauer PM, Schnermann J, Roberts DD, Isenberg JS. Thrombospondin-1 supports blood pressure by limiting eNOS activation and endothelial-dependent vasorelaxation. Cardiovasc Res 2010;88:471–81. https://doi.org/10.1093/cvr/cvq218.

[93] Isenberg JS, Hyodo F, Pappan LK, Abu-Asab M, Tsokos M, Krishna MC, Frazier WA, Roberts DD. Blocking thrombospondin-1/CD47 signaling alleviates deleterious effects of aging on tissue responses to ischemia. Arterioscler Thromb Vasc Biol 2007;27:2582–8.

[94] Florio T, Morini M, Villa V, Arena S, Corsaro A, Thellung S, Culler MD, Pfeffer U, Noonan DM, Schettini G, Albini A. Somatostatin inhibits tumor angiogenesis and growth via somatostatin receptor-3-mediated regulation of endothelial nitric oxide synthase and mitogen-activated protein kinase activities. Endocrinology 2003;144:1574–84.

[95] Lee SH, Nishino M, Mazumdar T, Garcia GE, Galfione M, Lee FL, Lee CL, Liang A, Kim J, Feng L, Eissa NT, Lin SH, Yu-Lee LY. 16-kDa prolactin down-regulates inducible nitric oxide synthase expression through inhibition of the signal transducer and activator of transcription 1/IFN regulatory factor-1 pathway. Cancer Res 2005;65(17):7984–92.

[96] Mauriz JL, Linares P, Macias RI, Jorquera F, Honrado E, Olcoz JL, González P, González-Gallego J. TNP-470 inhibits oxidative stress, nitric oxide production and nuclear factor kappa B activation in a rat model of hepatocellular carcinoma. Free Radic Res 2003; 37(8):841–948.

[97] Roodhart JM, Langenberg MH, Witteveen E, Voest EE. The molecular basis of class side effects due to treatment with inhibitors of the VEGF/VEGFR pathway. Curr Clin Pharmacol 2008; 3(2):132–43.

[98] Chiou WF, Chou CJ, Chen CF. Camptothecin suppresses nitric oxide biosynthesis in RAW 264.7 macrophages. Life Sci 2001;69:625–35.

[99] Yin MB, Li ZR, Toth K, Cao S, Durrani FA, Hapke G, Bhattacharya A, Azrak RG, Frank C, Rustum YM. Potentiation of irinotecan sensitivity by Se-methylselenocysteine in an in vivo tumor model is associated with downregulation of cyclooxygenase-2, inducible nitric oxide synthase, and hypoxia-inducible factor 1alpha expression, resulting in reduced angiogenesis. Oncogene 2006;25(17):2509–19.

[100] Tozer GM, Prise VE, Wilson J, Cemazar M, Shan S, Dewhirst MW, Barber PR, Vojnovic B, Chaplin DJ. Mechanisms associated with tumor vascular shut-down induced by combretastatin A-4 phosphate: intravital microscopy and measurement of vascular permeability. Cancer Res 2001;61(17):6413–22.

[101] de Wilt JH, Manusama ER, van Etten B, van Tiel ST, Jorna AS, Seynhaeve AL, ten Hagen TL, Eggermont AM. Nitric oxide synthase inhibition results in synergistic anti-tumour activity with melphalan and tumour necrosis factor alpha-based isolated limb perfusions. Br J Cancer 2000;83:1176–82.

[102] Dulak J, Jozkowicz A, Dembinska-Kiec A, Guevara I, Zdzienicka A, Zmudzinska-Grochot D, Florek I, Wojtowicz A, Szuba A, Cooke JP. Nitric oxide induces the synthesis of vascular endothelial growth factor by rat vascular smooth muscle cells. Arterioscler Thromb Vasc Biol 2000;20:659–66.

[103] Jozkowicz A, Cooke JP, Guevara I, Huk I, Funovics P, Pachinger O, Weidinger F, Dulak J. Genetic augmentation of nitric oxide synthase increases the vascular generation of VEGF. Cardiovasc Res 2001;51:773–83.

[104] Chin K, Kurashima Y, Ogura T, Tajiri H, Yoshida S, Esumi H. Induction of vascular endothelial growth factor by nitric oxide in human glioblastoma and hepatocellular carcinoma cells. Oncogene 1997;15(4):437–42.

[105] Milkiewicz M, Hudlicka O, Brown MD, Silgram H. Nitric oxide, VEGF, and VEGFR-2: interactions in activity-induced angiogenesis in rat skeletal muscle. Am J Physiol Heart Circ Physiol 2005;289:H336–43.

[106] Kimura H, Esumi H. Reciprocal regulation between nitric oxide and vascular endothelial growth factor in angiogenesis. Acta Biochem Pol 2003;50(1):49–59.

[107] Pichu S, Sathiyamoorthy J, Vimalraj S, Viswanathan V, Chatterjee S. Impact of lysyl oxidase (G473A) polymorphism on diabetic foot ulcers. Int J Biol Macromol 2017;103:242–7.

[108] Kimura H, Weisz A, Ogura T, Hitomi Y, Kurashima Y, Hashimoto K, D'Acquisto F, Makuuchi M, Esumi H. Identification of hypoxia-inducible factor 1 ancillary sequence and its function in vascular endothelial growth factor gene induction by hypoxia and nitric oxide. J Biol Chem 2001;276:2292–8.

[109] Kimura H, Weisz A, Kurashima Y, Hashimoto K, Ogura T, D'Acquisto F, Addeo R, Makuuchi M, Esumi H. Hypoxia response element of the human vascular endothelial growth factor gene mediates transcriptional regulation by nitric oxide: control of hypoxia-inducible factor-1 activity by nitric oxide. Blood 2000;95:189–97.

[110] Sandau KB, Zhou J, Kietzmann T, Brune B. Regulation of the hypoxia-inducible factor 1á by the inflammatory mediators nitric oxide and tumor necrosis factor-α in contrast to desferroxamine and phenylarsine oxide. J Biol Chem 2001;276:39805–11.

[111] Kasuno K, Takabuchi S, Fukuda K, Kizaka-Kondoh S, Yodoi J, Adachi T, Semenza GL, Hirota K. Nitric oxide induces hypoxia inducible factor 1 activation that is dependent on MAPK and phosphatidylinositol 3-kinase signaling. J Biol Chem 2004;279:2550–8.

[112] Metzen E, Zhou J, Jelkmann W, Fandrey J, Brune B. Nitric oxide impairs normoxic degradation of HIF-1α by inhibition of prolyl hydroxylases. Mol Biol Cell 2003;14:3470–81.

[113] Pae HO, Oh GS, Choi BM, Kim YM, Chung HT. A molecular cascade showing nitric oxide-heme oxygenase-1-vascular endothelial growth factorinterleukin-8 sequence in human endothelial cells. Endocrinology 2005;146:2229–38.

[114] Quintero M, Brennan PA, Thomas GJ, Moncada S. Nitric oxide is a factor in the stabilization of hypoxia-inducible factor-1α in cancer: role of free radical formation. Cancer Res 2006;66:770–4.

[115] Janaszak-Jasiecka A, Siekierzycka A, Bartoszewska S, Serocki M, Dobrucki LW, Collawn JF, Kalinowski L, Bartoszewski R. eNOS expression and NO release during hypoxia is inhibited by miR-200b in human endothelial cells. Angiogenesis 2018; https://doi.org/10.1007/s10456-018-9620-y.

[116] Sogawa K, Numayama-Tsuruta K, Ema M, Abe M, Abe H, Fujii-Kuriyama Y. Inhibition of hypoxia-inducible factor 1 activity by nitric oxide donors in hypoxia. Proc Natl Acad Sci U S A 1998;95:7368–73.

[117] Huang LE, Willmore WG, Gu J, Goldberg MA, Bunn F. Inhibition of hypoxia-inducible factor 1 activation by carbon monoxide and nitric oxide. Implications for oxygen sensing and signaling. J Biol Chem 1999;274:9038–44.

[118] Ciccone V, Monti M, Monzani E, Casella L, Morbidelli L. The metal-nonoate Ni(SalPip-NONO) inhibits in vitro tumor growth, invasiveness and angiogenesis. Oncotarget 2018;9(17):13353–65.

[119] Hagen T, Taylor CT, Lam F, Moncada S. Redistribution of intracellular oxygen in hypoxia by nitric oxide: effect on HIF1α. Science 2003;302:1975–8.

[120] Wang F, Sekine H, Kikuchi Y, Takasaki C, Miura C, Heiwa O, Shuin T, Fujii-Kuriyama Y, Sogawa K. HIF-1alpha-prolyl hydroxylase: molecular target of nitric oxide in the hypoxic signal transduction pathway. Biochem Biophys Res Commun 2000;295:657–62.

[121] Díez I, Calatayud S, Hernández C, Quintana E, O'Connor JE, Esplugues JV, Barrachina MD. Nitric oxide, derived from inducible nitric oxide synthase, decreases hypoxia inducible factor-1alpha in macrophages during aspirin-induced mesenteric inflammation. Br J Pharmacol 2010;159(8):1636–45.

[122] Sorensen AG, Batchelor TT, Zhang WT, Chen PJ, Yeo P, Wang M, Jennings D, Wen PY, Lahdenranta J, Ancukiewicz M, di Tomaso E, Duda DG, Jain RK. A "vascular normalization index" as potential mechanistic biomarker to predict survival after single dose of cediranib in recurrent glioblastoma patients. Cancer Res 2009;69:5296–300.

[123] Tsurumi Y, Murohara T, Krasinski K, Chen D, Witzenbichler B, Kearney M, Couffinhal T, Isner JM. Reciprocal relation between VEGF and NO in the regulation of endothelial integrity. Nat Med 1997;3:879–86.

[124] Thomas DD, Espey MG, Ridnour LA, Hofseth LJ, Mancardi D, Harris CC, Wink DA. Hypoxic inducible factor 1α, extracellular signal-regulated kinase, and p53 are regulated by distinct threshold concentrations of nitric oxide. Proc Natl Acad Sci U S A 2004;101:8894–9.

[125] Andrew PJ, Harant H, Lindley IJ. Nitric oxide regulates IL-8 expression in melanoma cells at the transcriptional level. Biochem Biophys Res Commun 1995;214(3):949–56.

[126] Xiong Q, Shi Q, Le X, Wang B, Xie K. Regulation of interleukin-8 expression by nitric oxide in human pancreatic adenocarcinoma. J Interferon Cytokine Res 2001;21(7):529–37.

[127] Cianchi F, Cortesini C, Fantappie O, Messerini L, Sardi I, Lasagna N, Perna F, Fabbroni V, Di Felice A, Perigli G, Mazzanti R, Masini E. Cyclooxygenase-2 activation mediates the proangiogenic effect of nitric oxide in colorectal cancer. Clin Cancer Res 2004;10(8):2694–704.

[128] Conelly L, Jacobs AT, Placious-Callender M, Moncada S, Hobbs AJ. Macrophage endothelial nitric oxide synthase autoregulates cellular activation and pro-inflammatory protein expression. J Biol Chem 2003;278:26480–7.

[129] Ziche M, Parenti A, Ledda F, Dell'Era P, Granger HJ, Maggi CA, Presta M. Nitric oxide promotes proliferation and plasminogen activator production by coronary venular endothelium through endogenous bFGF. Circ Res 1997;80:845–52.

[130] Lizarbe TR, García-Rama C, Tarín C, Saura M, Calvo E, López JA, López-Otín C, Folgueras AR, Lamas S, Zaragoza C. Nitric oxide elicits functional MMP-13 protein-tyrosine nitration during wound repair. FASEB J 2008;22(9):3207–15.

[131] Matsunaga T, Weihrauch DW, Moniz MC, Tessmer J, Warltier DC, Chilian WM. Angiostatin inhibits coronary angiogenesis during impaired production of nitric oxide. Circulation 2002;105:2185–91.

[132] Romagnani P, Lazzeri E, Lasagni L, Mavilia C, Beltrame C, Francalanci M, Rotondi M, Annunziato F, Maurenzig L, Cosmi L, Galli G, Salvadori M, Maggi E, Serio M. IP10 and Mig production by glomerular cells in human proliferative glomerulonephritis and regulation by nitric oxide. J Am Soc Nephrol 2002;13:53–64.

[133] Morbidelli L, Donnini S, Ziche M. Role of nitric oxide in the modulation of angiogenesis. Curr Pharm Design 2003;9:521–30.

[134] Kundu JK, Surh YJ. Inflammation: gearing the journey to cancer. Mutat Res 2008;659(1–2):15–30.

[135] Lala PK, Chakraborty C. Role of nitric oxide in carcinogenesis and tumor progression. Lancet Oncol 2001;(3):149–56.

[136] Pan J-W, Zhan R, Tong Y, Zhou Y, Zhang M. Expression of endothelial nitric oxide synthase and vascular endothelial growth factor in association with neovascularization in human primary astrocytoma. J Zhejiang Univ Sci B 2005;6(7):693–8.

[137] Lechner M, Lirk P, Rieder J. Inducible nitric oxide synthase (iNOS) in tumor biology: the two sides of the same coin. Sem Cancer Biology 2005;15:277–89.

[138] Fukumura D, Kashiwagi S, Jain RK. The role of nitric oxide in tumour progression. Nat Rev Cancer 2006;6(7):521–34.

[139] Terzuoli E, Finetti F, Costanza F, Giachetti A, Ziche M, Donnini S. Linking of mPGES-1 and iNOS activates stem-like phenotype in EGFR-driven epithelial tumor cells. Nitric Oxide 2017;66:17–29.

[140] Jenkins DC, Charles IG, Thomsen LL, Moss DW, Holmes LS, Baylis SA, Rhodes P, Westmore K, Emson PC, Moncada S. Roles of nitric oxide in tumour growth. Proc Natl Acad Sci U S A 1995;92:4392–6.

[141] Frank S, Stallmeyer B, Kampfer H, Schaffner C, Pfeilschifter J. Differential regulation of vascular endothelial growth factor and its receptor fms-like-tyrosine kinase is mediated by nitric oxide in rat renal mesangial cells. Biochem J 1999;338:367–74 Pt 2.

[142] Jadeski LC, Hum KO, Chakraborty C, Lala PK. Nitric oxide promotes murine mammary tumour growth and metastasis by stimulating tumour cell migration, invasiveness and angiogenesis. Int J Cancer 2000;86:30–9.

[143] Yang HT, Yan Z, Abraham JA, Terjung RL. VEGF(121)- and bFGF-induced increase in collateral blood flow requires normal nitric oxide production. Am J Physiol 2001;280(3):H1097–104.

[144] Andrade SP, Hart IR, Piper PJ. Inhibitors of nitric oxide synthase selectively reduce flow in tumor-associated neovasculature. Br J Pharmacol 1992;107(4):1092–5.

[145] Malone JM, Saed GM, Diamond MP, Sokol RJ, Munkarah AR. The effects of the inhibition of inducible nitric oxide synthase on angiogenesis of epithelial ovarian cancer. Am J Obstet Gynecol 2006;194(4):1110–8.

[146] Thomsen LL, Lawton FG, Knowles RG, Beesley JE, Riveros-Moreno V, Moncada S. Nitric oxide synthase activity in human gynaecological cancer. Cancer Res 1994;54:1352–4.

[147] Cobbs CS, Brenman JE, Aldape KD, Bredt DS, Israel MA. Expression of nitric oxide synthase in human central nervous system tumours. Cancer Res 1995;55:727–30.

[148] Gallo O, Masini E, Morbidelli L, Franchi A, Fini-Storchi I, Vergari WA, Ziche M. Role of nitric oxide in angiogenesis and tumour progression in head and neck cancer. J Natl Cancer Inst 1998;90:587–96.

[149] Kojima M, Morisaki T, Tsukahara Y, Uchiyama A, Matsunari Y, Mibu R, Tanaka M. Nitric oxide synthase expression and nitric oxide production in human colon carcinoma tissue. J Surg Oncol 1999;70(4):222–9.

[150] Swana HS, Smith SD, Perrotta PL, Saito N, Wheeler MA, Weiss RM. Inducible nitric oxide synthase with transitional cell carcinoma of the bladder. J Urol 1999;161(2):630–4.

[151] Marrogi AJ, Travis WD, Welsh JA, Khan MA, Rahim H, Tazelaar H, Pairolero P, Trastek V, Jett J, Caporaso NE, Liotta LA, Harris CC. Nitric oxide synthase, cyclooxygenase 2, and vascular endothelial growth factor in the angiogenesis of non-small cell lung carcinoma. Clin Cancer Res 2000;6(12):4739–44.

[152] Vakkala M, Kahlos K, Lakari E, Paakko P, Kinnula V, Soini Y. Inducible nitric oxide synthase expression, apoptosis, and angiogenesis in in situ and invasive breast carcinomas. Clin Cancer Res 2000;6(6):2408–16.

[153] Uotila P, Valve E, Martikainen P, Nevalainen M, Nurmi M, Harkonen P. Increased expression of cyclooxygenase-2 and nitric oxide synthase-2 in human prostate cancer. Urol Res 2001; 29(1):23–8.

[154] Aaltoma SH, Lipponen PK, Kosma VM. Inducible nitric oxide synthase (iNOS) expression and its prognostic value in prostate cancer. Anticancer Res 2001;21(4B):3101–6.

[155] Massi D, Franchi A, Sardi I, Magnelli L, Paglierani M, Borgognoni L, Maria Reali U, Santucci M. Inducible nitric oxide synthase expression in benign and malignant cutaneous melanocytic lesions. J Pathol 2001;194(2):194–200.

[156] Kostourou V, Robinson SP, Cartwright JE, Whitley GS. Dimethylarginine dimethylaminohydrolase I enhances tumour growth and angiogenesis. Br J Cancer 2002;87(6):673–80.

[157] Hulin JA, Tommasi S, Elliot D, Hu DG, Lewis BC, Mangoni AA. MiR-193b regulates breast cancer cell migration and vasculogenic mimicry by targeting dimethylarginine dimethylaminohydrolase 1. Sci Rep 2017;7(1):13996.

[158] Kostourou V, Robinson SP, Whitley GS, Griffiths JR. Effects of overexpression of dimethylarginine dimethylaminohydrolase on tumor angiogenesis assessed by susceptibility magnetic resonance imaging. Cancer Res 2003;63(16):4960–6.

[159] Kostourou V, Troy H, Murray JF, Cullis ER, Whitley GS, Griffiths JR, Robinson SP. Overexpression of dimethylarginine dimethylaminohydrolase enhances tumor hypoxia: an insight into the relationship of hypoxia and angiogenesis in vivo. Neoplasia 2004;6(4):401–11.

[160] Hasegawa K, Wakino S, Tanaka T, Kimoto M, Tatematsu S, Kanda T, Yoshioka K, Homma K, Sugano N, Kurabayashi M, Saruta T, Hayashi K. Dimethylarginine dimethylaminohydrolase 2 increases vascular endothelial growth factor expression through Sp1 transcription factor in endothelial cells. Arterioscler Thromb Vasc Biol 2006;26(7):1488–94.

[161] Smith CL, Birdsey GM, Anthony S, Arrigoni FI, Leiper JM, Vallance P. Dimethylarginine dimethylaminohydrolase activity modulates ADMA levels, VEGF expression, and cell phenotype. Biochem Biophys Res Commun 2003;308(4):984–9.

[162] Bonavida B, Khineche S, Huerta-Yepez S, Garbán H. Therapeutic potential of nitric oxide in cancer. Drug Resistance Uptodates 2006;9(3):157–73.

[163] Coulter JA, McCarthy HO, Roedl W, Wagner E, Robson T, Hirst DG. Nitric oxide-a novel therapeutic for cancer. Nitric Oxide 2008;19(2):192–8.

[164] Fitzpatrick B, Mehibel M, Cowen RL, Stratford IJ. iNOS as a therapeutic target for treatment of human tumors. Nitric Oxide 2008;19(2):217–24.

[165] Cheng H, Wang L, Mollica M, Re AT, Wu S, Zuo L. Nitric oxide in cancer metastasis. Cancer Lett 2014;353(1):1–7.

[166] Morbidelli L. Therapeutic potential of nitric oxide donors in cancer: focus on angiogenesis. Crit Rev Oncog 2016;21(5-6):447–58.

[167] Dong R, Wang X, Wang H, Liu Z, Liu J, Saavedra JE. Effects of JS-K, a novel anti cancer nitric oxide prodrug, on gene expression in human Hep3B cells. Biomed Pharmacother 2017;88:367–73.

[168] Parinandi NL, Sharma A, Eubank TD, Kaufman BF, Kutala VK, Marsh CB, Ignarro LJ, Kuppusamy P. Nitroaspirin (NCX-4016), an NO donor, is antiangiogenic through induction of loss of redox-dependent viability and cytoskeletal reorganization in endothelial cells. Antioxid Redox Signal 2007;9(11):1837–49.

[169] Kumar S, Singh RK, Bhardwaj TR. Therapeutic role of nitric oxide as emerging molecule. Biomed Pharmacother 2017;85:182–201.

[170] Herman AG, Moncada S. Therapeutic potential of nitric oxide donors in the prevention and treatment of atherosclerosis. Eur Heart J 2005;26(19):1945–55.

[171] Donnini S, Solito R, Giachetti A, Granger HJ, Ziche M, Morbidelli L. Fibroblast growth factor-2 mediates Angiotensin-converting enzyme inhibitor-induced angiogenesis in coronary endothelium. J Pharmacol Exp Ther 2006;319(2):515–22.

[172] Donnini S, Terzuoli E, Ziche M, Morbidelli L. Sulfhydryl angiotensin-converting enzyme inhibitor promotes endothelial cell survival through nitric-oxide synthase, fibroblast growth factor-2, and telomerase cross-talk. J Pharmacol Exp Ther 2010;332(3):776–84.

[173] Patarroyo Aponte MM, Francis GS. Effect of Angiotensin-converting enzyme inhibitors and Angiotensin receptor antagonists in atherosclerosis prevention. Curr Cardiol Rep 2012;14 (4):433–42.

[174] Estruch R, Ros E, Salas-Salvadó J, Covas MI, Corella D, Arós F, Gómez-Gracia E, Ruiz-Gutiérrez V, Fiol M, Lapetra J, Lamuela-Raventos RM, Serra-Majem L, Pintó X, Basora J, Muñoz MA, Sorlí JV, Martínez JA, Fitó M, Gea A, Hernán MA, Martínez-González MA, PREDIMED Study Investigators. Primary prevention of cardiovascular disease with a mediterranean diet supplemented with extra-virgin olive oil or nuts. N Engl J Med 2018;378(25):e34.

[175] Vasquez EC, Gava AL, Graceli JB, Balarini CM, Campagnaro BP, Pereira TM, Meyrelles SS. Novel therapeutic targets for phosphodiesterase 5 inhibitors: current state-of-the-art on systemic arterial hypertension and atherosclerosis. Curr Pharm Biotechnol 2016;17(4):347–64.

[176] Hashimoto A, Miyakoda G, Hirose Y, Mori T. Activation of endothelial nitric oxide synthase by cilostazol via a cAMP/protein kinase A- and phosphatidylinositol 3-kinase/Akt-dependent mechanism. Atherosclerosis 2006;189(2):350–7.

[177] Herath SC, Lion T, Klein M, Stenger D, Scheuer C, Holstein JH, Mörsdorf P, Rollmann MF, Pohlemann T, Menger MD, Histing T. Stimulation of angiogenesis by cilostazol accelerates fracture healing in mice. J Orthop Res 2015;33(12):1880–7.

[178] Sanada F, Kanbara Y, Taniyama Y, Otsu R, Carracedo M, Ikeda-Iwabu Y, Muratsu J, Sugimoto K, Yamamoto K, Rakugi H, Morishita R. Induction of angiogenesis by a Type III phosphodiesterase inhibitor, cilostazol, through activation of peroxisome proliferator-activated receptor-γ and cAMP pathways in vascular cells. Arterioscler Thromb Vasc Biol 2016;36(3):545–52.

[179] Tseng SY, Chao TH, Li YH, Cho CL. Cilostazol improves proangiogenesis functions in human early endothelial progenitor cells through the stromal cell-derived factor system and hybrid therapy provides a synergistic effect in vivo. Biomed Res Int 2016;2016:3639868.

[180] Cui J, Zhuang S, Qi S, Li L, Zhou J, Zhang W, Zhao Y, Qi N, Yin Y, Huang L. Hydrogen sulfide facilities production of nitric oxide via the Akt/endothelial nitric oxide synthases signaling pathway to protect human umbilical vein endothelial cells from injury by angiotensin II. Mol Med Rep 2017;16(5):6255–61.

[181] Bucci M, Papapetropoulos A, Vellecco V, Zhou Z, Pyriochou A, Roussos C, Roviezzo F, Brancaleone V, Cirino G. Hydrogen sulfide is an endogenous inhibitor of phosphodiesterase activity. Arterioscler Thromb Vasc Biol 2010;30(10):1998–2004.

[182] Monti M, Hyseni I, Pacini A, Monzani E, Casella L, Morbidelli L. Cross-talk between endogenous H2S and NO accounts for vascular protective activity of the metal-nonoate Zn(PipNONO)Cl. Biochem Pharmacol 2018;152:143–52.

[183] Terzuoli E, Monti M, Vellecco V, Bucci M, Cirino G, Ziche M, Morbidelli L. Characterization of zofenoprilat as an inducer of functional angiogenesis through increased H2 S availability. Br J Pharmacol 2015;172(12):2961–73.

[184] Luna-Vázquez FJ, Ibarra-Alvarado C, Rojas-Molina A, Romo-Mancillas A, López-Vallejo FH, Solís-Gutiérrez M, Rojas-Molina JI, Rivero-Cruz F. Role of nitric oxide and hydrogen sulfide in the vasodilator effect of ursolic acid and uvaol from black cherry Prunus serotina fruits. Molecules 2016;21(1):78.

[185] Bradley JM, Organ CL, Lefer DJ. Garlic-derived organic polysulfides and myocardial protection. J Nutr 2016;146(2):403S–409S.

[186] Williams G, Shi-Wen X, Abraham D, Selvakumar S, Baker DM, Tsui JC. Nitric oxide manipulation: a therapeutic target for peripheral arterial disease? Cardiol Res Pract 2012;2012:656247.

[187] Sharma R, Joubert J, Malan SF. Recent developments in drug design of NO-donor hybrid compounds. Mini Rev Med Chem 2018;18(14):1175–98.

[188] Abdellatif KR, Abdelall EK, Bakr RB. Nitric oxide-NASIDS donor prodrugs as hybrid safe anti-inflammatory agents. Curr Top Med Chem 2017;17(8):941–55.

[189] Wu D, Hu Q, Xiong Y, Zhu D, Mao Y, Zhu YZ. Novel H2S-NO hybrid molecule (ZYZ-803) promoted synergistic effects against heart failure. Redox Biol 2018;15:243–52.

[190] Li Y, Lin J, Zhi X, Li P, Jiang X, Yuan J. Triple stimuli-responsive keratin nanoparticles as carriers for drug and potential nitric oxide release. Korean J Couns Psychother 2018;91:606–14.

[191] Pinto ML, Rocha J, Gomes JR, Pires J. Slow release of NO by microporous titanosilicate ETS-4. J Am Chem Soc 2011;133(16):6396–402.

[192] Haikal RR, Hua C, Perry 4th JJ, O'Nolan D, Syed I, Kumar A, Chester AH, Zaworotko MJ, MH3 Y, Alkordi MH. Controlling the uptake and regulating the release of nitric oxide in microporous solids. ACS Appl Mater Interfaces 2017;9(50):43520–8.

[193] Wan X, Wang Y, Jin X, Li P, Yuan J, Shen J. Heparinized PCL/keratin mats for vascular tissue engineering scaffold with potential of catalytic nitric oxide generation. J Biomater Sci Polym Ed 2018;23:1–19.

[194] Hopkins SP, Pant J, Goudie MJ, Schmiedt C, Handa H. Achieving long-term biocompatible silicone via covalently immobilized S-nitroso-N-acetylpenicillamine (SNAP) that exhibits 4 months of sustained nitric oxide release. ACS Appl Mater Interfaces 2018; https://doi.org/10.1021/acsami.8b08647.

[195] Albaghdadi MS, Yang J, Brown JH, Mansukhani NA, Ameer GA, Kibbe MR. A tailorable in-situ light-activated biodegradable vascular scaffold. Adv Mater Technol 2017;2(4). pii: 1600243.

CHAPTER 5

Nitric Oxide and Cancer: To Inhibit or To Induce iNOS: That Is the Question?

Khosrow Kashfi*,†, Federica Vannini‡

*Department of Molecular, Cellular and Biomedical Sciences, Sophie Davis School of Biomedical Education, City University of New York School of Medicine, New York, NY, United States

†Avicenna Therapeutics, Inc, Raleigh, NC, United States

‡Institute of Life Sciences, Scuola Superiore Sant'Anna, Pisa, Italy

ABSTRACT

The role of NO in cancer has been quite perplexing, as both tumor promotion and inflammatory activities as well as antitumor and antiinflammatory properties have been described. These paradoxes have been explained in terms of a dual or a biphasic effect that is dependent on its local flux. In many tumors, iNOS expression is high, and this has correlated relatively well with poor patient outcomes, so much that some have suggested its expression may be used as a biomarker of poor patient's prognosis and perhaps survival. Many studies also suggest that a high iNOS expression results in tumor regression. In this review, we have evaluated iNOS inhibition and induction as two opposing strategies for controlling cancer and suggest that ultimately it is the tumor microenvironment that is critical to the outcome. We highlight that high NO levels produced by iNOS help to transfer the tumor-associated microglia/macrophages from a tumor-promoting and immunosuppressive M2-like state to a tumoricidal M1-like milieu.

ABBREVIATIONS

1400W	*N*-[[3-(aminomethyl)phenyl]methyl]-ethanimidamide
AG	aminoguanidine
AOAA	aminooxyacetic acid
Bcl-2	B-cell lymphoma 2
Cdk2	cyclin-dependent kinase 2
cGMP	cyclic guanosine monophosphate
EGFR	epidermal growth factor receptor

Therapeutic Application of Nitric Oxide in Cancer and Inflammatory Disorders. https://doi.org/10.1016/B978-0-12-816545-4.00005-0

GAPDH	glyceraldehyde-3-phosphate dehydrogenase
HIF-1α	hypoxia-inducible factor-1
iNOS	inducible nitric oxide synthase
L-NAME	N5-[imino(nitroamino)methyl]-L-ornithine methyl ester
L-NMMA	tilarginine (N^G-monomethyl-L-arginine)
MAPK	mitogen-activated protein kinase
MKP-1	mitogen-activated protein kinase phosphatase 1
NF-κB	nuclear factor kappa-light-chain-enhancer of activated B cells
nNOS	neuronal nitric oxide synthase
NO	nitric oxide
NOS	nitric oxide synthase
$P21^{Cip-1/Waf-1}$	cyclin-dependent kinase inhibitor 1
$P21^{RAS}$	RAS member
P53	oncoprotein
P-Akt	phosphorylated protein kinase B
P-ERK:	phosphorylated extracellular signal-regulated kinase
PKC	protein kinase C
P-p53	phosphorylated p53
PTP	protein tyrosine phosphatase
sGC	soluble guanylate cyclase
STAT3	signal transducer and activator of transcription 3

Conflict of Interest

No potential conflicts of interest were disclosed.

NO: ITS GENERATION AND SIGNALING

Nitric oxide (NO) is a free-radical gaseous molecule that is generated from L-arginine by three distinct isoforms of nitric oxide synthases (NOSs, EC1.14.13.39): the inducible form (iNOS or NOS2) and two constitutive forms, neuronal (nNOS or NOS1) and endothelial (eNOS or NOS3) (Fig. 1) [1]. nNOS and eNOS are calcium-dependent forms of the enzyme that are regulated by negative feedback [2], phosphorylation [3], and interaction with regulatory molecules [4]. iNOS is calcium-independent, is regulated transcriptionally, and is induced by inflammatory cytokines, oxidative stress, hypoxia, and various endotoxins [5, 6]. While historically NO has been recognized as being quite toxic, it is now recognized to be one of the three gasotransmitters that have multiple roles in normal physiology. The other two being hydrogen sulfide and carbon monoxide [7].

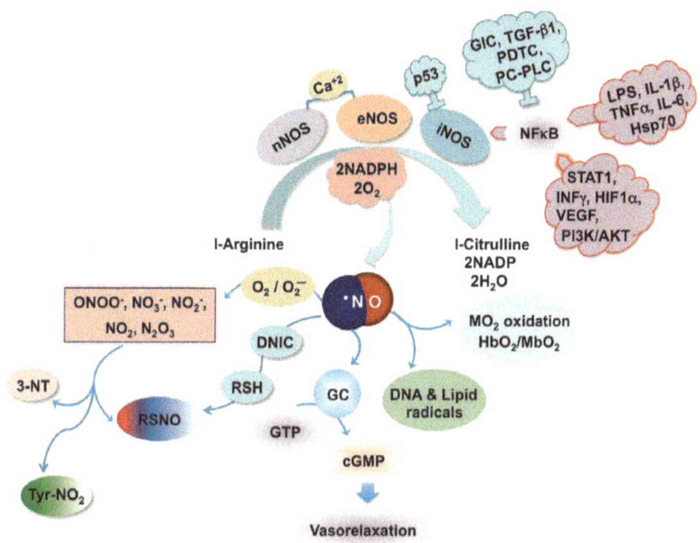

FIG. 1

Biosynthesis of nitric oxide and its mechanism of action. NO is produced by three nitric oxide synthase (NOS) isoforms: neuronal, endothelial, and inducible (nNOS, eNOS, and iNOS) that catalyze the oxidation of L-arginine to L-citrulline. The transcription factor NF-κB is central to iNOS regulation. LPS, IL-1β, TNF-α, IL-6, and Hsp70 have been shown to induce iNOS, whereas glucocorticoids (GIC), transforming growth factor-β1 (TGFβ1), antioxidants (e.g., pyrrolidine dithiocarbamate, PDTC), and inhibitors of phosphatidylcholine-specific phospholipase (PC-PLC) have been shown to inhibit iNOS expression by inhibiting NF-κB activation. iNOS expression is also directly induced by STAT1, INFα, HIF1-α, VEGF, and PI3K/AKT and inhibited by p53.

The biological effects of NO include direct effects where NO directly reacts with target molecules such as metal centers, DNA, and lipid radicals. Indirect effects are those where NO first reacts with other molecules such as O_2 or O_2^- to form nitrogen oxides that in turn go on to react with target molecules. Peroxynitrite ($ONOO^-$) can react with cysteine and/or tyrosine residues to form *S*-nitrosothiols (RSNO) and 3-nitrotyrosine (3-NT). RSNO can also be formed via N_2O_3 or through dinitrosyliron complex (DNIC), which requires chelatable iron. NO reacts with the active site of soluble guanylate cyclase (sGC) and produces cyclic GMP leading to vasorelaxation.

Two major pathways are at play regarding signaling mechanisms of NO. One is cGMP-dependent (NO-sGC-cGMP pathway), and the other is cGMP-independent, which is also referred to as the NO oxidative pathway. For the NO-sGC-cGMP signaling pathway, NO reacts with the active site of soluble guanylate cyclase (sGC) and produces cyclic GMP (cGMP). cGMP activates cGMP-dependent protein kinase G (PKG), which phosphorylates multiple substrates [8]. The other two major downstream elements that may also be activated are cGMP-dependent gated ion channels and cGMP-dependent phosphodiesterase [8]. In platelets, activation of the cGMP-dependent kinase phosphorylates a variety of substrates and is involved in platelet adhesion and aggregation [9].

In general, an increase in cGMP leads to smooth muscle relaxation (Fig. 1), vasorelaxation, and decrease of platelet aggregation [9, 10], reviewed in Ref. [11].

The cGMP-independent pathway is mainly through protein modification by *S*-nitrosylation of the cysteine residues [12–14] (Fig. 1). This type of posttranslational modification can affect transcriptional activity by changes in DNA binding of the protein; for example, such modifications to NF-κB results in its loss of binding to DNA [15, 16]. Hypoxia-inducible factor-1 (HIF-1) and estrogen receptor are also regulated by *S*-nitrosylation [17]. Signaling events regulated by *S*-nitrosylation may lead to either progression or inhibition of cancer. *S*-nitrosylation of NF-κB and matrix metalloproteinase 9 (MMP9) promotes cell death, whereas *S*-nitrosylation of caspase-3, caspase-9, and c-Jun N-terminal kinase prevents activity and inhibits apoptosis [16]. Death receptors were also found to be regulated by nitrosylation; for example, nitrosylation of the death receptor DR4 and Fas promotes apoptosis via the death-inducing signal complex [18, 19].

NO reacts with oxygen radicals to form peroxynitrite, which is a strong oxidative and nitrosative agent, resulting in direct nitrosylation or nitrosation or nitration of signaling proteins [11]. αNitrosative signaling in cancer cells is known to contribute to more proliferation and metastasis and resistance to therapy and, therefore, poor outcome of treatment [20]. On the other hand, apoptotic and necrotic cell deaths can occur by peroxynitrite via lipid peroxidation, cysteine oxidation, and also protein nitrosylation [21, 22].

NO AND CANCER

NO has a dichotomous role in cancer biology, with some reports suggesting that NO possesses antitumor and antiinflammatory properties, while others implicate NO in tumor promotion and inflammation. For example, the production of endogenous NO is associated with apoptosis in cancer cells [23] characterized by the upregulation of the tumor suppressor p53, changes in the expression of pro- and antiapoptotic Bcl-2 family members, activation of caspases, chromatin condensation, and DNA fragmentation [24]. On the other hand, a large body of experimental and clinical data suggests a promoting role of NO in tumor progression and metastasis.

Since the activation of iNOS is Ca^{2+}/calmodulin-independent, the amount of NO that can be generated by this enzyme is much larger and significantly more prolonged than that synthetized by the constitutive forms. It can reach micromolar concentrations for relatively long periods of time (days) compared with the nanomolar concentrations that are produced by the constitutive NOSs, usually no more than a few minutes at best [25]. This is a key feature of NOS that affects the NO-dependent biological effects. In fact, for decades, a concentration-dependent role of NO was observed in many physiological functions and pathological events, including cancer [11]. At low concentrations (<100 nM), NO regulates physiological processes such as blood flow, iron homeostasis, and neurotransmission, but it can also promote cellular proliferation and angiogenesis (Fig. 2). Notably, high levels of NO (<500 nM) generate DNA damage and oxidative/nitrosative stress and promote cytotoxic effects and apoptosis. Intermediate levels of NO are able to induce cytoprotection as well as invasiveness and metastasis [26]. Thus, NO shows both pro- and antitumorigenic properties depending on the concentration and duration of exposure. High levels of NO activate the tumor suppressor protein p53, leading to apoptosis, whereas low concentrations are necessary to stimulate the oncoprotein Akt via the antiapoptotic Ras signaling [27]. Other NO-mediated proapoptotic mechanisms include the activation of mitogen-activated protein kinase (MAPK)-phosphatase-1 (MKP-1) and the inhibition of NF-κB.

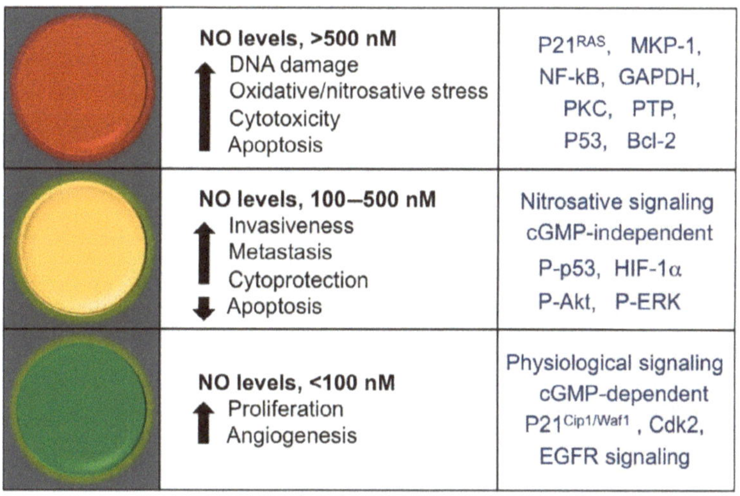

FIG. 2

Concentration-dependent effects of NO in the pathophysiological cell processes. At low concentrations (<100 nM), NO stimulates cell proliferation and angiogenesis, modulating cyclic guanosine monophosphate (cGMP)-dependent signaling, whose main targets are cyclin-dependent kinase inhibitor 1 ($P21^{Cip-1/Waf-1}$), cyclin-dependent kinase 2 (Cdk2), and epidermal growth factor receptor (EGFR). EGFR phosphorylation can also be directly inhibited by NO. At high concentrations (>500 nM), oxidative and nitrosative stress activate P21 (RAS member), mitogen-activated protein kinase phosphatase 1 (MKP)-1, nuclear factor kappa B (NF-κB), glyceraldehyde-3-phosphate dehydrogenase (GAPDH), protein kinase C (PKC), and protein tyrosine phosphatase (PTP). Cytotoxicity and apoptosis are increased by the oncoprotein P53 and B-cell lymphoma 2 (Bcl-2) modulation. At intermediate levels (100–500 nM), apoptosis is inhibited by P53 phosphorylation and invasiveness/metastasis are promoted by protein kinase B (PKB or AKT) and extracellular signal-regulated kinase (ERK) phosphorylation. NO also affects hypoxia-inducible factor-1 (HIF-1α) activation.

The dual role of iNOS in cancer must also be attributed to the tumor microenvironment, the intrinsic sensitivity of cells to NO, the bioavailability of NO substrates, and the biodistribution. The nitrosylation/nitration reaction, methylation, and ribosylation are some of the posttranslational modifications induced by NO to target proteins that can affect the cellular functions. For example, NO-mediated nitrosylation of the DNA alkyl-transferase inhibits the DNA repair capacity. Apoptosis can be inhibited by caspase and cGMP nitrosylation, but instead, it is promoted by S-nitrosylation of NF-κB, glyceraldehydes-3-phosphate dehydrogenase, Fas receptor, and Bcl-2 [28].

The redox state within the cells is the cause of the cellular sensitivity to NO-mediated cytotoxicity. The local oxygen concentration is dependent on the rate of delivery from the vasculature and the tissue distribution of iNOS isoforms, but NO itself can also affect the O_2 consumption by damaging the mitochondrial electron transport chain [29]. This is important for tumors that grow in the hypoxic microenvironment that are subjected to low levels of NO, whose chronic exposure can lead to NO adaptation and activation of tumorigenic pathways. On the other side, therapeutic strategies that increase NO levels to promote the cytotoxic effect in tumors that are not NO-dependent have proved effective. Singh et al. demonstrated that by exposing HT1080 and MDA231 cells, engineered to constitutively overexpress the iNOS gene, to anoxic conditions, NO generation was not observed. As O_2 availability was increased, NO production was detected, and it was able to potentiate the radiosensitivity of cells [30].

What about the tumors that are NO-dependent for their growth? A number of factors affecting the NO antitumorigenic properties make the design of a specific therapy challenging. Here, we review two additional applied approaches that use iNOS as a therapeutic target. iNOS can be either increased to promote cytotoxicity and apoptosis or inhibited to reduce inflammation, angiogenesis, and tumor proliferation.

INOS INHIBITION STRATEGIES

All NOS isoforms metabolize L-arginine to L-citrulline and NO via two consecutive NADPH-dependent monooxygenation reactions [7]. Two identical monomers constitute the active NOS, each having a C-terminal reductase and an N-terminal oxygenase domain. In mammals, all three isoforms contain highly conserved catalytic cores, which historically was an obstacle when attempting to find a specific inhibitor of a particular isoform [31]. The X-ray crystal structures of iNOS [32, 33] and eNOS [34] became available in the late 1990s and that of nNOS in 2002 [35]. These allowed for computer-based drug design of each isoform [36]. Potential therapeutic and challenges in developing NOS inhibitors have been reviewed in Refs. [37, 38].

In many clinical cancer studies, high iNOS expression has correlated relatively well with poor patient survival, reviewed in Ref. [7]. These include colon, breast, gastric, hepatocellular carcinoma, melanoma, ovarian, leukemia, gastric, prostate, esophageal, and cervical cancer [39, 40]. It has thus been suggested that iNOS expression may be used as a biomarker of poor patient prognosis and perhaps survival [41, 42].

Treatment of triple-negative breast cancer (TNBC) cell lines, which express high levels of iNOS with 1400W, a highly selective iNOS inhibitor; or L-NAME, a relatively selective eNOS inhibitor; or L-NMMA, a pan-NOS inhibitor, reduced cell proliferation, migration, and mammosphere formation [43]. In xenograft mouse models of TNBC, treatment with L-NAME and L-NMMA significantly reduced tumor growth. However, L-NMMA causes an acute elevation in blood pressure through inhibition of eNOS. To counteract this, amlodipine that is an antihypertensive agent was also incorporated. Administration of aminoguanidine, another iNOS specific inhibitor instead of 1400W that has shown toxicity at high doses [44] to nude mice bearing TNBC xenografts, abated tumor growth and metastatic burden [45]. Further, growth of glioma [46] or melanoma [47] cells in xenografts was significantly reduced when iNOS was silenced in these cells before they were implanted. The overarching data from all of these studies are the observation that the enhanced growth of the iNOS-overexpressing tumors appears to be due to enhanced angiogenesis.

Tumorigenesis induced by APC/β-catenin signaling was found to be inhibited by AG in the small intestine but not in the large intestine of Apc(Min/+) mice [48]. Genetic inhibition of iNOS led to fewer development of adenomas in the large intestine compared with Apc(Min/+) iNOS(+/+) mice. Cianchi and colleagues demonstrated a positive correlation between iNOS and COX2 expression in cancer cells as well as NO and PGE2 production, suggesting a novel therapeutic strategy based on selective COX2 and iNOS inhibitors [49]. The combination of the COX2 inhibitor celecoxib and the iNOS inhibitor SC-51 was reported to reduce the crypt formation in rats that were previously treated with the colon-specific carcinogen azoxymethane [50].

INOS INDUCTION STRATEGIES

L-arginine supplementation can reduce tumor occurrence and development in some tumor models, including chemically induced solid tumors, transplantable solid tumors, and human tumors. For example, rats subcutaneously injected with 1,2-dimethylhydrazine (DMH), a colonic procarcinogen, showed significantly reduced tumor incidence and tumor burden when they received L-arginine in their drinking water [51]. In an in vitro study using SW480 human colon cancer cells, L-arginine administration at similar or higher than physiological concentrations reduced the progression of colon cancer, with iNOS having a role in reducing angiogenesis [52]. Xenograft studies with SW480 cells showed that L-arginine supplementation significantly lowered MMP-2, MMP-9, and VEGF receptor levels in tumors, whereas tumor NO levels and spleen natural killer (NK) cell activities were significantly higher compared with the control untreated group. Thus, L-arginine treatment inhibited the progression of colon cancer in this model by increasing NO levels and consequently enhancing NK cell activity.

In a randomized, double-blind, placebo-controlled protocol consisting of 69 males and 51 females divided into two arms each of 60 patients with colorectal carcinoma (CRC) and 60 patients with colorectal adenoma (CRA), the subjects underwent colonoscopy with biopsy specimens taken from the tumor, paratumor mucosa (within 5 cm of the tumor), and normal mucosa (>10 cm distance from the tumor). Each group was randomly divided further into two groups of 30 patients each, a test group that was administered 30 g (120 mL) of freshly prepared L-arginine orally once a day for 3 days and a control group where the L-arginine was substituted with 5% glucose (120 mL). All subjects with CRC underwent radical surgical resection 3 days after treatment, whereas those with CRA underwent tumor resection by colonoscopy; tissue samples were obtained during the procedures [53]. The results showed that eNOS and nNOS were expressed in the normal and paratumor mucosa but not in the tumor cells, whereas the glandular cells in the normal and paratumor mucosa did not express iNOS. The tumor cells showed moderate expression for iNOS, which was significantly increased after treatment. Before treatment, serum NO levels in the patients with CRC were similar to those in the patients with CRA; however, after treatment, the values significantly increased in the CRC group with no significant changes in the CRA group. Thus, L-arginine treatment blocked the formation and development of colorectal tumors; this effect correlated with the increased serum NO levels through iNOS induction, reviewed in Ref. [7].

High iNOS expression has been associated with favorable prognoses in ovarian (10-year prognoses) [54] and non-small-cell lung cancers [55]. Overexpression of iNOS by gene transfer in prostate and colon cancer cells produced enhanced amounts of NO and increased the sensitivity of these cells to cisplatin-induced cell death [56] or to radiation therapy [57]. In a mouse model of thyroid cancer, iNOS overexpression inhibited tumorigenesis [58]. iNOS expressing pancreatic cancer cells did not produce tumors or metastases in xenograft mouse models due to NO upregulation and apoptosis [59]. Overexpressing iNOS by other means has also produced anticancer effects. Microencapsulation of iNOS expressing cells and delivering them to human colon and ovarian cancer xenografts led to increases in apoptosis [60]. Retroviral delivery of iNOS in mice produced slowly progressing tumors and fewer lung metastases than control mice transfected with iNOS-negative retrovirus [61]. Thus, therapeutic overexpression of iNOS within the tumor through gene therapy or other means may be a fruitful approach in treating some cancers. Retroviral and adenoviral vectors are used to transfer iNOS cDNA into cancer cells, but this gene therapy is usually performed with high doses of viral vector to increase efficacy, increasing the risk of developing uncontrolled reactions with other targets that may present a danger to the host [62].

NO donors, capable of producing a sustained release with a wide range of half-lives, are attractive as sensitizers of cancer cells due to NO-mediated cytotoxicity [63]. However, development of such agents has not led to clinical applications due to potential adverse effects of NO donors in vivo, such as toxic accumulation of metabolites [64]; thus, the focus has somewhat shifted toward developing hybrid NO-releasing compounds. One such collection of agents has been the NO-releasing nonsteroidal antiinflammatory drugs (NO-NSAIDs) [65, 66]. NO-aspirin, NO-sulindac, and NO-naproxen exerted colon cancer growth inhibition in vitro and in vivo, interacting with the main proapoptotic signaling cascade [67]. This class of compounds and their molecular targets have been thoroughly reviewed elsewhere [65, 66, 68].

From the combination of the NO-NSAIDs' antitumorigenic activity and the hydrogen sulfide-dependent gastrointestinal defense mechanisms/anticancer properties, novel nitric oxide-hydrogen sulfide (NOSH) hybrid compounds were synthetized [69, 70]. NOSH-sulindac, NOSH-aspirin, and NOSH-naproxen were all able to inhibit tumor cell proliferation and induce tumor cell death both in vitro and in xenograft mouse models, showing no signs of gastrointestinal damage [71–74].

DOES EXPOSURE TO NO CAUSE HUMAN CANCER?

This is an overarching question, recently reviewed in Ref. [7]. The major sources of human exposure to NO are nitrovasodilators, which include nitroglycerine (glyceryl trinitrate), amyl nitrite, isosorbide mono- and dinitrate, erythrityl tetranitrate, and sodium nitroprusside [75]. These medications are taken sublingually, orally, or subcutaneously for the treatment of angina pectoris and other coronary artery diseases. Nitroglycerine has been used effectively for over 100 years, and the other organic medicinal nitrates have been available since the 1930s. In theory, these medications can inhibit or promote the development of cancer. For example, glyceryl trinitrate induced apoptosis through the activation of caspases in three different colon cancer cell lines [76] and inhibited tumorigenesis in murine skin [77]. Further, isosorbide mononitrate and isosorbide dinitrate were shown to inhibit angiogenesis, tumor growth, and metastasis in mice [78]. In contrast, feeding glyceryl trinitrate to F344 rats induced hepatocellular carcinomas that were characterized by K-ras point mutations [79]. Consistent with this finding are data showing elevated blood levels of nitrates and nitrites in patients with hepatocellular carcinoma [80].

Given the clinical importance and extensive use of NO-releasing vasodilators, the association between the commonly used NO-releasing nitrates and colorectal cancer risk was undertaken by using the data from the Framingham Heart and Offspring studies [81]. The Framingham Heart study is a long-term, multigenerational study, designed to identify genetic and environmental factors influencing the development of cardiovascular and other diseases [82]. Examination and testing of 5209 residents of Framingham, Massachusetts, was initiated in 1948 and with approximately 200 members of the original cohort currently alive and under follow-up. An offspring cohort and a third-generation cohort were subsequently added, followed more recently by two minority cohorts (Omni Group 1 and Omni Group 2) [83]. Using these data, it was concluded that nitrovasodilators do not increase the risk of colorectal cancer [81], reviewed in Ref. [75].

Statins are cholesterol-lowering drugs that are widely used in both men and women. As a class, they competitively and reversibly inhibit the enzyme 3-hydroxy-3-methylglutaryl-CoA (HMG-CoA) reductase, which is the rate-limiting enzyme for cholesterol biosynthesis in the liver; this leads to decreases in serum low-density lipoprotein cholesterol (LDL-C), triglyceride, and cholesterol levels [84]. It has been proposed that statins exert both LDL-C-dependent and LDL-C-independent or pleiotropic effects [85, 86]. One of the pleiotropic effects of statins is the upregulation of eNOS resulting in higher production of endothelial NO, which is important for vasodilation, platelet aggregation, vascular smooth muscle proliferation, and endothelial-leukocyte interactions [86, 87]. Statins upregulate eNOS activity through multiple mechanisms. One pathway involves Rho/ROCK signaling. In vitro studies show that Rho inhibition increases eNOS expression [88]. Statins also increase eNOS activity by post-translational activation of the phosphatidylinositol 3-kinase/protein kinase Akt pathway as eNOS is phosphorylated by Akt [3]. Other pleiotropic effects of statins, which increase eNOS activity, include effects on caveolin 1 and transcription factor Kruppel-like factor-2 and mobilization of endothelial progenitor cells [86].

Emerging research suggests that statins may prevent various types of cancers, including colon [89–91] and breast [92–94] as well as ovarian, pancreatic, lung, and lymphomas, reviewed in Ref. [95]. Statins inhibit cancer cell proliferation by arresting the cell cycle at the G_1-S phase and inducing apoptosis [96, 97]. Fluvastatin inhibits epidermal growth factor-induced invasion of pancreatic cancer cells by inhibiting the translocation of RhoA from the cytosol to the membrane [98]. Furthermore, statins have been shown to inhibit in vivo tumor growth at clinically relevant doses and to diminish metastasis in animal models [99, 100]. Of interest, simvastatin and fluvastatin have been shown to enhance NO levels through increases in iNOS mRNA and protein expressions. Statin-induced NO and tumor cell cytotoxicity were inhibited by 1400W [*N*-(3-aminomethyl)benzylacetamidine], a relatively specific iNOS inhibitor [101].

It is important to note that these agents have been used for more than 30 years. They induce iNOS and/or eNOS, yet there are no data to suggest that people who have been using these drugs, essentially for long period of times, are at a greater risk of developing any kind of cancer [7].

THE TUMOR MICROENVIRONMENT: INOS, INNATE IMMUNITY, AND CANCER

NO is an effector of the innate immune system, which includes the macrophages, neutrophils, and natural killer cells [102]. These use pattern recognition receptors to recognize molecular patterns associated with pathogens [103]. Activated macrophages then inhibit pathogen replication by releasing a variety of effector molecules, including NO.

Resting immune cells lack iNOS; however, a variety of extracellular stimuli can activate a number of signaling pathways that converge to induce iNOS expression. The innate immune pathway can activate signaling pathways including the mitogen-activated protein kinase (MAPK) and the NF-κB. These pathways converge to activate *iNOS* transcription [102]. Released cytokines from host cells can also activate NO production, including tumor necrosis factor-α (TNF-α) and interleukin-1β (IL-1β). Interferon-γ (IFN-γ) produced by activated immune cells can activate iNOS expression and can synergistically activate iNOS expression in combination with other agents such as lipopolysaccharide (LPS). IFN-γ interacts with the interferon receptor 1 (IFNR1) and IFNR2 complex, which activates Jak family kinases and signal transducers and activators of transcription (STAT) pathways [104]. The interferon-signaling pathway can also activate iNOS transcription.

High levels of NO can lead to the repolarization of tumor-associated microglia/macrophages (TAMs) from the tumor-promoting and immunosuppressive M2-like state to the M1 state. It is important to note that innate immune cells like microglia and macrophages are the first line of defense against pathogens and tumors [105]. Among the host of immune cells, microglia, in their proinflammatory M1 state, can kill tumors directly and indirectly by functioning as specialized antigen-presenting cells and via the activation and recruitment of other tumoricidal innate immune cells like natural killer (NK) cells and peripheral

M1-type macrophages [106–109]. In some tumors, the major portion of the tumor mass is composed of the M2-type TAM [106, 110, 111].

Recently, it was reported that mice harboring glioblastoma (GBM) tumors that died when they were vehicle-treated expressed high levels of tumor-promoting arginase1 (ARG1) and low levels of iNOS and M2-like TAM. In sharp contrast, the mice rescued from GBM by antibody-linked curcumin (CC) and phytosomal curcumin (CCP) contained low levels of ARG1 and high levels of iNOS and M1-like TAM, in the scar tissue [106]. The authors also demonstrated that CCP induced the transcription factor STAT1, which is known to trigger iNOS and IL12 syntheses in the TAM [112, 113]. Recruited NK cells can cause suppression of the M2-type markers, such as P-STAT3, IL10, and ARG1 in the TAM. Simultaneously, the NK cells can also increase the tumoricidal M1-type markers, such as P-STAT1 and IL12 in the TAM. The consequent IL12-evoked NK activation may amplify and prolong the M1 polarization of TAM via IFN-γ released by activated NK cells [112, 114], thereby causing significant increases in iNOS, which would catalyze the formation of cytotoxic levels of NO inside the tumors. Thus, repolarization of TAM from a tumor-promoting M2-like population to a tumoricidal M1-like milieu can be achieved in part by the induction of iNOS (Fig. 3).

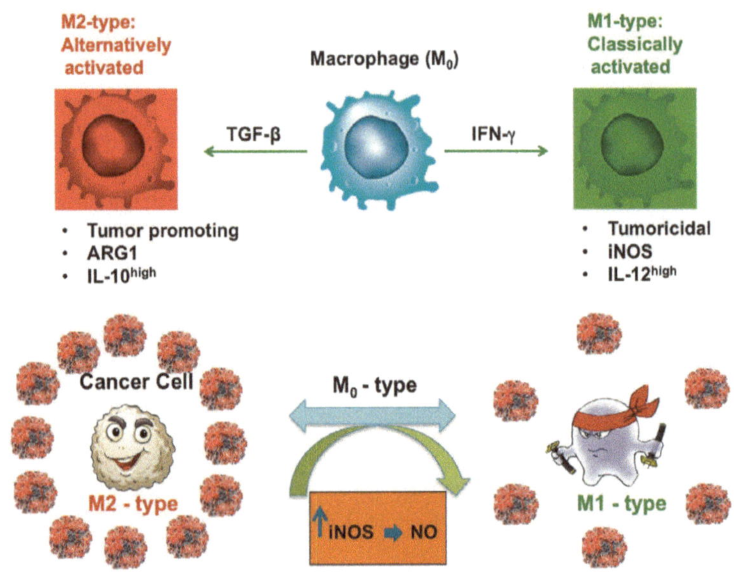

FIG. 3

Role of tumor-associated macrophages (M2/M1) in tumor progression. Induction of iNOS produces high levels of NO, which leads to the repolarization of TAM from a tumor-promoting M2-like population to a tumoricidal M1-like population.

FINAL THOUGHTS AND FUTURE DIRECTIONS

The effect of NO donors or inhibitors in vitro may not appropriately represent the pathophysiological environment in vivo. Side effects associated with non-selective drug distribution can lead to uncontrolled NO levels, which could also be affected by different oxidant species and NO substrate availability within the tumor microenvironment. Le et al. demonstrated that NO overexpressed multiple angiogenic molecules in vitro but suppressed tumor growth and metastases in ectopic and orthotopic xenograft nude mouse models, in favor of apoptosis [59].

In recent years, nanoparticle-based systems have been developed to deliver sustained NO-releasing compounds to a specific target site, improving their biodistribution and increasing their levels in the tumor tissue, preserving their antitumorigenic efficacy [115]. The search for the best biocompatible systems and the more effective strategies to induce a controlled delivery is continuing.

The observations that high levels of iNOS are expressed in various human cancers may ultimately mean that the tumor microenvironment in its "wisdom" is switching from a tumor-promoting M2-like population of TAMs to a tumoricidal M1-like milieu. It is well appreciated that this view may not be acceptable in some circles; nevertheless, it does give a plausible explanation for the observations and does provide some food for thought.

References

[1] Alderton WK, Cooper CE, Knowles RG. Nitric oxide synthases: structure, function and inhibition. Biochem J 2001;357:593–615.

[2] Stuehr DJ. Structure-function aspects in the nitric oxide synthases. Annu Rev Pharmacol Toxicol 1997;37:339–59.

[3] Dimmeler S, Fleming I, Fisslthaler B, Hermann C, Busse R, Zeiher AM. Activation of nitric oxide synthase in endothelial cells by Akt-dependent phosphorylation. Nature 1999;399:601–5.

[4] Sessa WC. eNOS at a glance. J Cell Sci 2004;117:2427–9.

[5] Nathan C, Xie QW. Nitric oxide synthases: roles, tolls, and controls. Cell 1994;78:915–8.

[6] Kleinert H, Schwarz PM, Forstermann U. Regulation of the expression of inducible nitric oxide synthase. Biol Chem 2003;384:1343–64.

[7] Kashfi K. The dichotomous role of H2S in cancer cell biology? Deja vu all over again. Biochem Pharmacol 2018;149:205–23.

[8] Friebe A, Koesling D. Regulation of nitric oxide-sensitive guanylyl cyclase. Circ Res 2003;93:96–105.

[9] Dangel O, Mergia E, Karlisch K, Groneberg D, Koesling D, Friebe A. Nitric oxide-sensitive guanylyl cyclase is the only nitric oxide receptor mediating platelet inhibition. J Thromb Haemost 2010;8:1343–52.

[10] Lincoln T, Cornwell T, Komalavilas P, MacMillan-Crow LA, Boerth NJ. The nitric oxide-cyclic GMP signaling system. In: Barany M, editor. Biochemistry of smooth muscle contraction. San Diego: Academic Press; 1996. p. 257–68.

[11] Vannini F, Kashfi K, Nath N. The dual role of iNOS in cancer. Redox Biol 2015;6:334–43.

[12] Lipton SA, Choi YB, Pan ZH, Lei SZ, Chen HS, Sucher NJ, et al. A redox-based mechanism for the neuroprotective and neurodestructive effects of nitric oxide and related nitroso-compounds. Nature 1993;364:626–32.

[13] Stamler JS. Redox signaling: nitrosylation and related target interactions of nitric oxide. Cell 1994;78:931–6.

[14] Stamler JS, Singel DJ, Loscalzo J. Biochemistry of nitric oxide and its redox-activated forms. Science (New York, NY) 1992;258:1898–902.

[15] Broillet MC. S-nitrosylation of proteins. Cell Mol Life Sci 1999;55:1036–42.

[16] Hess DT, Matsumoto A, Kim SO, Marshall HE, Stamler JS. Protein S-nitrosylation: purview and parameters. Nat Rev Mol Cell Biol 2005;6:150–66.

[17] Liu H, Colavitti R, Rovira II, Finkel T. Redox-dependent transcriptional regulation. Circ Res 2005;97:967–74.

[18] Leon-Bollotte L, Subramaniam S, Cauvard O, Plenchette-Colas S, Paul C, Godard C, et al. S-nitrosylation of the death receptor fas promotes fas ligand-mediated apoptosis in cancer cells. Gastroenterology 2011;140: 2009–18, 18 e1–4.

[19] Tang Z, Bauer JA, Morrison B, Lindner DJ. Nitrosylcobalamin promotes cell death via S nitrosylation of Apo2L/TRAIL receptor DR4. Mol Cell Biol 2006;26:5588–94.

[20] Glynn SA, Boersma BJ, Dorsey TH, Yi M, Yfantis HG, Ridnour LA, et al. Increased NOS2 predicts poor survival in estrogen receptor-negative breast cancer patients. J Clin Invest 2010;120:3843–54.

[21] Ducrocq C, Blanchard B, Pignatelli B, Ohshima H. Peroxynitrite: an endogenous oxidizing and nitrating agent. Cell Mol Life Sci 1999;55:1068–77.

[22] Szabo C, Ischiropoulos H, Radi R. Peroxynitrite: biochemistry, pathophysiology and development of therapeutics. Nat Rev Drug Discov 2007;6:662–80.

[23] Cui S, Reichner JS, Mateo RB, Albina JE. Activated murine macrophages induce apoptosis in tumor cells through nitric oxide-dependent or -independent mechanisms. Cancer Res 1994;54:2462–7.

[24] Brune B, von Knethen A, Sandau KB. Nitric oxide (NO): an effector of apoptosis. Cell Death Differ 1999;6:969–75.

[25] Vahora H, Khan MA, Alalami U, Hussain A. The potential role of nitric oxide in halting cancer progression through chemoprevention. J Cancer Prevention 2016;21:1–12.

[26] Burke AJ, Sullivan FJ, Giles FJ, Glynn SA. The yin and yang of nitric oxide in cancer progression. Carcinogenesis 2013;34:503–12.

[27] Janakiram NB, Rao CV. iNOS-selective inhibitors for cancer prevention: promise and progress. Future Med Chem 2012;4:2193–204.

[28] Mocellin S. Nitric oxide: cancer target or anticancer agent? Curr Cancer Drug Targets 2009;9:214–36.

[29] Thomas DD, Liu X, Kantrow SP, Lancaster Jr. JR. The biological lifetime of nitric oxide: implications for the perivascular dynamics of NO and O2. Proc Natl Acad Sci U S A 2001;98: 355–60.

[30] Singh S, Cowen RL, Chinje EC, Stratford IJ. The impact of intracellular generation of nitric oxide on the radiation response of human tumor cells. Radiat Res 2009;171:572–80.

[31] Daff S. NO synthase: structures and mechanisms. Nitric Oxide Biol Chem 2010;23:1–11.

[32] Crane BR, Arvai AS, Gachhui R, Wu C, Ghosh DK, Getzoff ED, et al. The structure of nitric oxide synthase oxygenase domain and inhibitor complexes. Science (New York, NY) 1997;278:425–31.

[33] Fischmann TO, Hruza A, Niu XD, Fossetta JD, Lunn CA, Dolphin E, et al. Structural characterization of nitric oxide synthase isoforms reveals striking active-site conservation. Nat Struct Biol 1999;6:233–42.

[34] Raman CS, Li H, Martasek P, Kral V, Masters BS, Poulos TL. Crystal structure of constitutive endothelial nitric oxide synthase: a paradigm for pterin function involving a novel metal center. Cell 1998;95:939–50.

[35] Li H, Shimizu H, Flinspach M, Jamal J, Yang W, Xian M, et al. The novel binding mode of N-alkyl-N′-hydroxyguanidine to neuronal nitric oxide synthase provides mechanistic insights into NO biosynthesis. Biochemistry 2002;41:13868–75.

[36] Ji H, Stanton BZ, Igarashi J, Li H, Martasek P, Roman LJ, et al. Minimal pharmacophoric elements and fragment hopping, an approach directed at molecular diversity and isozyme selectivity. Design of selective neuronal nitric oxide synthase inhibitors. J Am Chem Soc 2008;130:3900–14.

[37] Vitecek J, Lojek A, Valacchi G, Kubala L. Arginine-based inhibitors of nitric oxide synthase: therapeutic potential and challenges. Mediators Inflamm 2012;2012:318087.

[38] Mukherjee P, Cinelli MA, Kang S, Silverman RB. Development of nitric oxide synthase inhibitors for neurodegeneration and neuropathic pain. Chem Soc Rev 2014;43:6814–38.

[39] Cheng R, Ridnour LA, Glynn SA, Switzer CH, Flores-Santana W, Hussain P, et al. Nitric oxide and cancer: an overview. In: Bonavida B, editor. Nitric oxide (NO) and cancer: prognosis, prevention, and therapy. Springer; 2010. p. 3–20.

[40] Saied EM, El-Etreby NM. The role and prognostic value of inducible nitric oxide synthase (iNOS) and interleukin-33 (IL-33) in serous and mucinous epithelial ovarian tumours. Ann Diagn Pathol 2017;27:62–8.

[41] de Oliveira GA, Cheng RYS, Ridnour LA, Basudhar D, Somasundaram V, McVicar DW, et al. Inducible nitric oxide synthase in the carcinogenesis of gastrointestinal cancers. Antioxid Redox Signal 2017;26:1059–77.

[42] Switzer CH, Glynn SA, Ridnour LA, Cheng RY, Vitek MP, Ambs S, et al. Nitric oxide and protein phosphatase 2A provide novel therapeutic opportunities in ER-negative breast cancer. Trends Pharmacol Sci 2011;32:644–51.

[43] Granados-Principal S, Liu Y, Guevara ML, Blanco E, Choi DS, Qian W, et al. Inhibition of iNOS as a novel effective targeted therapy against triple-negative breast cancer. Breast Cancer Res 2015;17:25.

[44] Alderton WK, Angell AD, Craig C, Dawson J, Garvey E, Moncada S, et al. GW274150 and GW273629 are potent and highly selective inhibitors of inducible nitric oxide synthase in vitro and in vivo. Br J Pharmacol 2005;145:301–12.

[45] Heinecke JL, Ridnour LA, Cheng RY, Switzer CH, Lizardo MM, Khanna C, et al. Tumor microenvironment-based feed-forward regulation of NOS2 in breast cancer progression. Proc Natl Acad Sci U S A 2014;111:6323–8.

[46] Kostourou V, Cartwright JE, Johnstone AP, Boult JK, Cullis ER, Whitley G, et al. The role of tumour-derived iNOS in tumour progression and angiogenesis. Br J Cancer 2011;104:83–90.

[47] Sikora AG, Gelbard A, Davies MA, Sano D, Ekmekcioglu S, Kwon J, et al. Targeted inhibition of inducible nitric oxide synthase inhibits growth of human melanoma in vivo and synergizes with chemotherapy. Clin Cancer Res 2010;16:1834–44.

[48] Ahn B, Ohshima H. Suppression of intestinal polyposis in Apc(Min/+) mice by inhibiting nitric oxide production. Cancer Res 2001;61:8357–60.

[49] Cianchi F, Cortesini C, Fantappie O, Messerini L, Schiavone N, Vannacci A, et al. Inducible nitric oxide synthase expression in human colorectal cancer: correlation with tumor angiogenesis. Am J Pathol 2003;162:793–801.

[50] Rao CV, Indranie C, Simi B, Manning PT, Connor JR, Reddy BS. Chemopreventive properties of a selective inducible nitric oxide synthase inhibitor in colon carcinogenesis, administered alone or in combination with celecoxib, a selective cyclooxygenase-2 inhibitor. Cancer Res 2002;62:165–70.

[51] Ma Q, Hoper M, Anderson N, Rowlands BJ. Effect of supplemental L-arginine in a chemical-induced model of colorectal cancer. World J Surg 1996;20:1087–91.

[52] Yeh CL, Pai MH, Li CC, Tsai YL, Yeh SL. Effect of arginine on angiogenesis induced by human colon cancer: in vitro and in vivo studies. J Nutr Biochem 2010;21:538–43.

[53] Ma Q, Wang Y, Gao X, Ma Z, Song Z. L-arginine reduces cell proliferation and ornithine decarboxylase activity in patients with colorectal adenoma and adenocarcinoma. Clin Cancer Res 2007;13:7407–12.

[54] Anttila MA, Voutilainen K, Merivalo S, Saarikoski S, Kosma VM. Prognostic significance of iNOS in epithelial ovarian cancer. Gynecol Oncol 2007;105:97–103.

[55] Puhakka A, Kinnula V, Napankangas U, Saily M, Koistinen P, Paakko P, et al. High expression of nitric oxide synthases is a favorable prognostic sign in non-small cell lung carcinoma. APMIS 2003;111:1137–46.

[56] Adams C, McCarthy HO, Coulter JA, Worthington J, Murphy C, Robson T, et al. Nitric oxide synthase gene therapy enhances the toxicity of cisplatin in cancer cells. J Gene Med 2009;11:160–8.

[57] Chung P, Cook T, Liu K, Vodovotz Y, Zamora R, Finkelstein S, et al. Overexpression of the human inducible nitric oxide synthase gene enhances radiation-induced apoptosis in colorectal cancer cells via a caspase-dependent mechanism. Nitric oxide 2003;8:119–26.

[58] Soler MN, Bobe P, Benihoud K, Lemaire G, Roos BA, Lausson S. Gene therapy of rat medullary thyroid cancer by naked nitric oxide synthase II DNA injection. J Gene Med 2000;2:344–52.

[59] Le X, Wei D, Huang S, Lancaster Jr. JR, Xie K. Nitric oxide synthase II suppresses the growth and metastasis of human cancer regardless of its up-regulation of protumor factors. Proc Natl Acad Sci U S A 2005;102:8758–63.

[60] Xu W, Liu LZ, Loizidou M, Ahmed M, Charles IG. The role of nitric oxide in cancer. Cell Res 2002;12:311–20.

[61] Juang SH, Xie K, Xu L, Wang Y, Yoneda J, Fidler IJ. Use of retroviral vectors encoding murine inducible nitric oxide synthase gene to suppress tumorigenicity and cancer metastasis of murine melanoma. Cancer Biother Radiopharm 1997;12:167–75.

[62] Lehrman S. Virus treatment questioned after gene therapy death. Nature 1999;401:517–8.

[63] Huerta S, Chilka S, Bonavida B. Nitric oxide donors: novel cancer therapeutics (review). Int J Oncol 2008;33:909–27.

[64] Lee DH, Pfeifer GP. Mutagenesis induced by the nitric oxide donor sodium nitroprusside in mouse cells. Mutagenesis 2007;22:63–7.

[65] Kashfi K. Anti-inflammatory agents as cancer therapeutics. Adv Pharmacol 2009;57:31–89.

[66] Kashfi K. Nitric oxide-releasing hybrid drugs target cellular processes through S-nitrosylation. Forum Immunopathol Dis Therapeut 2012;3:97–108.

[67] Rigas B, Kashfi K. Nitric-oxide-donating NSAIDs as agents for cancer prevention. Trends Mol Med 2004;10:324–30.

[68] Kashfi K, Esmaili M. NO-H2S-releasing chimeras as a multifaceted approach to cancer therapy. In: Bonavida B, editor. Cancer sensitizing agents for chemotherapy; nitric oxide (donor/induced) in chemosensitization. Elsevier/Academic Press; 2017. p. 105–42.

[69] Kodela R, Chattopadhyay M, Kashfi K. Synthesis and biological activity of NOSH-naproxen (AVT-219) and NOSH-sulindac (AVT-18A) as potent anti-inflammatory agents with chemotherapeutic potential. MedChemComm 2013;4:1472–81.

[70] Kodela R, Chattopadhyay M, Kashfi K. NOSH-aspirin: a novel nitric oxide-hydrogen sulfide-releasing hybrid: a new class of anti-inflammatory pharmaceuticals. ACS Med Chem Lett 2012;3:257–62.

[71] Kashfi K, Chattopadhyay M, Kodela R. NOSH-sulindac (AVT-18A) is a novel nitric oxide- and hydrogen sulfide-releasing hybrid that is gastrointestinal safe and has potent anti-inflammatory, analgesic, antipyretic, anti-platelet, and anti-cancer properties. Redox Biol 2015;6:287–96.

[72] Vannini F, MacKessack-Leitch AC, Eschbach EK, Chattopadhyay M, Kodela R, Kashfi K. Synthesis and anti-cancer potential of the positional isomers of NOSH-aspirin (NBS-1120) a dual nitric oxide and hydrogen sulfide releasing hybrid. Bioorg Med Chem Lett 2015;25: 4677–82.

[73] Kodela R, Chattopadhyay M, Velazquez-Martinez CA, Kashfi K. NOSH-aspirin (NBS-1120), a novel nitric oxide- and hydrogen sulfide-releasing hybrid has enhanced chemo-preventive properties compared to aspirin, is gastrointestinal safe with all the classic therapeutic indications. Biochem Pharmacol 2015;98:564–72.

[74] Chattopadhyay M, Kodela R, Duvalsaint PL, Kashfi K. Gastrointestinal safety, chemotherapeutic potential, and classic pharmacological profile of NOSH-naproxen (AVT-219) a dual NO- and H2S-releasing hybrid. Pharmacol Res Perspect 2016;4:e00224.

[75] Kashfi K, Duvalsaint PL. Nitric oxide donors and therapeutic applications in cancer. In: Seabra AB, editor. Nitric oxide donors: novel biomedical applications. Oxford: Elsevier/ Academic Press; 2017. p. 75–119.

[76] Millet A, Bettaieb A, Renaud F, Prevotat L, Hammann A, Solary E, et al. Influence of the nitric oxide donor glyceryl trinitrate on apoptotic pathways in human colon cancer cells. Gastroenterology 2002;123:235–46.

[77] Trikha P, Sharma N, Athar M. Nitroglycerin: a NO donor inhibits TPA-mediated tumor promotion in murine skin. Carcinogenesis 2001;22:1207–11.

[78] Pipili-Synetos E, Papageorgiou A, Sakkoula E, Sotiropoulou G, Fotsis T, Karakiulakis G, et al. Inhibition of angiogenesis, tumour growth and metastasis by the NO-releasing vasodilators, isosorbide mononitrate and dinitrate. Br J Pharmacol 1995;116:1829–34.

[79] Tamano S, Ward JM, Diwan BA, Keefer LK, Weghorst CM, Calvert RJ, et al. Histogenesis and the role of p53 and K-ras mutations in hepatocarcinogenesis by glyceryl trinitrate (nitroglycerin) in male F344 rats. Carcinogenesis 1996;17:2477–86.

[80] Notas G, Xidakis C, Valatas V, Kouroumalis A, Kouroumalis E. Levels of circulating endothelin-1 and nitrates/nitrites in patients with virus-related hepatocellular carcinoma. J Viral Hepat 2001;8:63–9.

[81] Muscat JE, Dyer AM, Rosenbaum RE, Rigas B. Nitric oxide-releasing medications and colorectal cancer risk: the framingham study. Anticancer Res 2005;25:4471–4.

[82] Dawber TR, Kannel WB, Lyell LP. An approach to longitudinal studies in a community: the Framingham study. Ann N Y Acad Sci 1963;107:539–56.

[83] Wei G. The Framingham heart study (FHS). Project period: 9/1/1948–5/1/2015.

[84] Istvan ES, Deisenhofer J. Structural mechanism for statin inhibition of HMG-CoA reductase. Science (New York, NY) 2001;292:1160–4.

[85] Liao JK, Laufs U. Pleiotropic effects of statins. Annu Rev Pharmacol Toxicol 2005;45:89–118.

[86] Oesterle A, Laufs U, Liao JK. Pleiotropic effects of statins on the cardiovascular system. Circ Res 2017;120:229–43.

[87] Gauthier TW, Scalia R, Murohara T, Guo JP, Lefer AM. Nitric oxide protects against leukocyte-endothelium interactions in the early stages of hypercholesterolemia. Arterioscler Thromb Vasc Biol 1995;15:1652–9.

[88] Laufs U, Liao JK. Post-transcriptional regulation of endothelial nitric oxide synthase mRNA stability by rho GTPase. J Biol Chem 1998;273:24266–71.

[89] Poynter JN, Gruber SB, Higgins PD, Almog R, Bonner JD, Rennert HS, et al. Statins and the risk of colorectal cancer. N Engl J Med 2005;352:2184–92.

[90] Jung YS, Park CH, Eun CS, Park DI, Han DS. Statin use and the risk of colorectal adenoma: a meta-analysis. J Gastroenterol Hepatol 2016;31:1823–30.

[91] Pikoulis E, Margonis GA, Angelou A, Zografos GC, Antoniou E. Statins in the chemoprevention of colorectal cancer in established animal models of sporadic and colitis-associated cancer. Eur J Cancer Prevention 2016;25:102–8.

[92] Muck AO, Seeger H, Wallwiener D. Inhibitory effect of statins on the proliferation of human breast cancer cells. Int J Clin Pharmacol Ther 2004;42:695–700.

[93] Koyuturk M, Ersoz M, Altiok N. Simvastatin induces apoptosis in human breast cancer cells: p53 and estrogen receptor independent pathway requiring signalling through JNK. Cancer Lett 2007;250:220–8.

[94] Ji Y, Rounds T, Crocker A, Sussman B, Hovey RC, Kingsley F, et al. The effect of atorvastatin on breast cancer biomarkers in high-risk women. Cancer Prevention Res 2016;9:379–84.

[95] Vallianou NG, Kostantinou A, Kougias M, Kazazis C. Statins and cancer. Anticancer Agents Med Chem 2014;14:706–12.

[96] Sivaprasad U, Abbas T, Dutta A. Differential efficacy of 3-hydroxy-3-methylglutaryl CoA reductase inhibitors on the cell cycle of prostate cancer cells. Mol Cancer Ther 2006;5:2310–6.

[97] Campbell MJ, Esserman LJ, Zhou Y, Shoemaker M, Lobo M, Borman E, et al. Breast cancer growth prevention by statins. Cancer Res 2006;66:8707–14.

[98] Kusama T, Mukai M, Iwasaki T, Tatsuta M, Matsumoto Y, Akedo H, et al. Inhibition of epidermal growth factor-induced RhoA translocation and invasion of human pancreatic cancer cells by 3-hydroxy-3-methylglutaryl-coenzyme a reductase inhibitors. Cancer Res 2001;61:4885–91.

[99] Kusama T, Mukai M, Iwasaki T, Tatsuta M, Matsumoto Y, Akedo H, et al. 3-Hydroxy-3-methylglutaryl-coenzyme a reductase inhibitors reduce human pancreatic cancer cell invasion and metastasis. Gastroenterology 2002;122:308–17.

[100] Paragh G, Kertai P, Kovacs P, Paragh Jr. G, Fulop P, Foris G. HMG CoA reductase inhibitor fluvastatin arrests the development of implanted hepatocarcinoma in rats. Anticancer Res 2003;23:3949–54.

[101] Kotamraju S, Williams CL, Kalyanaraman B. Statin-induced breast cancer cell death: role of inducible nitric oxide and arginase-dependent pathways. Cancer Res 2007;67:7386–94.

[102] Lowenstein CJ, Padalko E. iNOS (NOS2) at a glance. J Cell Sci 2004;117:2865–7.

[103] Medzhitov R. Toll-like receptors and innate immunity. Nat Rev Immunol 2001;1:135–45.

[104] Darnell Jr. JE, Kerr IM, Stark GR. Jak-STAT pathways and transcriptional activation in response to IFNs and other extracellular signaling proteins. Science (New York, NY) 1994;264:1415–21.

[105] Kreutzberg GW. Microglia, the first line of defence in brain pathologies. Arzneimittelforschung 1995;45:357–60.

[106] Mukherjee S, Fried A, Hussaini R, White R, Baidoo J, Yalamanchi S, et al. Phytosomal curcumin causes natural killer cell-dependent repolarization of glioblastoma (GBM) tumor-associated microglia/macrophages and elimination of GBM and GBM stem cells. J Exp Clin Cancer Res 2018;37:168.

[107] Li W, Graeber MB. The molecular profile of microglia under the influence of glioma. Neuro Oncol 2012;14:958–78.

[108] Brantley EC, Guo L, Zhang C, Lin Q, Yokoi K, Langley RR, et al. Nitric oxide-mediated tumoricidal activity of murine microglial cells. Transl Oncol 2010;3:380–8.

[109] Zhang X, Tian W, Cai X, Wang X, Dang W, Tang H, et al. Hydrazinocurcumin Encapsuled nanoparticles "re-educate" tumor-associated macrophages and exhibit anti-tumor effects on breast cancer following STAT3 suppression. PloS One 2013;8:e65896.

[110] Mukherjee S, Baidoo JNE, Sampat S, Mancuso A, David L, Cohen LS, et al. Liposomal TriCurin, a synergistic combination of curcumin, epicatechin gallate and resveratrol, repolarizes tumor-associated microglia/macrophages, and eliminates glioblastoma (GBM) and GBM stem cells. Molecules (Basel, Switzerland) 2013;23:201–21.

[111] Kostianovsky AM, Maier LM, Anderson RC, Bruce JN, Anderson DE. Astrocytic regulation of human monocytic/microglial activation. J Immunol 2008;181:5425–32.

[112] Mukherjee S, Hussaini R, White R, Atwi D, Fried A, Sampat S, et al. TriCurin, a synergistic formulation of curcumin, resveratrol, and epicatechin gallate, repolarizes tumor-associated macrophages and triggers an immune response to cause suppression of HPV+ tumors. Cancer Immunol Immunother 2018;67:761–74.

[113] Hagemann T, Biswas SK, Lawrence T, Sica A, Lewis CE. Regulation of macrophage function in tumors: the multifaceted role of NF-kappaB. Blood 2009;113:3139–46.

[114] Bellora F, Castriconi R, Dondero A, Reggiardo G, Moretta L, Mantovani A, et al. The interaction of human natural killer cells with either unpolarized or polarized macrophages results in different functional outcomes. Proc Natl Acad Sci U S A 2010;107:21659–64.

[115] Quinn JF, Whittaker MR, Davis TP. Delivering nitric oxide with nanoparticles. J Control Release 2015;205:190–205.

CHAPTER 6

Arginine Metabolism Regulates Nitric Oxide Production in Melanoma Tumor Microenvironment to Provide Survival Advantage to Tumor Cells

Sun-Hee Kim, Dai Ogata, Jason Roszik, Elizabeth A. Grimm, Suhendan Ekmekcioglu

Department of Melanoma Medical Oncology, The University of Texas MD Anderson Cancer Center, Houston, TX, United States

ABSTRACT

Over the last decade, remarkable progress has been made in the areas of targeted therapy and immunotherapy for the treatment of cancer patients. However, durable clinical benefit has occurred only in a subset of responding patients. In our field of melanoma, we believe that the arginine metabolism leads to nitric oxide (NO) production, resulting in marker and functional changes in both melanoma and the microenvironmental cells, including potential effector T lymphocytes, concurrently suppressing the local immune response. Arginine metabolism is dependent on the activity of the families of nitric oxide synthase (NOS) and arginase (ARG) enzymes. NOS oxidizes arginine to citrulline and NO, and ARG hydrolyzes arginine into ornithine and urea. NO can react with oxides to form peroxynitrite ($NO3^-$) that rapidly forms two posttranslational modifications (PTMs) on proteins, nitrosylation of thiols (SNO) and nitration of tyrosines (NT). We and others have shown the expression of the inducible isoform of NOS enzyme (iNOS) in many human tumors, including melanoma. ARG1 activity has also been shown to be associated with the M2-polarized, protumoral tumor-associated macrophages (TAMs). Collectively, ARG and NOS enzymes metabolize arginine and are crucial components of immune suppression pathways, and the metabolic products of these enzymes are important moderators of T-cell function in cancer.

Therapeutic Application of Nitric Oxide in Cancer and Inflammatory Disorders. https://doi.org/10.1016/B978-0-12-816545-4.00006-2

ABBREVIATIONS

ARG	arginase
COX-2	cyclooxygenase-2
MDSCs	myeloid-derived suppressor cells
mPGES1	microsomal PGE2 synthase
NO	nitric oxide
NOS	nitric oxide synthase
NT	nitrotyrosine
PGE2	prostaglandin E2
SNO	*S*-nitrosothiol
TAM	tumor-associated macrophages
TIL	tumor-infiltrating lymphocytes

Conflict of Interest

No potential conflicts of interest were disclosed.

INTRODUCTION

Arginine-catabolizing enzymes, mainly nitric oxide synthases (NOS) and arginases (ARGs), are united for the control of the immune response. Since immunotherapy became a major treatment modality for many cancers, arginine metabolism has emerged as our focus due to its role as a dynamic regulator of immune responses, both innate and adaptive immune responses. The major arginine-catabolizing enzymes involved in inflammatory immune responses are the three isoforms of NOS (NOS1, NOS2, and NOS3) and ARGs (ARG1 and ARG2). All NOS enzymes metabolize arginine to produce nitric oxide (NO), which is a relatively stable intracellular gas that is known to diffuse through lipid bilayers and persist for at least few seconds [1]. NO can react with oxides to form peroxynitrite ($NO3^-$) that rapidly forms two posttranslational modifications (PTMs) on proteins, nitrosylation of thiols (SNO) and nitration of tyrosines (NT) [2, 3]. NT is considered a surrogate marker of NO to detect adducts in the human tissue by various methods including immunohistochemistry and provides a useful measure of the severity of tissue exposure to reactive nitrogen species as an indicator of inflammatory disease severity and progression. A second arginine metabolism via ARG leads to ornithine and urea production. Ornithine is the precursor for the production of polyamines and proline. Polyamines fundamentally regulate cell proliferation and differentiation, and proline is critical for the collagen synthesis, which is a primarily efficient protein in wound healing [4]. While we discuss the functional consequences of the regulated arginine metabolism in the tumor microenvironment, it is important to highlight that our research focus is the NOS-produced NO and its functional role on subsequent effects. Thus, our chapter will specifically address the role of NO in the melanoma tumor microenvironment that is regulated by the arginine metabolism to provide a survival advantage to tumor cells.

The importance of the arginine metabolism as a novel field of investigation includes that of arginine depletion retarding the growth of some cancers [5], whereas others report that arginine supplementation enhances antitumor effects, possibly by enhancing immune function [6, 7]. Arginine supplementation has been discovered to stimulate T-cell and NK-cell activities and promotes the production of proinflammatory cytokines [8, 9]. The roles of arginine now need to be understood in the context of NOS and ARG activities and tumor-derived NO. Additional preliminary data from our lab indicate that NO leads to the subsequent activation of COX-2 and other inflammatory mediators also leading to a prooxidant microenvironment that supports cancer cell growth and suppresses antitumor immunity. Cyclooxygenase-2 (COX-2) and microsomal PGE2 synthase (mPGES1) and their product prostaglandin E2 (PGE2) are reported to be upregulated in melanoma, to enhance carcinogenesis and tumor progression, and to support immunosuppression [10, 11]. Specifically, in melanoma, tumor-derived COX activity plays a role as the key suppressor of type I IFN-mediated and T-cell-mediated tumor elimination and as an inducer of an inflammatory signature associated with progression in a melanoma mouse model [11]. Another potential source of NO from the arginine metabolism in cancers is from myeloid-derived suppressor cells (MDSCs), which are a heterogeneous population of cells that often infiltrate cancers and sites of inflammation and infection and that have a remarkable ability to suppress T-cell responses [12]. Therapeutic strategies targeting NO production can effectively reverse MDSC-mediated immunosuppression by interfering with the inflammation-driven MDSC accrual in the tumor microenvironment, and it is a highly active area for translational approaches, currently. However, MDSC trafficking into the tumor or the strategies to regulate this move for therapeutic use are out of our focus for this review.

NOS2 (INOS) AND NT EXPRESSION IN MELANOMA TUMOR ASSOCIATES WITH DISEASE PROGRESSION

We and others have demonstrated that many human tumors, including melanoma, express iNOS [13–18]. iNOS expression and NO production have been shown to promote many cancers' growth, survival, and resistance to therapy. Molecular analysis also demonstrated that NO production regulates resistance to apoptosis shown by experiments in which chemical quenching of NO resulted in G2 growth arrest followed by a gain of cisplatin-induced apoptosis [19]. In addition, inhibition of NO production acts synergistically with cisplatin to reduce tumor development in experimental melanoma [20]. Clearly, the ground of regulatory mechanisms, which are fundamental for these interactions, is multifactorial and complex. However, there is a consensus on the proinflammatory factors' role in NO concentrations in the tumor microenvironment that decides on the outcome. NO produced by immune and cancer cells at high concentrations could cause nitrosative stress, which may lead to tumor death [21]. In contrast, constitutive expression of iNOS results in the production of low levels of NO [22], which helps the growth and progression of tumor cells [20, 23].

Our original study suggested that NT expression, along with iNOS expression, by melanoma metastases may be predictive of a poor outcome, information that may be of clinical value when making therapeutic decisions for individual patients [24]. This early study reported by us was followed with many other cancer types, including colon [25] and prostate [26]. Overall, these studies strengthen the findings in inflammation and cancer interactions by showing that iNOS expression and tyrosine nitration may be indicators of cancer development and progression. Besides iNOS protein expression being a predictor of poor outcome in multiple cancers, it has also been associated with inflammatory and immunosuppressive progressions in the tumor microenvironment. We have recently identified iNOS as a potential mediator of immune suppression in pancreatic ductal adenocarcinoma (PDAC) [27]. Expression of iNOS protein by tumor cells negatively impacts the antitumor immune response by mediation of immune suppression. The functional roles of iNOS in antitumor immunity may comprise recruitment and activation of MDSCs, tumor-associated macrophages (TAMs), and Th2 lymphocytes. Therefore, understanding the regulation of the immune cell response and the functions of tumor-expressed iNOS will provide significant opportunities for therapeutic interventions.

NT EXPRESSION IN MELANOMA ASSOCIATES WITH REDUCED TUMOR INFILTRATING LYMPHOCYTES (TIL) GROWTH

We have previously reported that iNOS expression in melanoma significantly associates with resistance to apoptosis and shorter patient survival [13, 19]. We have recently observed a global tumor NT expression in human stages IIIc and IV melanoma. NT is a marker of NO reaction with reactive oxygen species (ROS) to form peroxynitrite that irreversibly nitrates tyrosines and that in expression in tissues associated with poor TIL growth, as well as a poorer clinical response to TIL therapy (unpublished data). Based on this preliminary association data, we now hypothesize that the arginine metabolism is likely to result in TIL growth inhibition, activity, and/or function that is currently being tested. NT has been identified as a surrogate marker of in situ inflammation, but its role in cellular immune response remains unclear. An intriguing study showed that oxidation of tyrosine to NT on either the TCR or the MHC molecules can affect recognition of the MHC class I-restricted epitope of lymphocytic choriomeningitis virus glycoprotein 33 by CD8 T Cells [28]. This study further demonstrates that specific recognition of nitrated epitopes by T cells is not restricted to MHC class II epitopes, but that the conversion of tyrosine to nitrotyrosine can also profoundly affect recognition of MHC class I-restricted epitopes. If CD8 T cells (as previously observed in CD4 T cells) can significantly recognize epitopes containing an NT posttranslational modification as antigenic, we believe this would affect immune cell activation in melanoma patients.

TUMOR NT EXPRESSION CORRELATES WITH M2 MACROPHAGE TYPES IN TUMOR MICROENVIRONMENT

Polarization of macrophages from the classical phenotype (M1) to the alternative phenotype (M2) happens during inflammation in the tissue. This phenomenon is, somewhat, underappreciated for its role in T-cell activation, proliferation, signaling, and apoptosis. M2 macrophages upregulate certain MHC class II molecules, but at the same time, they are not capable of an efficient antigen presentation [29]. Cytokines, such as IL-10, and many chemokines in the tumor microenvironment not only induce immune suppression but also may help tumor progression due to the activation of many subsets of the M2 macrophages. Most relevant to this current topic, metabolic alterations present in tumor cells produce signals that can further modify these polarizations. The M2 phenotype is marked by an increased overall cell redox potential and upregulated immunomodulatory cytokines. M2 macrophages express high levels of ARG, and switching the environment to boost M1 phenotype could be an alternative approach to control tumor growth. In general, ARG and NOS enzymes' roles in macrophage activation in rodents are well established. However, it is not that clear whether they have similar roles in macrophages from humans. Despite of contradicting results from diverse scientific communities that embrace a variety of viewpoints on their specific role in macrophages, it is more clear that both enzymes have an important role in inflammation-related cancers [15]. Our finding of a highly significant ($p < 0.0004$) association between tumor NT positivity and M2-type macrophage phenotype (Fig. 1) supports our hypothesis that ARG activity enhances

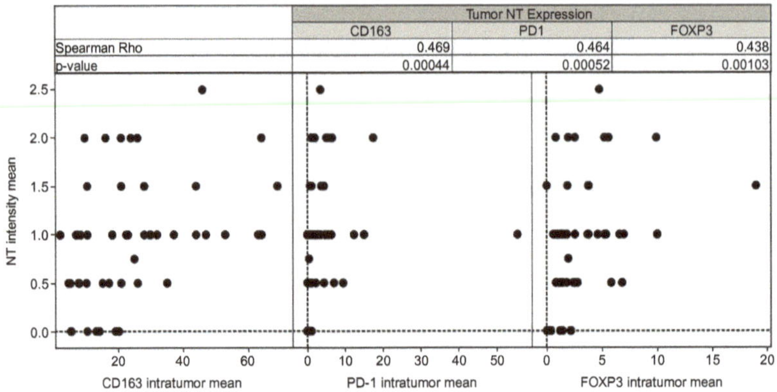

FIG. 1

Association between tumor NT positivity and M2-type (CD163) macrophage phenotype.

immunosuppressive and inflammatory features in the tumor microenvironment, via the PTM NO-mediated tyrosine nitration. We have also observed a significant association between tumor NT and PD-1 expression ($p < 0.0005$) and also FOXP3 expression ($p < 0.001$), which could indicate Treg recruitment, as further regulation of negative signals for the immune response.

EXPRESSION OF ARGININE TRANSPORTERS IN HUMAN MELANOMA CELLS

Arginine mainly enters mammalian cells through membrane-bound transporters, CAT1 and CAT2B [30]. From the Cancer Cell Line Encyclopedia (CCLE) database, we found that the expression CAT2B is various in human melanoma cells and a subset of human melanoma cells have its high level although the expression of CAT1 is similar in those cell lines (Fig. 2). It has been reported that CAT1 is constitutively expressed in most tissues, while CAT2B is cytokine-inducible [30]. Thus, the diverse expression of CAT2B may be associated with inflammatory factors expressed in human melanoma cells.

INOS/NO POSITIVELY REGULATES MPGES1 AND PGE2 PRODUCTION

It has been shown that COX-2 and inducible nitric oxide synthase (iNOS) can be produced simultaneously in the same tissues in several inflammatory models [31]. In our previous study, stage III melanoma TMA analysis showed that about 90% of mPGES1-positive samples also stained positive for the iNOS antibody. In addition, transient iNOS expression significantly enhanced PGE2

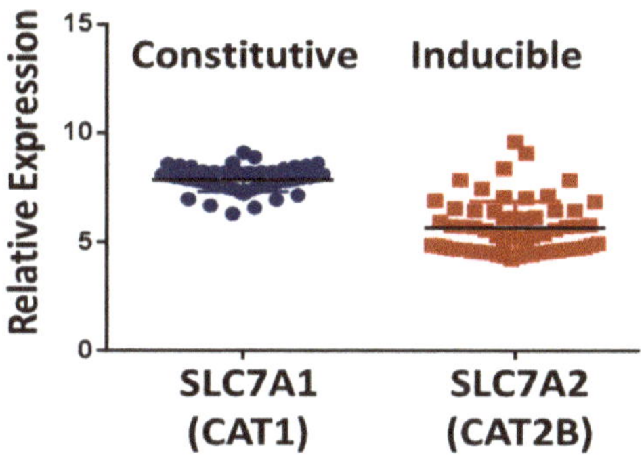

FIG. 2
Arginine transporter levels in human melanoma cells from CCLE database.

production. Moreover, NO donors, GSNO, and SNAP significantly enhanced PGE2 production. These results suggest that iNOS and NO are upstream of PGE2 biosynthesis in melanoma cells. We found that mPGES1 was tyrosine-nitrated as a function of iNOS expression. Our data imply that iNOS regulates mPGES1 activity through a posttranslational NT modification, which ultimately enhances PGE2 production [10].

HIGH MICROSOMAL PGE2 SYNTHASE-1 LEVELS ASSOCIATE WITH LOW CD8 T CELLS AND POORER MELANOMA PATIENT SURVIVAL

In melanoma and other cancers, COX-2 and its product PGE2 are reported to enhance carcinogenesis and tumor progression and to support immunosuppression. Specifically, in melanoma, tumor-derived COX activity plays a role as the key suppressor of type I IFN-mediated and T-cell-mediated tumor elimination and as an inducer of an inflammatory signature associated with progression [11]. We have also found that tumors in a stage III melanoma TMA analysis demonstrated high mPGES1 staining intensity and significantly associated with lowest CD8 levels (unpublished data) ($r = -0.2722$, $p = 0.0128$). Importantly, stage III melanoma patients with high mPGES1 and low CD8 signature had significantly increased risk of death when compared with patients with low mPGES1/low CD8 and low mPGES1/high CD8. This result supports that the COX-2/mPGES1 pathway contributes to immune evasion.

CONCLUSION

An effective cancer immunotherapy needs to exploit differences of tumor cells versus normal cells to enable an effective immune response and to overcome immunosuppressive mechanisms at the tumor site. An effective T-cell-mediated antitumor immune response, as opposed to immunosuppressive mechanisms, is a progressive process, and the subsequent effective immune response or active immune suppression can produce an antitumor immune reaction as opposed to immune escape and tumor growth. Therefore, improving any of these steps or controlling immune suppression has the potential to promote a more effective antitumor immune response.

Overall, the research from our group confirms that NT-positive tumors also contain more immunosuppressive macrophages, the M2 subclass. Moreover, the iNOS/NO pathway upregulates another major immunosuppressive pathway, that of COX-2 and mPGES1, so that a complex and connected arginine metabolism pathway appears likely to regulate the T-cell infiltration and function that results in tumor growth and potential immune suppression. The genetic ablation of this course may cause an increased T-cell infiltration, so this pathway is possibly amenable to an experimental pharmacological intervention.

References

[1] Habib S, Ali A. Biochemistry of nitric oxide. Indian J Clin Biochem 2011;26(1):3–17.

[2] Kovacs I, Lindermayr C. Nitric oxide-based protein modification: formation and site-specificity of protein S-nitrosylation. Front Plant Sci 2013;4:137.

[3] Yakovlev VA, Mikkelsen RB. Protein tyrosine nitration in cellular signal transduction pathways. J Recept Signal Transduct Res 2010;30(6):420–9.

[4] Rodriguez PC, Ochoa AC, Al-Khami AA. Arginine metabolism in myeloid cells shapes innate and adaptive immunity. Front Immunol 2017;8:93.

[5] Lind DS. Arginine and cancer. J Nutr 2004;134:2837S–41S [discussion 2853S].

[6] Yoon JK, Frankel AE, Feun LG, Ekmekcioglu S, Kim KB. Arginine deprivation therapy for malignant melanoma. Clin Pharmacol 2013;5:11–9.

[7] Feun L, You M, Wu CJ, Kuo MT, Wangpaichitr M, Spector S, Savaraj N. Arginine deprivation as a targeted therapy for cancer. Curr Pharm Des 2008;14:1049–57.

[8] Rodriguez PC, Quiceno DG, Ochoa AC. L-arginine availability regulates T-lymphocyte cell-cycle progression. Blood 2017;109:1568–73.

[9] Lamas B, Vergnaud-Gauduchon J, Goncalves-Mendes N, Perche O, Rossary A, Vasson MP, Farges MC. Altered functions of natural killer cells in response to L-arginine availability. Cell Immunol 2012;280:182–90.

[10] Kim SH, Hashimoto Y, Cho SN, Roszik J, Milton DR, Dal F, Kim SF, Menter DG, Yang P, Ekmekcioglu S, Grimm EA. Microsomal PGE2 synthase-1 regulates melanoma cell survival and associates with melanoma disease progression. Pigment Cell Melanoma Res 2016;29:297–308.

[11] Zelenay S, van der Veen AG, Bottcher JP, Snelgrove KJ, Rogers N, Acton SE, Chakravarty P, Girotti MR, Marais R, Quezada SA, Sahai E, Reis e Sousa C. Cyclooxygenase-dependent tumor growth through evasion of immunity. Cell 2015;162:1257–70.

[12] Gabrilovich DI, Nagaraj S. Myeloid-derived suppressor cells as regulators of the immune system. Nat Rev Immunol 2009;9:162–74.

[13] Ekmekcioglu S, Ellerhorst JA, Prieto VG, Johnson MM, Broemeling LD, Grimm EA. Tumor iNOS predicts poor survival for stage III melanoma patients. Int J Cancer 2006;119:861–6.

[14] Johansson CC, Egyhazi S, Masucci G, Harlin H, Mougiakakos D, Poschke I, Nilsson B, Garberg L, Tuominen R, Linden D, Stolt MF, Hansson J, Kiessling R. Prognostic significance of tumor iNOS and COX-2 in stage III malignant cutaneous melanoma. Cancer Immunol Immunother 2009;58:1085–94.

[15] Grimm EA, Sikora AG, Ekmekcioglu S. Molecular pathways: inflammation-associated nitric-oxide production as a cancer-supporting redox mechanism and a potential therapeutic target. Clin Cancer Res 2013;19:5557–63.

[16] Heinecke JL, Ridnour LA, Cheng RY, Switzer CH, Lizardo MM, Khanna C, Glynn SA, Hussain SP, Young HA, Ambs S, Wink DA. Tumor microenvironment-based feed-forward regulation of NOS2 in breast cancer progression. Proc Natl Acad Sci U S A 2014; 111:6323–8.

[17] Tanese K, Grimm EA, Ekmekcioglu S. The role of melanoma tumor-derived nitric oxide in the tumor inflammatory microenvironment: its impact on the chemokine expression profile, including suppression of CXCL10. Int J Cancer 2012;131:891–901.

[18] Ekmekcioglu S, Grimm EA, Roszik J. Targeting iNOS to increase efficacy of immunotherapies. Hum Vaccin Immunother 2017;13:1105–8.

[19] Tang CH, Grimm EA. Depletion of endogenous nitric oxide enhances cisplatin-induced apoptosis in a p53-dependent manner in melanoma cell lines. J Biol Chem 2004;279: 288–98.

[20] Sikora AG, Gelbard A, Davies MA, Sano D, Ekmekcioglu S, Kwon J, Hailemichael Y, Jayaraman P, Myers JN, Grimm EA, Overwijk WW. Targeted inhibition of inducible nitric oxide synthase inhibits growth of human melanoma in vivo and synergizes with chemotherapy. Clin Cancer Res 2010;16(6):1834–44.

[21] Massi D, Franchi A, Sardi I, Magnelli L, Paglierani M, Borgognoni L, Maria Reali U, Santucci M. Inducible nitric oxide synthase expression in benign and malignant cutaneous melanocytic lesions. J Pathol 2001;194(2):194–200.

[22] Jeannin JF, Leon L, Cortier M, Sassi N, Paul C, Bettaieb A. Nitric oxide-induced resistance or sensitization to death in tumor cells. Nitric Oxide 2008;19(2):158–63.

[23] Brüne B. Nitric oxide: NO apoptosis or turning it ON? Cell Death Differ 2003;10(8):864–9.

[24] Ekmekcioglu S, Ellerhorst J, Smid CM, Prieto VG, Munsell M, Buzaid AC, Grimm EA. Inducible nitric oxide synthase and nitrotyrosine in human metastatic melanoma tumors correlate with poor survival. Clin Cancer Res 2000;6(12):4768–75.

[25] Gochman E, Mahajna J, Shenzer P, Dahan A, Blatt A, Elyakim R, Reznick AZ. The expression of iNOS and nitrotyrosine in colitis and colon cancer in humans. Acta Histochem 2012;114 (8):827–35.

[26] Thomas LN, Merrimen J, Bell DG, Rendon R, Too CK. Prolactin- and testosterone-induced carboxypeptidase-D correlates with increased nitrotyrosines and Ki67 in prostate cancer. Prostate 2015;75(15):1726–36.

[27] Bailey P, Chang DK, Forget MA, Lucas FA, Alvarez HA, Haymaker C, Chattopadhyay C, Kim SH, Ekmekcioglu S, Grimm EA, et al. Exploiting the neoantigen landscape for immunotherapy of pancreatic ductal adenocarcinoma. Sci Rep 2016;6:35848.

[28] Hardy LL, Wick DA, Webb JR. Conversion of tyrosine to the inflammation-associated analog 3′-nitrotyrosine at either TCR- or MHC-contact positions can profoundly affect recognition of the MHC class I-restricted epitope of lymphocytic choriomeningitis virus glycoprotein 33 by CD8 T cells. J Immunol 2008;180(9):5956–62.

[29] Pannell M, Labuz D, Celik MO, Keye J, Batra A, Siegmundn B, Halina Machelska H. Adoptive transfer of M2 macrophages reduces neuropathic pain via opioid peptides. J Neuroinflammation 2016;13:262.

[30] Visigalli R, Bussolati O, Sala R, Barilli A, Rotoli BM, Parolari A, Alamanni F, Gazzola GC, Dall'Asta V. The stimulation of arginine transport by TNFalpha in human endothelial cells depends on NF-kappaB activation. Biochim Biophys Acta 2004;1664:45–52.

[31] Cuzzocrea S, Salvemini D. Molecular mechanisms involved in the reciprocal regulation of cyclooxygenase and nitric oxide synthase enzymes. Kidney Int 2007;71:290–7.

CHAPTER 7

Cancer and Beyond: Discovery and Development of NO-Releasing Therapeutics

Jan Scicinski*, Khosrow Kashfi*,†

*Avicenna Therapeutics, Inc, Raleigh, NC, United States

†Department of Molecular, Cellular and Biomedical Sciences, Sophie Davis School of Biomedical Education, City University of New York School of Medicine, New York, NY, United States

ABSTRACT

Nitric oxide, the endogenous vasodilator, is a key signaling molecule that has been implicated in many biological processes and disease states. The manipulation of local nitric oxide concentration using drugs that deliver NO has been exploited for the treatment of many disease indications ranging from heart and circulatory disorders to the treatment and prevention of cancers. In this brief review, we focus on three drugs that have different mechanisms of NO release, eliciting distinct biological effects leading to contrasting potential applications: NBS-1120, an enhanced NSAID (eNSAID) based on the aspirin scaffold that releases nitric oxide and hydrogen sulfide; JS-K, an arylating NO donor; and RRx-001, an immune stimulatory agent that induces NO release indirectly. All three compounds affect the local concentrations of NO; however, their ultimate mechanisms of action are complex, arising from the multifaceted nature of nitric oxide and its context-specific activity. Here, we describe the known activity and safety of these NO donors, current hypotheses on their mechanism of actions, and their developmental status.

ABBREVIATIONS

5-FU	5-fluorouracil
AD	Alzheimer's disease
ADT-OH	anethole trithione
AML	acute myeloid leukemia
ARDS	acute respiratory distress syndrome

Therapeutic Application of Nitric Oxide in Cancer and Inflammatory Disorders. https://doi.org/10.1016/B978-0-12-816545-4.00007-4

cGMP	cyclic guanosine monophosphate
COX	cyclooxygenase
CT	computerized tomography
CV	cardiovascular
Cys	cysteine
DETA/NO	diethylenetriamine/nitric oxide adduct
DMSO	dimethyl sulfoxide
DNMT	DNA methyltransferase
eNSAID	enhanced nonsteroidal anti-inflammatory drug
FoxM1	forkhead box M1
G6PD	glucose-6-phosphate dehydrogenase
GI	gastrointestinal
GSH	glutathione
GST	glutathione *S*-transferase
H&E stain	hematoxylin and eosin stain
H2S	hydrogen sulfide
Hb	hemoglobin
HL-60	human promyelocytic leukemia cell line
HT-29	human colon cancer cell line
JNK	c-Jun N-terminal kinase
MAPK	mitogen-activated protein kinase
MM	multiple myeloma
NET	neuroendocrine tumor
NO	nitric oxide
NSAID	nonsteroidal antiinflammatory drug
NSCLC	non-small-cell lung cancer
PABA/NO	para-aminobenzoic acid/nitric oxide adduct
PPP	pentose phosphate pathway
RBC	red blood cell
RNS	reactive nitrogen species
RONs	reactive oxygen and nitrogen species
ROS	reactive oxygen species
RT-PCR, qPCR	reverse transcription polymerase chain reaction
SAHA	suberoylanilide hydroxamic acid; vorinostat. A histone deacetylase inhibitor
SCLC	small-cell lung cancer
SNAP	*S*-nitroso-*N*-acetyl-penicillamine
TAM	tumor-associated macrophage
TNAZ	trinitroazetidine

Conflict of Interest

No potential conflicts of interest were disclosed.

LOCATION, LOCATION, LOCATION

Nitric oxide, named the "molecule of the year" in 1992 [1], is a ubiquitous, endogenous vasodilating free radical that has been implicated in a growing list of biological functions and disease states. Due to its high reactivity and complex chemistry, nitric oxide is an unlikely biological mediator; however, an important consequence of its reactivity is its rapidity of signaling, spatially limited action, and fast inactivation. High reactivity enables nitric oxide to act as an exquisitely regulated autocrine and paracrine signaling agent, not requiring complex clearance pathways to turn off signaling effects [2]. Nitric oxide is cleared through dilution away from the secretory center and is rapidly oxidized to nitrate and nitrite and is sequestered by reduced protein thiols. These nitrosated species can be viewed as downstream metabolites that relay, modulate, and prolong the biological activity of NO [3].

Nitric oxide activity is also highly dependent on its concentration. Higher local concentrations favor chemical reactions that lead to highly reactive species such as peroxynitrite, nitrogen dioxide, and dinitrogen trioxide [4]. These species partake in radical reactions and can trigger cell death via oxidation and nitration reactions, serving as highly effective endogenous agents against pathogens and, indeed, tumors [5, 6]. In tumor biology, it is now understood that very low levels of NO secretion in the microenvironment produces an anticancer effect. Likewise, very high levels of NO lead to oxidative and nitrosative stress, pushing tumors over their threshold of free radical tolerance and into cell death [7]. Intermediate levels, however, have been shown to promote tumor growth [8] (Fig. 1).

Given the vasodilatory activity of nitric oxide, the development of NO-releasing drugs was a logical step for treating various cardiovascular indications. Inhaled nitric oxide itself is used extensively for numerous perioperative applications, typically to treat pulmonary and right ventricular failure in patients undergoing heart or lung transplantation and other cardiac surgery and to treat hypoxemia [9, 10]. Further indications such as traumatic brain injury, sickle cell disease, acute respiratory distress syndrome (ARDS), and sepsis are all subject of current or recent clinical trials. NO donors that include nitroglycerine, isosorbide dinitrate (Isordil), and isosorbide mononitrate have been used extensively to treat angina with sublingual and IV formulations; however, systemic dosing with NO-releasing agents presents a number of challenges. Systemic release of NO results in marked changes in blood pressure as vasodilation occurs

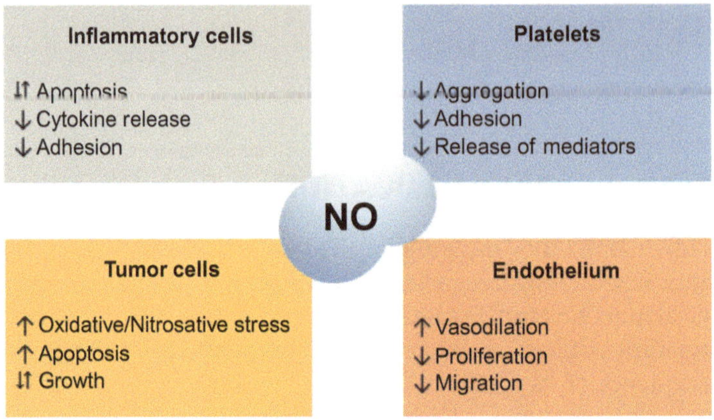

FIG. 1

Multifaceted action of nitric oxide *Adapted from Miller MR, Megson IL. Recent developments in nitric oxide donor drugs. Br J Pharmacol 2007;151(3):305–21. PubMed PMID: 17401442 (web archive link).*

throughout the body. Vasodilation in brain capillaries also results in migraine-like headaches, a sometimes severe yet ironic side effect of sildenafil arising from increased levels of nitric oxide [11]. Nitric oxide activity is therefore context-specific and location- and concentration-dependent.

The design of second-generation NO-based therapeutics acknowledged the observation that nitric oxide activity is highly context-specific [8] and the understanding that nitric oxide release is not the whole story. Therefore, other additional approaches that release NO at specific locations or release NO together with another active species that provides synergy have been investigated.

In this brief review, we focus on three drugs that have different mechanisms of NO release, eliciting distinct biological effects leading to contrasting potential applications: NBS-1120, an enhanced NSAID (eNSAID) based on the aspirin scaffold that releases nitric oxide and hydrogen sulfide; JS-K, an arylating NO donor and GSH depleter; and RRx-001, an immune stimulatory agent that induces NO release indirectly (Fig. 2). All three compounds affect the local

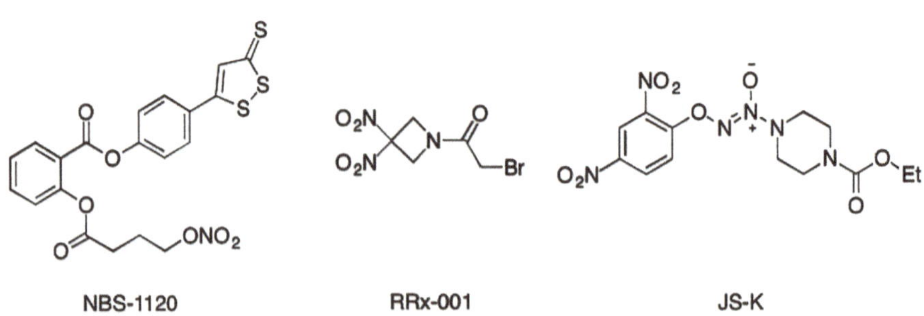

FIG. 2

Chemical structure of RRx-001, NBS-1120, and JS-K.

concentrations of NO; however, their ultimate mechanisms of action are complex, arising from the multifaceted nature of nitric oxide and its context-specific activity. Here, we describe the known activity and safety of these NO donors, current hypotheses on their mechanism of action, and their developmental status. In addition to their anticancer applications, we explore potential indications beyond cancer.

JS-K AND *N*-DIAZENIUMDIOLATES

N-Diazeniumdiolates are a class of molecule that readily fragment to release two molecules of nitric oxide. Consequently, many structural variations of these "NONOates" have been studied as a convenient and stable source of nitric oxide that is released under biological conditions [12, 13]. While these compounds are stable as solids, dissolution and small pH changes result in spontaneous fragmentation. Although this is not necessarily an issue when systemic exposure to NO is desired, it does present some developmental challenges and does not enable targeted delivery of nitric oxide to the tumor. In addition, rapid exposure to systemic NO can lead to sudden drops in blood pressure through the systemic vasodilatory effect of NO (Fig. 3).

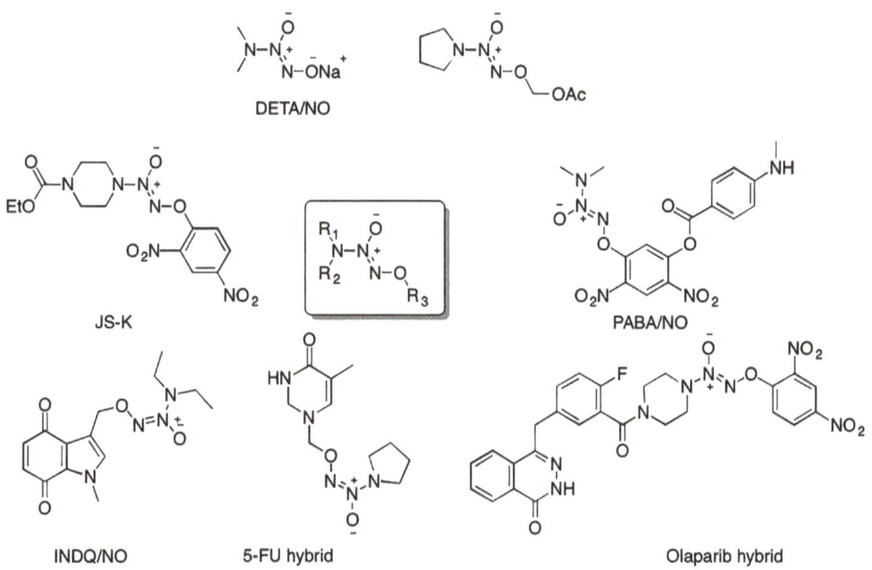

FIG. 3

Diazeniumdiolate NO donors. DETA/NO has been used extensively as a NO donor in biological experiments. PABA/NO was derived from JS-K as a better substrate of the GST isoform [14]. INeDQ/NO is an example of a differentially activated NO prodrug, in this case by DT-diaphorase [15]. A number of cytotoxic and enzyme inhibitor/NO donor hybrids have been synthesized, here exemplified by 5-FU [16] and olaparib [17] hybrids, respectively.

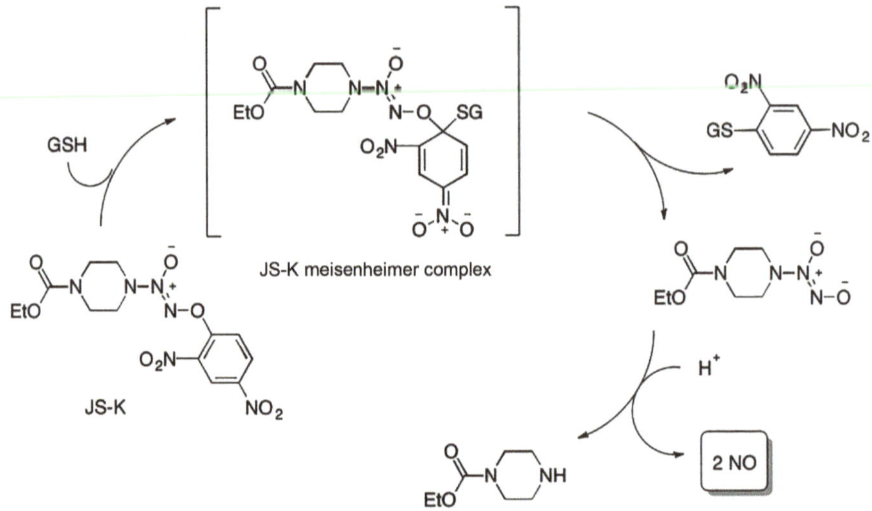

FIG. 4
JS-K activation by glutathione (GSH) and release of nitric oxide.

The rational design of JS-K sprung from early work on arylated diazeniumdiolates [14, 18], which demonstrated that stabilized diazeniumdiolates could be activated to release the diazeniumdiolate ion and therefore nitric oxide, by the influence of nucleophiles. Screening, structural optimization, and activity confirmation in a human leukemia cell line in vivo identified JS-K, designed to be activated by the endogenous nucleophile, glutathione, as the most potent and promising candidate [19]. JS-K contains the diazeniumdiolate moiety functionalized with a stabilizing 2,4-dinitrophenyl group that is sensitive to cleavage in the presence of sulfur nucleophiles such as the endogenous tripeptide, glutathione (GSH) (Fig. 4). The overexpression of glutathione *S*-transferase (GST) in tumor cells catalyzes the conjugation of cytotoxic agents leading to their detoxification [20, 21]. Here, GST catalyzes the sequestration of tumor GSH by JS-K resulting in the fragmentation of JS-K and the release of two molecules of nitric oxide per JS-K molecule resulting in an increase in the oxidative and nitrosative stress at the point of fragmentation, i.e., inside the tumor cell.

Leukemia cells were initially chosen as a target for JS-K due to their particular sensitivity to nitric oxide [22, 23] and reactive oxygen species in general [24]. Further studies, however, demonstrated a broad range of activities in prostate [19, 25–27], colon [28], non-small-cell lung [29], breast [30, 31], glioma [32–34], hepatocellular [35], and renal [36] cancer cell lines. The most dramatic effects were, however, observed in leukemia and multiple myeloma. An IC_{50} of 0.2–0.5 μM was observed in the HL-60 human myeloid leukemia line, together with apoptosis induction in a concentration- and caspase-dependent manner. An in vivo experiment in subcutaneously implanted HL-60 tumors in NOS-SCID mice showed a growth reduction of ~50% when JS-K was dosed IV three times a week at a dose low enough to not induce hypotension. Excision and examination of the tumors revealed extensive central necrosis, common with other ROS- and RNS-mediated drugs [19]. JS-K also exhibited synergistic activity against AML (HL-60 cell line) with cytarabine (ARA-C), though not, interestingly, daunorubicin or etoposide [37]. Microarray and real-time RT-PCR analysis on JS-K treated HL-60 cells was used to identify upregulated or downregulated genes to provide further insight into potential molecular targets. Gene expression analysis demonstrated that multiple molecular targets were implicated in the mechanism of action of JS-K and confirmed that the induction of apoptosis through apoptosis-related gene expression was an important factor [38].

JS-K was active in multiple myeloma (MM) cell lines and in patient-derived MM cells, inducing extrinsic and intrinsic apoptotic pathways via the JNK pathway, DNA double-strand breaks, and the induction intratumoral oxidative stress. These effects were shown to be mediated via nitric oxide. Importantly, JS-K was active in the survival-inducing presence of IL-6 and IGF-1 and in multiple myeloma cells that adhered to bone marrow stromal cells. Combination with bortezomib sensitized multiple myeloma cells to JS-K [39]. JS-K inhibited angiogenesis [40], prolonged survival, and inhibited tumor growth in a mouse model of multiple myeloma [39].

Development of JS-K as an intravenously dosed drug required good aqueous solubility and stability. The poor solubility and tendancy to unpredictably precipitate [19], together with reactivity with whole blood components [41], prompted formulation work that would enhance the solubility profile while also stabilizing the molecule in whole blood enabling tumor-specific activation and release of NO. A micellular formulation using Pluronic P123 polymer increased solubility and stability in saline, whole blood, and plasma [42] and increased accumulation into the nucleus compared with free JS-K [41]. Pluronic P123 formulated JS-K had equivalent in vitro activity against HL-60 cells compared with free JS-K, while the micellular formulation had a somewhat higher activity in vivo [42], potentially attributed to higher nuclear concentrations and enhanced stability in systemic circulation as evidenced by the absence of NO-induced hypotension.

Mechanistically, JS-K induces apoptosis and DNA double-strand breaks and resulting in DNA damage and the activation of MAPK stress signaling pathways [43, 44]. JS-K also induces cell death by mitotic catastrophe as part of necrosis in glioblastoma multiforme [34]. In common with other NO-releasing drugs, these effects are most likely mediated by the generation of reactive oxygen species (ROS) [26, 29, 45]. The action of reactive oxygen and nitrogen species

(RONs) changes the redox potential of the tumor microenvironment [44] and leads to a plethora of downstream actions including knockdown of thioredoxin 1 [28], impact on WNT signaling [25, 46], and reduction in tumor cell invasiveness through the inhibition of TIMP-2 [31]. In addition to the activation and release of NO from JS-K by GSH, an alternative chemical pathway to JS-K activity exists through direct reaction with exposed reduced cysteine residues on proteins [47]. Capping these residues may lead to protein inactivation.

The preclinical results in AML and MM provided a rationale for the development of JS-K as a treatment for AML and multiple myeloma, and Orphan Drug Designation was granted by the FDA for these indications. JS-K is currently undergoing IND-enabling studies.

NOSH-ASPIRIN

The link between inflammatory states and cancer has been well established [48], and therefore the reduction of inflammation by common nonsteroidal antiinflammatory drugs (NSAIDs) such as aspirin could potentially reduce the incidence of cancer. A number of comparative studies [49, 50], facilitated in part by the widespread use of low-dose aspirin, enabled meta-analyses to be performed to determine whether chronic dosing with NSAIDs is beneficial in the prevention of cancer [51, 52]. In one meta-analysis, Veetil [53] was able to demonstrate the benefit of chronic administration of low-dose aspirin on recurrence of colorectal adenomas, an observation supported by other systemic reviews on colorectal cancer incidence and metastasis [54, 55]. Similar analyses were carried out in esophageal [56], gastric, biliary, and breast cancer [57].

The mechanisms by which NSAIDs exert chemoprotection are complex. In addition to the impact of COX-2 [51] that contributes to tumor growth and invasion in epithelial ovarian cancer through the action of PI3K/AKT kinase [58], aspirin has been shown to inhibit expression of the oncoprotein c-Myc through its action on blood platelets [59], reducing the tumor-promoting activity of platelets in colon and pancreatic tumor cells [60]. Overexpression of c-Myc has important consequences in the maintenance of cancer stem cells and chemoresistance [61]. A recent report on the inhibition of c-Myc by RRx-001 (see below) [62] suggests a general c-Myc inhibitory role for prooxidant NO-releasing drugs.

Although NSAIDs significantly reduce the incidence of cancer in the clinic, long-term dosing has been associated with serious GI, CV, and renal side effects, limiting the effectiveness of the drugs and making the long-term use of NSAIDs not feasible [63]. Moreover, although aspirin, naproxen, sulindac, and other NSAIDs have been shown to have cytotoxic effects in in vitro preclinical studies, the doses required to translate these effects in vivo are too toxic to be feasible

[52]. Alternative antiinflammatory drugs to NSAIDs such as the COX-2 inhibitors were promising, well funded, but ultimately unsuccessful due to serious CV issues [64].

Nitric oxide possess antiinflammatory activity, acting by inhibiting NF-κB activation [65] and reducing expression of key mediators of inflammation in vascular endothelium, mast cells, and macrophages [66–68]. Nitric oxide-releasing NSAIDs (NO-NSAIDs) were designed to combine the antiinflammatory activities of NO and the parent NSAID. NO-releasing groups were integrated into the NSAID scaffold through the addition of short-chain ester-linked nitroxybutyl or nitrosothiol groups [63, 69]. These hybrid molecules, in particular those incorporating nitrate esters as the NO donor of choice, have a long clinical history in which antiinflammatory and analgesic effects have been demonstrated in a wide range of disease settings [70, 71]; however, none have received regulatory marketing approval to date. Notably, these compounds have been shown to possess markedly reduced GI, CV, and renal toxicity, compared with the parent NSAIDs [72], arising from the similar properties of NO to prostaglandins resulting in the prevention of mucosal injury [73, 74]. These compounds have shown some promising anticancer activity with NCX4016, a NO-releasing derivative of aspirin, inhibiting cell cycle progression and proliferation in a number of different human cancer cell lines two to three orders of magnitude greater than aspirin itself [75, 76].

Hydrogen sulfide is another key, though less well-known gasotransmitter [77]. Both hydrogen sulfide and nitric oxide are produced in the vasculature, and their interaction strongly influences vasorelaxation (Fig. 5) [78]. Signaling of NO and H2S to produce vasorelaxation occurs through cyclic guanosine monophosphate (cGMP) through the tandem action of NO activation of cGMP and H2S raising the levels of cGMP [79, 80]. In a similar way to nitric oxide, H2S activity is context-specific where different local concentrations appear to result in much varied activity. In the context of cancer, low or endogenous levels of H2S can promote tumor growth, while sustained high concentrations have an anticancer effect [81]. Consequently, hydrogen sulfide is both pro- and antiinflammatory [82, 83] with both anticancer and tumor-promoting activities [84, 85].

Hydrogen sulfide-donating and hydrogen sulfide-releasing molecules have been shown to possess anticancer activity in preclinical in vitro and in vivo studies [80]. Hydrogen sulfide-releasing NSAIDs have been prepared and shown to possess anticancer activity. H2S-releasing naproxen induced apoptosis of HT-29 colon cancer cells in vitro, while hydrogen sulfide-releasing aspirin reduced growth of estrogen receptor-negative breast cancer cells in vitro and in vivo, both inducing G0/G1 cell cycle arrest and acting through the downregulation of NF-κB signaling [86, 87]. Interestingly, ATB-346, a H2S-donating

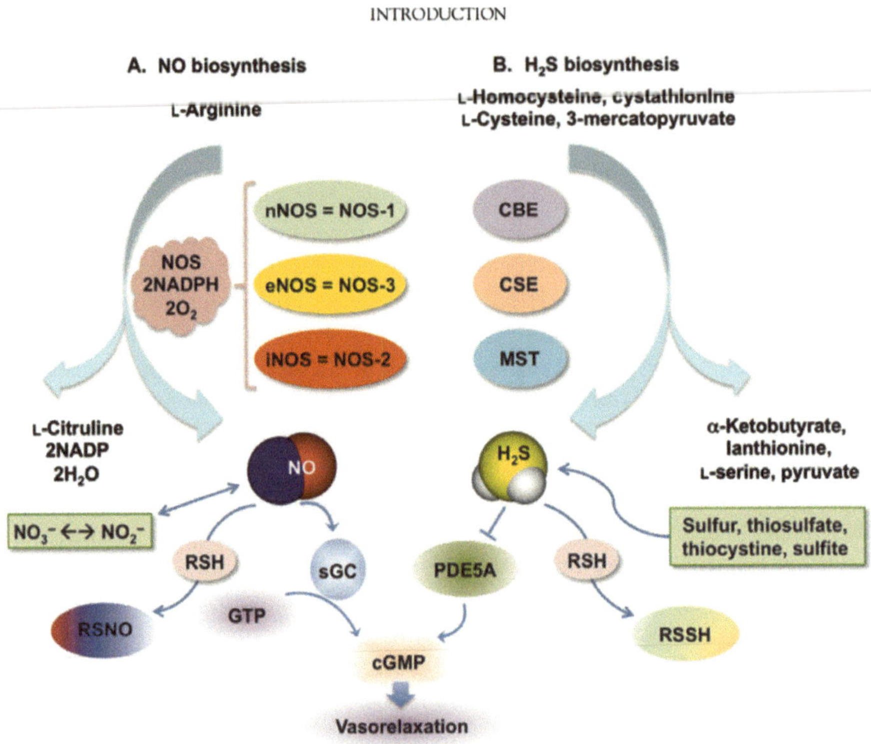

FIG. 5

Biosynthesis and interactions between nitric oxide and hydrogen sulfide. *From Kashfi K, M. E. NO-H2S-releasing chimeras as a multifaceted approach to cancer therapy. Nitric oxide (donor/induced) in chemosensitization. 1st ed. Elsevier; 2017. pp. 105–42.*

hybrid of naproxen, suppressed the expression of oncoproteins such as c-Myc and beta-catenin, while naproxen did not possess this activity [88]. In addition, ATB-346 showed gastroprotective activity compared with naproxen itself [89]. ATB-346 is being developed by Antibe Therapeutics for the treatment of osteoarthritis [90, 91].

As both nitric oxide and hydrogen sulfide potentiate the anticancer activity of NSAIDs, while mitigating gastric toxicity, it was hypothesized that NSAID molecules that donated both gasotransmitters would result in potent anticancer agents with an enhanced safety profile [80]. A number of these NOSH-NSAIDs that are based on the aspirin [92, 93], naproxen [94], and sulindac [95, 96] scaffolds have been described (Fig. 6). All three representative NOSH-NSAIDs described in Fig. 6 were devoid of any GI toxicity while possessing NSAID activity on par with their parent NSAIDs [80]. Herein, we focus on the activity of NBS-1120, NOSH-aspirin.

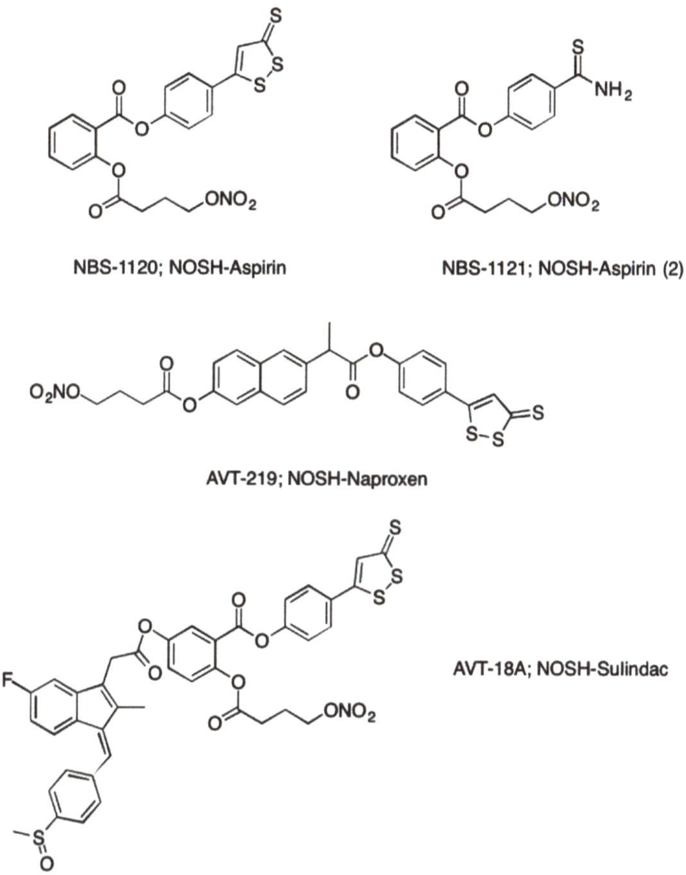

FIG. 6

Representative NOSH-NSAIDs. AVT-18A is a NO and H2S hybrid of sulindac and aspirin.

NBS-1120, first described in a seminal paper in 2012, is a nitric oxide- and hydrogen sulfide-releasing hybrid based on the aspirin scaffold [93]. Nitric oxide donation is effected through a nitrate ester functionality, while hydrogen sulfide donation occurs through the anethole trithione (ADT-OH) [97] subunit (Fig. 7).

NBS-1120 showed a time-dependent release of NO as measured using a nitrate/nitrite colorimetric assay, while H2S release was found to be greater when incubated in tissue compared with the medium suggesting intracellular release or formation of H2S was a key mediator of activity [92, 93].

The anticancer activity of NBS-1120 was found to be significantly greater than aspirin and in the submicromolar range in a panel of different human tumor

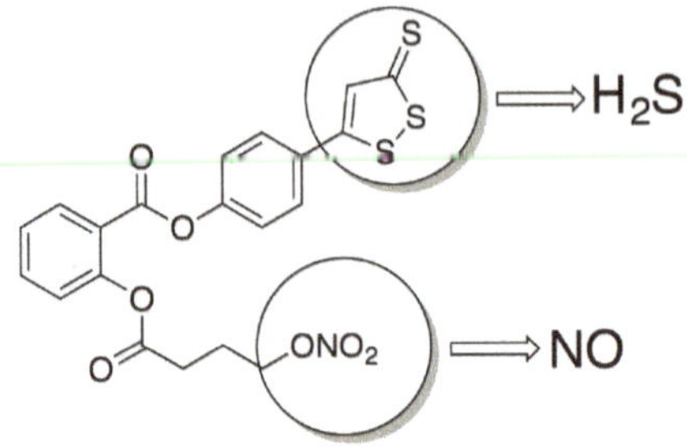

FIG. 7

NBS-1120: NO- and H2S-releasing subunits.

cell lines (Fig. 8). Colon HT-29 and pancreatic BxPC3 cells express both COX-1 and COX-2, whereas colon HCT15 and pancreatic MIA PaCa2 cells are COX-null, suggesting that the effects observed are COX-independent. In HT-29 colon cancer cells, IC_{50} ranged from 46 nM at 24 h to 8 nM at 72 h. Interestingly, unlike JS-K, NBS-1120 activity was not greatest in a leukemia cell line. Of the four NOSH-NSAIDs illustrated in Fig. 6, NBS-1120 exhibited the highest potency in these cell lines [93–95].

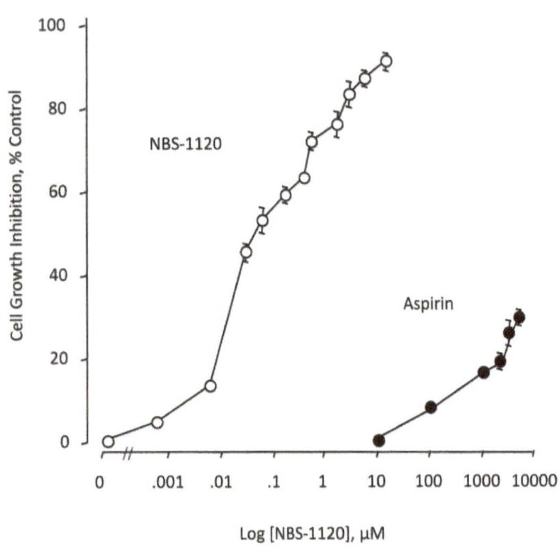

Tumor Type	Cell line	NBS-1120 IC_{50} at 24h, nM	ASA IC_{50} at 24h, nM
Colon	HT29	48 ± 3	>5,000,000
	HCT15	50 ± 5	
	SW480	60 ± 4	
Breast	MDA-MB231	100 ± 1	
	SK BR3	75 ± 5	
	MCF7	280 ± 16	
Pancreas	MIA PaCa2	47 ± 5	
	Bx PC3	57 ± 4	
Lung	A540	50 ± 7	
Prostate	LNCAP	88 ± 8	
Leukemia	Jurkat	100 ± 8	

FIG. 8

In vitro activity of NBS-1120 against a panel of human tumor cell lines compared with aspirin (ASA). *Adapted from Chattopadhyay M, Kodela R, Olson KR, Kashfi K. NOSH-aspirin (NBS-1120), a novel nitric oxide- and hydrogen sulfide-releasing hybrid is a potent inhibitor of colon cancer cell growth in vitro and in a xenograft mouse model. Biochem Biophys Res Commun 2012;419(3):523–8. PubMed PMID: 22366248 (web archive link).*

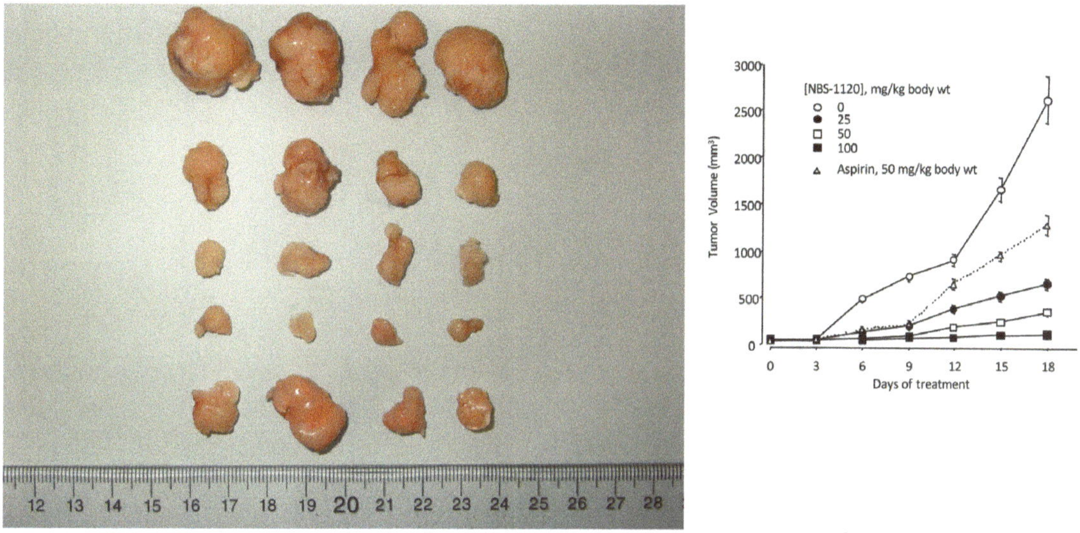

FIG. 9

NBS-1120 prevents growth of HT-29 human colon cancer cells in mouse model. *Left*: Tumor growth delay curves. *Right*: Images of excised tumors showing fivefold difference in potency between NBS-1120 and aspirin.

In vivo NBS-1120 activity was evaluated in athymic nude tumor-bearing mice using human colon cancer xenografts (HT-29 cells) [98], human estrogen receptor-negative (MDA-MB-231) and positive (MCF-7) breast cancer cells, and human pancreatic cancer cells (MIA PaCa2) [99]. After tumors were established, the mice were dosed with NBS-1120 (25, 50, and 100 mg/kg), vehicle (1% methylcellulose), or aspirin (50 mg/kg) daily for 18 days. NBS-1120 inhibited tumor growth and mass and was found to be approximately fivefold potent than aspirin (Fig. 9)[98]. In a similar experiment conducted in the same HT-29 tumor model, NBS-1120 exhibited synergy when dosed with 5-FU [99].

A key part of these experiments was the assessment of GI toxicity. While inspection of stomachs excised from mice treated with NBS-1120 showed no adverse observations at all, GI erosions, an indication of ulcer development, were noted in the stomachs of mice treated with aspirin [92, 100]. In a similar chemopreventative study with AVT-219 and its parent, naproxen, in the HT-29 xenograft colorectal cancer model, at equimolar dosing, all naproxen-dosed animals died within 14 days, exhibiting signs of severe gastric bleeding together with liver and spleen injury. In contrast, the AVT-219-treated mice exhibited no signs of overt toxicity for the 30-day duration of the experiment together with observation of marked tumor growth delay vs untreated controls [94]. In a separate experiment to confirm the gastrointestinal safety of NBS-1120, equimolar acute doses of NBS-1120 (477 mg/kg) or aspirin (180 mg/kg) were administered by gavage to rats and the stomachs excised 6 h post dose. The administration of

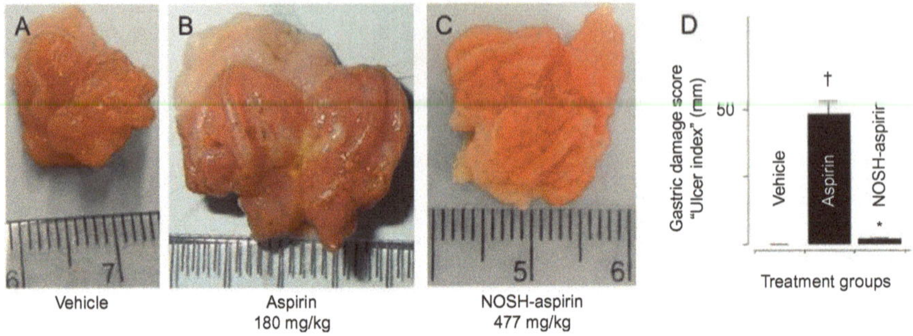

FIG. 10

NBS-1120 does not cause gastric damage: (A) Vehicle-treated rat stomach; (B) Aspirin-treated rat stomach showing ulceration and bleeding; (C) NBS-1120-treated rat stomach showing no ulceration or mucosal damage; (D) Comparison of "Ulcer Index," a measure of gastric damage for the three groups. *From Kodela R, Chattopadhyay M, Velazquez-Martinez CA, Kashfi K. NOSH-aspirin (NBS-1120), a novel nitric oxide- and hydrogen sulfide-releasing hybrid has enhanced chemo-preventive properties compared to aspirin, is gastrointestinal safe with all the classic therapeutic indications. Biochem Pharmacol 2015;98(4):564–72. PubMed PMID: 26394025 (web archive link).*

aspirin resulted in extensive damage of the mucosa (Fig. 10), while rats receiving NBS-1120 did not result in any damage [98].

To investigate the contributions to the activity of NBS-1120 of its various components, a structure-activity and reconstitution experiment was carried out in HT-29 cells using aspirin, *S*-nitroso-*N*-acetyl-penicillamine (SNAP), as the NO donor and ADT-OH for hydrogen sulfide. The results (Table 1) showed that the activity of the combined individual components of NBS-1120 did

Table 1 IC50 values for cell growth inhibition by various components of NOSH-aspirin or other agents that release NO or H2S alone or in combination

Compound	IC_{50} (μM): 24 h
Aspirin	>1000
SNAP	530±45
ADT-OH	26±3
Aspirin+SNAP	710±35
Aspirin+ADT-OH	380±45
Aspirin+ADT-OH+SNAP	450±35
NBS-1120	0.05±0.005

From Chattopadhyay M, Kodela R, Olson KR, Kashfi K. NOSH-aspirin (NBS-1120), a novel nitric oxide- and hydrogen sulfide-releasing hybrid is a potent inhibitor of colon cancer cell growth in vitro and in a xenograft mouse model. Biochem Biophys Res Commun 2012;419(3):523–8. PubMed PMID: 22366248.

not equal the biological activity of intact NBS-1120. Thus, the sum of the parts does not equal the whole [92]. Although there are no data on the relative kinetics of the release of the two gasotransmitters, it is possible that the superior activity of NBS-1120 stems from the complex interaction of the released components with the NO and H2S signaling pathways and the potential formation of the highly reactive and short-lived species, HSNO [101–103]. Positional isomers of NBS-1120 were investigated to determine whether ring substitution in NBS-1120 conferred the optimum activity. Although all positional isomers possessed the same biological activity, the parent NBS-1120 had the most pronounced anticancer effect [104].

In common with other NO-releasing agents described in this review, NBS-1120 possesses a broad mechanism of action that impacts a plethora of molecular targets. NBS-1120 inhibits proliferation through cell cycle arrest and the induction of apoptosis and inhibition of beta-catenin partially through the induction of oxidative and nitrosative stress. Downstream, these effects impact the regulation of NF-kB, FoxM1, and p53 via protein *S*-nitrosylation and *S*-sulfhydration [105–107].

The inhibition of Forkhead box M1 (FoxM1) is of particular interest. FoxM1 is a nucleus-located protein that plays a key role in normal cell cycle function, particularly during G1/S-, G2-, and M-phase transitions [108]. Overexpression of FoxM1 by tumors promotes rapid cell cycle turnover and therefore aggressive tumor growth, and its activity has been linked to DNA damage signaling, resistance to cancer drugs, aggressive tumors, and poor clinical outcomes [109]. FoxM1 upregulation has been found in most solid tumors with high expression levels correlating negatively with enhanced survival [110–117]. Inhibition of FoxM1 has been linked to the reversal of resistance in ovarian [109], lung [118, 119], and gastric [120] cancers as well as glioblastoma [121]. These observations suggest that NBS-1120 may have a utility in reversing acquired resistance enabling patients to revisit therapy that was previously effective. These clinical approaches are discussed in more detail with RRx-001 below.

NBS-1120 is a promising anticancer agent that exhibits good activity in different models of human solid tumors both as a single agent and in combination. Although not tested, as a NO-releasing agent, NBS-1120 is anticipated to be a potent radiosensitizer [122]. NBS-1120 is currently under active pre-IND development.

RRX-001

RRx-001 is a novel, first-in-class small molecule with redox-active epi-immunotherapeutic anticancer properties due to enter phase III clinical trials this year. The multiple mechanisms of action of RRx-001 that include NO release and are described in more detail below combine to sensitize tumors to conventional chemo- and radiotherapies. The discovery of RRx-001 arose through an innovative collaborative partnership between ATK Thiokol (now Northrop Grumman Innovation Systems), and Stanford University under the auspices of InterWest Partners that provided funding for the program. The goal was to explore the potential of energetic materials as radio- and chemosensitizers for therapeutic purposes, with a focus on the treatment of cancer [123].

Chemical compounds that have a high amount of stored chemical energy are known as energetic molecules. The stored chemical energy is often in the form of highly strained ring systems or the presence of multiple nitrogen- and oxygen-containing functional groups. In both cases, copious quantities of energy can be released either by ring fragmentation or the release of gaseous nitrogen- and oxygen-containing fragments. These highly reactive intermediates, formed as free radicals, were thought to function as DNA repair inhibitors, performing a similar role to that of oxygen in radiotherapy [124]. Of the initial compounds evaluated, trinitroazetidine (TNAZ) [125, 126] was found to be the most potent. Optimization and exploration of structure-activity relationships yielded RRx-001, a dinitroazetidine derivative with single-agent, radio-, and chemosensitization activity without the drawback of bulk explosive properties (Fig. 2), as the lead compound [127]. RRx-001 encompasses the structural units of the dinitroazetidine, derived from TNAZ and an α-bromoacetamide. These units were combined to enable assembly of antibody and prodrug derivatives and to introduce an atom that reduced the overall number of nitrogen and oxygen atoms in the molecule, thus reducing its energetic liability (Fig. 11). Investigation of structure-activity relationships of close analogs revealed that both subunits were critical to the activity of RRx-001 [128].

The activity of RRx-001 against a panel of cancer cell lines revealed a micromolar activity profile. RRx-001 induced reactive oxygen species, double-strand DNA breaks, and apoptosis and increased tumor blood flow [129] and was shown to be a potent radiosensitizer. In vitro potency of RRx-001 was enhanced

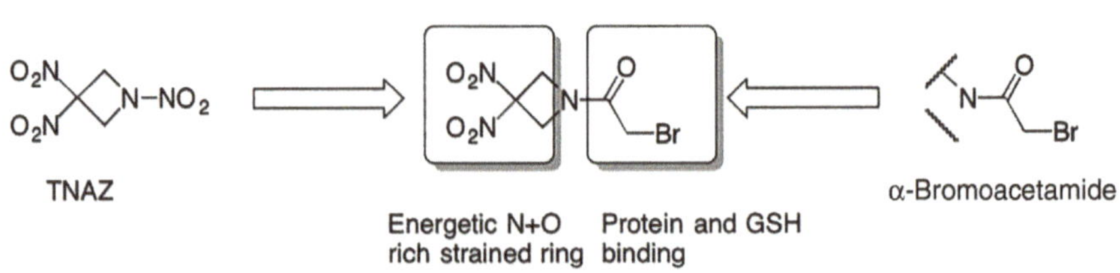

FIG. 11

Design of RRx-001. Combination of energetic nitrogen- and oxygen-rich strained ring system with a derivatizing center.

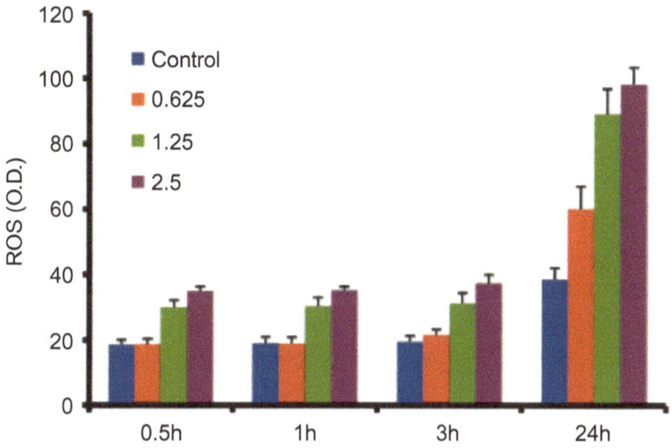

FIG. 12

RRx-001 increases ROS in a dose- and time-dependent manner in MM cell lines [130].

under hypoxic conditions [127]. In vivo studies confirmed single-agent activity on par with cisplatin and strong radiosensitizing effects in the murine syngeneic SCCVII tumor model [129] (Fig. 12).

Similar to JS-K, potent activity against the leukemia and lymphoma cell lines, HL60 and Daudi cells, suggested the potential of RRx-001 in the treatment of multiple myeloma. RRx-001 decreased the viability of resistant multiple myeloma cell lines and patient-derived primary cell lines and inhibited multiple myeloma proliferation in the presence of bone marrow stromal cells. RRx-001 was active as a single agent and exhibited synergy when dosed with pomalidomide, bortezomib, or SAHA [130] (Fig. 13).

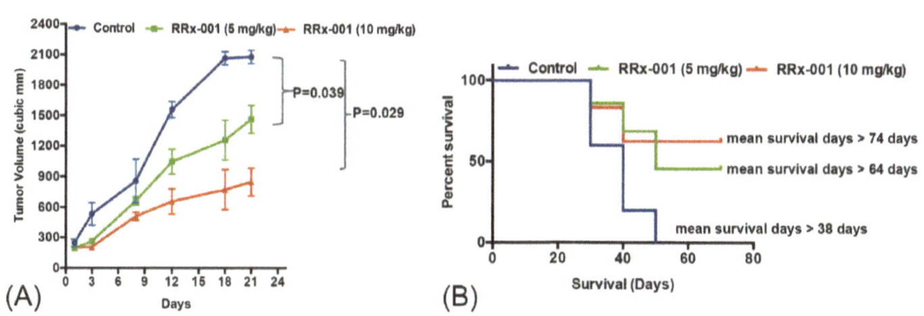

FIG. 13

RRx-001 inhibits multiple myeloma tumor growth and prolongs survival. *From Das DS, Ray A, Das A, Song Y, Tian Z, Oronsky B, et al. A novel hypoxia-selective epigenetic agent RRx-001 triggers apoptosis and overcomes drug resistance in multiple myeloma cells. Leukemia 2016;30(11):2187–97. PubMed PMID: 27118403 (web archive link).*

RRx-001 metabolism is closely linked to its activity. On infusion, RRx-001 rapidly permeates red blood cells (RBCs) and binds covalently and irreversibly to glutathione and cysteine and to a conserved residue on hemoglobin, β-Cys-93 [131], that is implicated in detoxification of electrophilic toxicants and nitric oxide transport [132, 133]. Intracellular depletion of glutathione increases oxidative stress inside the RBCs. Furthermore, binding to β-Cys-93 greatly potentiates the nitrite reductase activity of hemoglobin, catalyzing the reduction of nitrite to nitric oxide only in the presence of hypoxia [134]. Thus, the oxidatively stressed RBCs begin producing nitric oxide once they enter the highly hypoxic tumor microenvironment. The current mechanistic hypotheses for the action of RRx-001 rest on the fate of this small subunit of modified RBCs. A consequence of RBC oxidative stress is membrane scrambling. Thus, phosphatidyl serine (PS), a lipid normally located on the interior membrane, is externalized [135]. In addition to being a universal "eat me" signal, RBCs with externalized PS can bind to tumor vasculature that expresses PS receptors [136] and become internalized [137]. Additionally, the higher local concentration of PS attracts tumor-associated macrophages (TAMs) that internalize and digest the RBCs. Similar to iron overload triggering a M2-to-M1 conversion in a number of pathologies [138], the high content of free heme and iron in the RRx-001-stressed RBCs [137] could be responsible for the observed M2-to-M1 phenotype transition in tumor-associated macrophages (TAMs) and resulting antitumor activity (Fig. 14).

As described above and in addition to the RRx-001-induced activity of TAMs, RRx-001 increases the reactive oxygen and nitrogen species (RONs) inside tumors, which induce the observed apoptosis and DNA damage. The increase in RONs, manifested by the upregulation and nuclear translocation of nrf2 [139], impact epigenetic processes, inhibiting DNA methyltransferases, histone deacetylases, and lysine demethylases [140]. In addition, inhibition of DNMT1

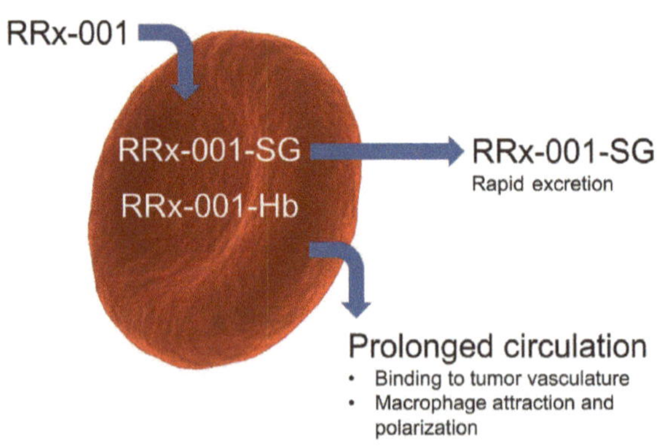

FIG. 14
RRx-001 metabolism and binding to RBCs.

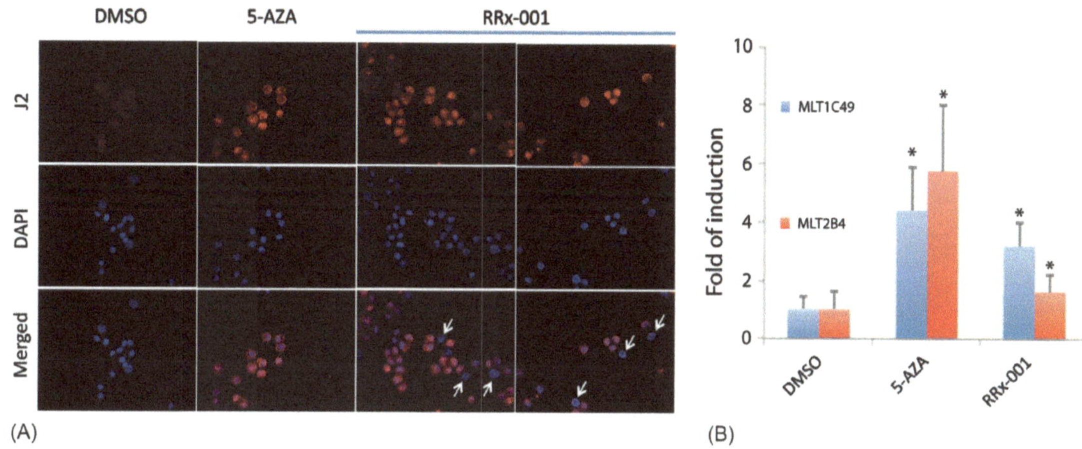

FIG. 15

RRx-001 induction of an interferon response through viral mimicry. (A) Immunofluorescence staining of viral double-strand RNA using J2 antibody. DAPI nuclear staining as control for RRx-001, 5-azacytidine (5-AZA), and DMSO control. White arrows point to cells that were negative for dsRNA. (B) Transcript levels of two endogenous retroviruses were upregulated by 5-AZA and RRx-001 compared with DMSO as determined by qPCR. *From Zhao H, Ning S, Nolley R, Scicinski J, Oronsky B, Knox SJ, et al. The immunomodulatory anticancer agent, RRx-001, induces an interferon response through epigenetic induction of viral mimicry. Clin Epigenetics 2017;9:4. PubMed PMID: 28149332 (web archive link).*

leads to the induction of an interferon response in tumors through viral mimicry [141] (Fig. 15).

Similar to the mechanism of chemoprevention described for NSAIDs above, RRx-001 has been shown to inhibit both c-Myc [62] and platelets [142], both important factors controlling the maintenance of cancer stem cells and thus chemoresistance. As described above, it is possible that RRx-001 also inhibits FoxM1, and these mechanisms together, with the epigenetic and immune effects, could contribute to the observed clinical reversal of resistance [143].

In a phase 1 clinical trial [144], RRx-001 was well tolerated with pain and vasodilation on infusion the only notable drug-related adverse event over 25 patients in six dose cohorts. The transient injection site pain and vasodilation was attributed to the release of nitric oxide on binding of RRx-001 to the β-Cys-93 of hemoglobin [145]. Consequently, the incidence of local infusion pain was mitigated by a change in the administration process: coadministration with 10 mL of autologous blood allowed sufficient time for nitric oxide release and resequestration enabling fast infusion times.

Evidence for the reversal of resistance or resensitization activity of RRx-001 began to surface during the phase 1 clinical trial [144, 146]. In particular, four patients who were resistant to chemotherapy were rechallenged with a previously effective regimen post RRx-001 were found to respond well.

Resensitization was explored in a number of clinical trials: Of these, QUADRUPLE THREAT (NCT02489903) explored resensitization to platinum-based therapies in patients with refractory/resistant small-cell lung cancer (SCLC), EGFR+ non-SCLC, high-grade neuroendocrine (NET), and resistant/refractory gynecologic cancers including ovarian and malignant mixed Müllerian tumors. Patients received RRx-001 4 mg weekly followed by the appropriate platinum doublet therapy.

Although full data from this trial have not yet been published, a number of case reports describing patient responses have appeared in the literature. Four case reports in SCLC patients [147–150] describe partial responses on chemotherapy after priming with RRx-001. Three case reports in NSCLC describing similar results have been published [151–153]. In the NET patient cohort, a number of case studies were reported [154–158]. One patient with very rare, high-grade neuroendocrine prostate cancer reported a complete metabolic response having received RRx-001 followed by reintroduced carboplatin/etoposide [155]. Another patient with a very rare recurrent platinum refractory ovarian cancer exhibited a partial response on PD-1 therapy after priming with RRx-001, suggesting an increase in tumor antigens post treatment [154].

Observations in phase 1 and 2 clinical trials of RRx-001 suggested a reduced incidence of toxicities attributed to platinum therapy once the patients had been primed with RRx-001. These observations prompted preclinical research to examine this effect. After pretreating mice with RRx-001 (5 mg/kg every other day for 3 days), cisplatin was administered and typical clinical labs were taken to assess toxicity (Fig. 16). Pretreatment with RRx-001 decreased blood urea

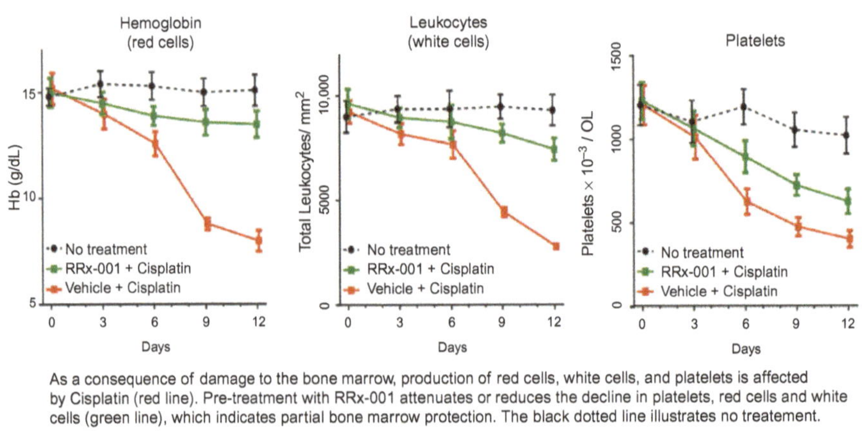

FIG. 16

Pretreatment with RRx-001 attenuates or reduces the decline in platelets, red cells, and white cells after cisplatin treatment.

nitrogen and creatinine levels and reduced chromosome aberration compared with the cisplatin only group [159]. Similar experiments suggested RRx-001 protected against doxorubicin cardiotoxicity [145].

RRx-001 is currently in phase II development and start of phase III clinical trials is expected this year.

BEYOND CANCER

Due to the short life expectancy of patients, paucity of treatments, poor outcomes, and a high diversity of cancer types, the development of new cancer medication has been in the spotlight. In particular, the "war on cancer" sharpened this focus, providing greater investment for research and development. In addition, regulatory agencies have looked favorably on new and novel cancer therapeutics often providing incentives in the form of streamlined development pathways to speed the time to approval. Consequently, the regulatory pathway to approval for anticancer agents is somewhat less onerous than for many less serious, chronic conditions. As a result, anticancer agents with a plethora of additional activity in different indications are often developed for cancer indications first, exploring other indications in clinical trials post initial approval.

As a result of the pleiotropic nature of nitric oxide, the three agents discussed in this review have activity beyond cancer indications. Indeed, the multiple mechanisms of action and many molecular targets are an advantage, conferring robust activity with a lower risk of acquired tolerance and resistance.

NBS-1120 has been shown to be an effective antiinflammatory agent, exhibiting strong antiinflammatory, antipyretic, analgesic, and antiplatelet activity (Fig. 17) [93, 98]. In some cases, NBS-1120 showed superior activity to aspirin in models of inflammatory pain: In an inflammatory hyperalgesia model (Freund's adjuvant), NBS-1120 produced a greater and longer-lasting antinociceptive effect compared with aspirin. Similarly, NBS-1120 activity was superior to aspirin in the acetic acid-induced writhing response and the carrageenan-induced inflammatory hyperalgesia models. Mechanistic investigations revealed that, unlike aspirin, NBS-1120 reduced prostaglandin E2-induced hyperalgesia and reduced IL-1β in the carrageenan-induced paw inflammation model [160].

In Alzheimer's disease (AD), the deposition of β-amyloid is associated with chronic, local inflammation [161], and once deposited, the β-amyloid plaques can induce glial cells to secrete reactive oxygen species and proinflammatory agents that may lead to further β-amyloid accumulation [162]. Although it is uncertain whether AD is a consequence of inflammation, the strong antiinflammatory activity of NBS-1120 promoted the investigation of the activity of this

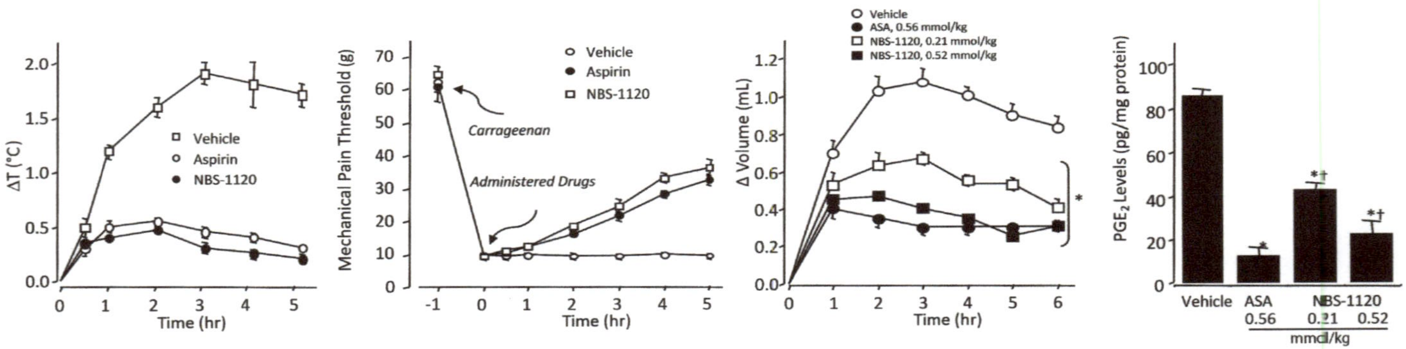

Equipotent fever reduction compared to aspirin in rats

Pain reduction comparable to aspirin in rats

Comparable anti-inflammatory activity to aspirin in rat paw edema model

FIG. 17
NBS-1120 retains NSAID activity.

compound in various models of AD. Since NSAIDs have proved to be anti-AD agents by reducing local inflammation, Drochioiu et al. hypothesized that NBS-1120 would be beneficial in neurodegenerative diseases due to the synergistic activities of the NBS-1120 components acting as independent anti-inflammatory agents, ROS scavengers, and metal binders [163]. In a study comparing the effect of NBS-1120, NO- and H2S-releasing aspirin on neuronal cell death induced by factors such as LPS and interferon-γ secreted from glial cells, Lee et al. found that NBS-1120 was the most potent of the three compounds, promoting neuronal cell survival dose dependently. The investigators' conclusion was that the powerful impact on cell survival by NBS-1120 warranted further work to explore the potential of NBS-1120 in neurodegenerative diseases [164]. Indeed, a NO- and H2S-releasing hybrid of *N*-butyl phthalide (NBP) was found to be effective an ischemia and reperfusion mouse model [165]. NBP is an antioxidant, antiischemic drug that was approved in China for stroke in 2002 [166].

Due to the release of NO under hypoxic conditions, it was hypothesized that treatment with RRx-001 would result in enhanced vasodilation under hypoxic conditions, allowing faster reoxygenation and preventing reperfusion injury. To investigate this hypothesis, RRx-001 was studied in animal models of hemorrhagic shock [167] and ischemia reperfusion injury [168].

Hemorrhagic shock arises from severe loss of blood resulting in a precipitous drop in blood pressure with insufficient delivery of oxygen and nutrients to vital organs. Acute hemorrhage in hamsters, fitted with a dorsal skinfold window chamber, was induced by the rapid removal of approximately 50% of total blood volume and, after 1 h, the animals were resuscitated with "blood" (fresh packed RBCs), "nitrite" (fresh packed RBCs with nitrite), "RRx-001" (fresh packed RBCs with 2 mg/ml of RRx-001), and "RRx-001 + N" (fresh packed RBCs with 2 mg/mL of RRx-001 with nitrite). Microvascular observations were taken through the dorsal window chamber. In all hemodynamic parameters and in the assessment of the number of apoptotic and necrotic cells (Fig. 18), resuscitation with RRx-001 treated blood was superior to that with blood alone. The presence of nitrite supplement, added to provide a feedstock for NO generation, did not significantly improve the parameters, but did result in methemoglobinemia [167].

Sickle cell disease, a hereditary blood disorder, is characterized by a single-gene mutation that leads to an abnormality in hemoglobin. During periods of ischemia, RBCs become rigid as a result of hemoglobin polymerization, contracting to the characteristic shape of a sickle moon. A particularly painful feature of the disease is the vaso-occlusive crises that arise when rigid, sickled RBCs become trapped in organ microvasculature. Strategies for treatment of this disorder have focused on stabilization of fetal hemoglobin with hydroxyurea and

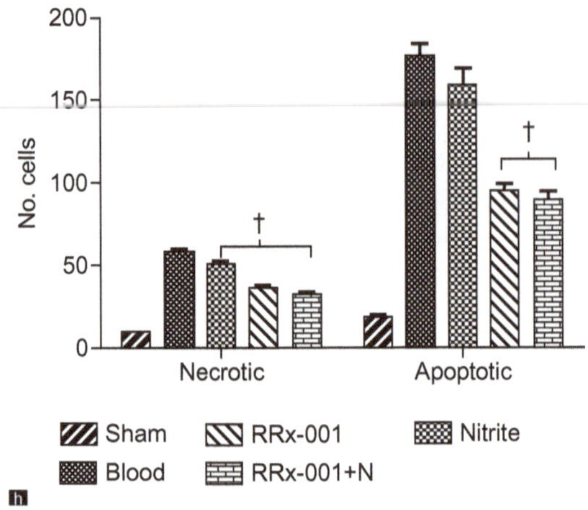

FIG. 18

Number of apoptotic and necrotic cells at 8 h following resuscitation.

increasing the oxygen affinity of sickle cell hemoglobin to prevent hypoxia-induced hemoglobin polymerization [169]. In addition, since cell-free hemoglobin scavenges NO faster than normal RBCs, low NO is thought to contribute to sickling [170]. Inhaled NO has been explored clinically as a potential anti-sickling agent, no significant differences in any of the measures compared with placebo [171] were observed. In addition to potentiating the nitrite reductase capacity of hemoglobin to form NO, RRx-001 has been previously reported to increase hemoglobin oxygen affinity [172]. RRx-001 was found to greatly potentiate the release of NO from sickle cell hemoglobin derived from patient's blood under hypoxia [173] and to prevent and reverse sickling [145] in vitro.

The sickle cell mutation that results in hemoglobin abnormality arose from evolutionary pressure of malarial infection. RRx-001 binding to hemoglobin suggested potential antimalarial activity [174]. Therefore, RRx-001 was investigated in an animal model of cerebral malaria. Cerebral malaria is a serious complication of *Plasmodium falciparum infection*. The parasite depends on enhanced glucose flux through the pentose phosphate pathway (PPP) for survival; therefore, inhibition of this key metabolic pathway could lead to reduction in parasitemia and improved survival [175]. Rather than using *P. falciparum*, infection of C57BL/6 mice with *Plasmodium berghei* is a well-established experimental model of cerebral malaria. RRx-001 was found to be a potent inhibitor of G6PD, a key enzyme in the PPP [176], and was found to significantly prolong survival, reduce parasitemia, and improve motor score both as a single agent and in combination with artemether [177], an artemisinin derivative (Fig. 19).

These results suggest that RRx-001 could find utility in treating malaria, potentiating the activity of artemisinin against resistant strains of *P. falciparum*.

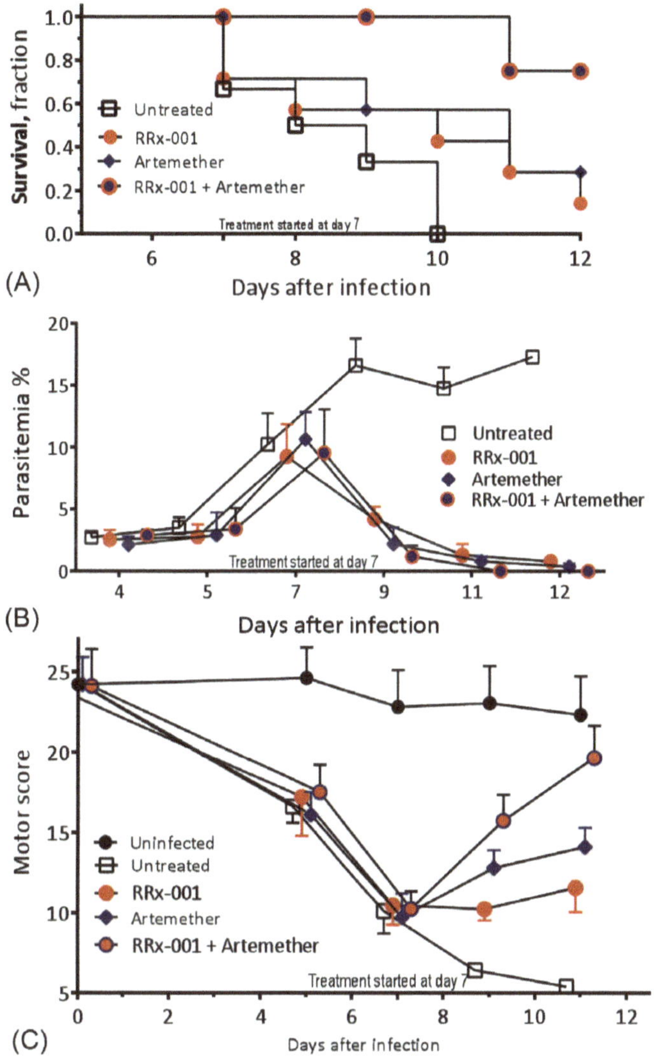

FIG. 19

(A) RRx-001 significantly prolongs survival in combination with artemether compared with the corresponding single agents or control. (B) RRx-001 alone or in combination with artemether reduces parasitemia. (C) RRx-001 in combination with arthemeter RRx-001 significantly improved motor score. *From Yalcin O, Oronsky B, Carvalho LJ, Kuypers FA, Scicinski J, Cabrales P. From METS to malaria: RRx-001, a multi-faceted anticancer agent with activity in cerebral malaria. Malar J 2015;14:218. PubMed PMID: 26017006 (web archive link).*

OUTLOOK

Second-generation NO donors, such as the three compounds that are the focus of this review, release nitric oxide in very different ways. All three compounds contribute additional activity beyond that associated just with NO donation. Nevertheless, these compounds do have much in common. Both JS-K and RRx-001 deplete glutathione while producing nitric oxide, leading to an increase in tumor-associated oxidative and nitrosative stress. With the pleiotropic activity of NO, it is no surprise that the mechanisms of action for these three drug candidates are difficult to establish. While JS-K releases NO on reaction with GSH, the released metabolite possesses a 2,4-dinitrophenol subunit that could contribute to the activity of the molecule through the inhibition of glycolysis and the Warburg effect leading to enhanced cell death. NBS-1120 releases the parent NSAID, NO, and H2S. Interaction of H2S and NO may lead to many useful biological functions for NBS-1120 including potential for resensitization of patients.

The three compounds that are the focus of this review are currently in active development against cancer indications, with JS-K and NBS-1120 still at an early stage and RRx-001 entering phase III. Not surprisingly, no noncancer clinical trials have been initiated with these as yet, and their potential for treatment of diseases beyond cancer is yet to be fully explored. Nevertheless, current and future clinical trials can inform development decisions and may justify the investment and development of these compounds for noncancer indications.

Acknowledgments

The authors would like to thank Dr. Joseph F. Nowoslawski for providing updated information on the development of JS-K.

References

[1] Culotta E, Koshland Jr. DE. NO news is good news, Science 1992;258(5090):1862–5. 1361684.

[2] Miller MR, Megson IL. Recent developments in nitric oxide donor drugs, Br J Pharmacol 2007;151(3):305–21. 17401442.

[3] Yang Y, Loscalzo J. CHAPTER 10-S-nitrosated proteins: formation, metabolism, and function. In: Van Faassen E, Fyodorovich Vanin A, editors. Radicals for life. Amsterdam: Elsevier; 2007. p. 201–21.

[4] Radi R. Peroxynitrite, a stealthy biological oxidant, J Biol Chem 2013;288(37):26464–72. 23861390.

[5] Hirst DG, Robson T. Nitrosative stress in cancer therapy, Front Biosci 2007;12:3406–18. 17485309.

[6] Hirst DG, Robson T. Nitrosative stress as a mediator of apoptosis: implications for cancer therapy, Curr Pharm Des 2010;16(1):45–55. 20214617.

[7] Schulze A, Harris AL. How cancer metabolism is tuned for proliferation and vulnerable to disruption, Nature 2012;491(7424):364–73. 23151579.

[8] Oronsky B, Fanger GR, Oronsky N, Knox S, Scicinski J. The implications of hyponitroxia in cancer, Transl Oncol 2014;7(2):167–73. 24731473.

[9] George I, Xydas S, Topkara VK, Ferdinando C, Barnwell EC, Gableman L, et al. Clinical indication for use and outcomes after inhaled nitric oxide therapy, Ann Thorac Surg 2006;82 (6):2161–9. 17126129.

[10] Oz MC, Ardehali A. Collective review: perioperative uses of inhaled nitric oxide in adults, Heart Surg Forum 2004;7(6):E584–9. 15769692.

[11] Lin GY, Lee JT, Peng GS, Yang FC. Sildenafil can induce the onset of a cluster headache bout, Can Urol Assoc J 2014;8(5–6):E378–80. 24940471.

[12] Davies KM, Wink DA, Saavedra JE, Keefer LK. Chemistry of the diazeniumdiolates. 2. Kinetics and mechanism of dissociation to nitric oxide in aqueous solution, J Am Chem Soc 2001;123 (23):5473–81. 11389629.

[13] Diodati JG, Quyyumi AA, Keefer LK. Complexes of nitric oxide with nucleophiles as agents for the controlled biological release of nitric oxide: hemodynamic effect in the rabbit, J Cardiovasc Pharmacol 1993;22(2):287–92. 7692171.

[14] Keefer LK. Behalf of the JSKC. Broad-spectrum anti-cancer activity of o-arylated diazeniumdiolates, Immunopathol Dis Therap 2010;1(3):205–18. 21949595.

[15] Sharma K, Iyer A, Sengupta K, Chakrapani H. INDQ/NO, a bioreductively activated nitric oxide prodrug, Org Lett 2013;15(11):2636–9. 23659457.

[16] Cai TB, Tang X, Nagorski J, Brauschweiger PG, Wang PG. Synthesis and cytotoxicity of 5-fluorouracil/diazeniumdiolate conjugates, Bioorg Med Chem 2003;11(23):4971–5. 14604659.

[17] Maciag AE, Holland RJ, Kim Y, Kumari V, Luthers CE, Sehareen WS, et al. Nitric oxide (NO) releasing poly ADP-ribose polymerase 1 (PARP-1) inhibitors targeted to glutathione S-transferase P1-overexpressing cancer cells, J Med Chem 2014;57(6):2292–302. 24521039.

[18] Saavedra JE, Srinivasan A, Bonifant CL, Chu J, Shanklin AP, Flippen-Anderson JL, et al. The secondary amine/nitric oxide complex ion R(2)N[N(O)NO](-) as nucleophile and leaving group in (S9N)Ar reactions, J Org Chem 2001;66(9):3090–8. 11325274.

[19] Shami PJ, Saavedra JE, Wang LY, Bonifant CL, Diwan BA, Singh SV, et al. JS-K, a glutathione/glutathione S-transferase-activated nitric oxide donor of the diazeniumdiolate class with potent antineoplastic activity, Mol Cancer Ther 2003;2(4):409–17. 12700285.

[20] Townsend DM, Tew KD. The role of glutathione-S-transferase in anti-cancer drug resistance, Oncogene 2003;22(47):7369–75. 14576844.

[21] Gate L, Tew KD. Glutathione S-transferases as emerging therapeutic targets, Expert Opin Ther Targets 2001;5(4):477–89. 12540261.

[22] Magrinat G, Mason SN, Shami PJ, Weinberg JB. Nitric oxide modulation of human leukemia cell differentiation and gene expression, Blood 1992;80(8):1880–4. 1382708.

[23] Shami PJ, Sauls DL, Weinberg JB. Schedule and concentration-dependent induction of apoptosis in leukemia cells by nitric oxide, Leukemia 1998;12(9):1461–6. 9737697.

[24] Gonzales EA, Ledesma RJ, McAllister DJ, Perry SM, Dyer CA, Maye JP. Effects of guided imagery on postoperative outcomes in patients undergoing same-day surgical procedures: a randomized, single-blind study, AANA J 2010;78(3):181–8. 20572403.

[25] Laschak M, Spindler KD, Schrader AJ, Hessenauer A, Streicher W, Schrader M, et al. JS-K, a glutathione/glutathione S-transferase-activated nitric oxide releasing prodrug inhibits androgen receptor and WNT-signaling in prostate cancer cells, BMC Cancer 2012;12:130. 22462810.

[26] Qiu M, Chen L, Tan G, Ke L, Zhang S, Chen H, et al. JS-K promotes apoptosis by inducing ROS production in human prostate cancer cells, Oncol Lett 2017;13(3):1137–42. 28454225.

[27] Tan G, Qiu M, Chen L, Zhang S, Ke L, Liu J. JS-K, a nitric oxide pro-drug, regulates growth and apoptosis through the ubiquitin-proteasome pathway in prostate cancer cells, BMC Cancer 2017;17(1):376. 28549433.

[28] Edes K, Cassidy P, Shami PJ, Moos PJ. JS-K, a nitric oxide prodrug, has enhanced cytotoxicity in colon cancer cells with knockdown of thioredoxin reductase 1, PLoS One 2010;5(1): e8786. 20098717.

[29] Maciag AE, Chakrapani H, Saavedra JE, Morris NL, Holland RJ, Kosak KM, et al. The nitric oxide prodrug JS-K is effective against non-small-cell lung cancer cells in vitro and in vivo: involvement of reactive oxygen species, J Pharmacol Exp Ther 2011;336(2):313–20. 20962031.

[30] McMurtry V, Saavedra JE, Nieves-Alicea R, Simeone AM, Keefer LK, Tari AM. JS-K, a nitric oxide-releasing prodrug, induces breast cancer cell death while sparing normal mammary epithelial cells, Int J Oncol 2011;38(4):963–71. 21271218.

[31] Simeone AM, McMurtry V, Nieves-Alicea R, Saavedra JE, Keefer LK, Johnson MM, et al. TIMP-2 mediates the anti-invasive effects of the nitric oxide-releasing prodrug JS-K in breast cancer cells, Breast Cancer Res 2008;10(3):R44. 18474097.

[32] Weyerbrock A, Osterberg N, Psarras N, Baumer B, Kogias E, Werres A, et al. JS-K, a glutathione S-transferase-activated nitric oxide donor with antineoplastic activity in malignant gliomas, Neurosurgery 2012;70(2):497–510. 21849924.

[33] Weidensteiner C, Reichardt W, Shami PJ, Saavedra JE, Keefer LK, Baumer B, et al. Effects of the nitric oxide donor JS-K on the blood-tumor barrier and on orthotopic U87 rat gliomas assessed by MRI, Nitric Oxide 2013;30:17–25. 23370169.

[34] Gunzle J, Osterberg N, Saavedra JE, Weyerbrock A. Nitric oxide released from JS-K induces cell death by mitotic catastrophe as part of necrosis in glioblastoma multiforme, Cell Death Dis 2016;7(9)e234927584787.

[35] Liu L, Wang D, Wang J, Wang S. The nitric oxide prodrug JS-K induces Ca(2+)-mediated apoptosis in human hepatocellular carcinoma HepG2 cells, J Biochem Mol Toxicol 2016; 30(4):192–9. 26616367.

[36] Qiu M, Ke L, Zhang S, Zeng X, Fang Z, Liu J. JS-K, a GST-activated nitric oxide donor prodrug, enhances chemo-sensitivity in renal carcinoma cells and prevents cardiac myocytes toxicity induced by doxorubicin, Cancer Chemother Pharmacol 2017;80(2):275–86. 28608259.

[37] Shami PJ, Maciag AE, Eddington JK, Udupi V, Kosak KM, Saavedra JE, et al. JS-K, an arylating nitric oxide (NO) donor, has synergistic anti-leukemic activity with cytarabine (ARA-C), Leuk Res 2009;33(11):1525–9. 19193435.

[38] Liu J, Malavya S, Wang X, Saavedra JE, Keefer LK, Tokar E, et al. Gene expression profiling for nitric oxide prodrug JS-K to kill HL-60 myeloid leukemia cells, Genomics 2009;94(1):32–8. 19348908.

[39] Kiziltepe T, Hideshima T, Ishitsuka K, Ocio EM, Raje N, Catley L, et al. JS-K, a GST-activated nitric oxide generator, induces DNA double-strand breaks, activates DNA damage response pathways, and induces apoptosis in vitro and in vivo in human multiple myeloma cells, Blood 2007;110(2):709–18. 17384201.

[40] Kiziltepe T, Anderson KC, Kutok JL, Jia L, Boucher KM, Saavedra JE, et al. JS-K has potent anti-angiogenic activity in vitro and inhibits tumour angiogenesis in a multiple myeloma model in vivo, J Pharm Pharmacol 2010;62(1):145–51. 20723011.

[41] Kaur I, Terrazas M, Kosak KM, Kern SE, Boucher KM, Shami PJ. Cellular distribution studies of the nitric oxide-generating antineoplastic prodrug O(2)-(2,4-dinitrophenyl)1-((4-ethoxy-carbonyl)piperazin-1-yl)diazen-1-ium-1,2-diolate formulated in Pluronic P123 micelles, J Pharm Pharmacol 2013;65(9):1329–36. 23927471.

[42] Kaur I, Kosak KM, Terrazas M, Herron JN, Kern SE, Boucher KM, et al. Effect of a Pluronic ((R)) P123 formulation on the nitric oxide-generating drug JS-K, Pharm Res 2015;32 (4):1395–406. 25330743.

[43] Chakrapani H, Kalathur RC, Maciag AE, Citro ML, Ji X, Keefer LK, et al. Synthesis, mechanistic studies, and anti-proliferative activity of glutathione/glutathione S-transferase-activated nitric oxide prodrugs, Bioorg Med Chem 2008;16(22):9764–71. 18930407.

[44] Maciag AE, Holland RJ, Robert Cheng YS, Rodriguez LG, Saavedra JE, Anderson LM, et al. Nitric oxide-releasing prodrug triggers cancer cell death through deregulation of cellular redox balance, Redox Biol 2013;1:115–24. 24024144.

[45] Qiu M, Chen L, Tan G, Ke L, Zhang S, Chen H, et al. A reactive oxygen species activation mechanism contributes to JS-K-induced apoptosis in human bladder cancer cells, Sci Rep 2015;5:15104. 26458509.

[46] Nath N, Chattopadhyay M, Pospishil L, Cieciura LZ, Goswami S, Kodela R, et al. JS-K, a nitric oxide-releasing prodrug, modulates ss-catenin/TCF signaling in leukemic Jurkat cells: evidence of an S-nitrosylated mechanism, Biochem Pharmacol 2010;80(11):1641–9. 20797387.

[47] Shami PJ, Saavedra JE, Bonifant CL, Chu J, Udupi V, Malaviya S, et al. Antitumor activity of JS-K [O2-(2,4-dinitrophenyl) 1-[(4-ethoxycarbonyl)piperazin-1-yl]diazen-1-ium-1,2-diolate] and related O2-aryl diazeniumdiolates in vitro and in vivo, J Med Chem 2006;49 (14):4356–66. 16821795.

[48] Rayburn ER, Ezell SJ, Zhang R. Anti-inflammatory agents for cancer therapy, Mol Cell Pharmacol 2009;1(1):29–43. 20333321.

[49] Sandler RS, Halabi S, Baron JA, Budinger S, Paskett E, Keresztes R, et al. A randomized trial of aspirin to prevent colorectal adenomas in patients with previous colorectal cancer, N Engl J Med 2003;348(10):883–90. 12621132.

[50] Yang L, Lv Z, Xia W, Zhang W, Xin Y, Yuan H, et al. The effect of aspirin on circulating tumor cells in metastatic colorectal and breast cancer patients: a phase II trial study, Clin Transl Oncol 2018;20(7):912–21. 29243075.

[51] Kashfi K, Rigas B. Non-COX-2 targets and cancer: expanding the molecular target repertoire of chemoprevention, Biochem Pharmacol 2005;70(7):969–86. 15949789.

[52] Langley RE. Clinical evidence for the use of aspirin in the treatment of cancer, Ecancermedicalscience 2013;7:297. 23589726.

[53] Veettil SK, Lim KG, Ching SM, Saokaew S, Phisalprapa P, Chaiyakunapruk N. Effects of aspirin and non-aspirin nonsteroidal anti-inflammatory drugs on the incidence of recurrent colorectal adenomas: a systematic review with meta-analysis and trial sequential analysis of randomized clinical trials, BMC Cancer 2017;17(1):763. 29137605.

[54] Cooper K, Squires H, Carroll C, Papaioannou D, Booth A, Logan RF, et al. Chemoprevention of colorectal cancer: systematic review and economic evaluation, Health Technol Assess 2010;14(32):1–206. 20594533.

[55] Kune GA, Kune S, Watson LF. Colorectal cancer risk, chronic illnesses, operations, and medications: case control results from the Melbourne Colorectal Cancer Study, Cancer Res 1988;48(15):4399–404. 3390835.

[56] Hao W, Shen Y, Feng M, Wang H, Lin M, Fang Y, et al. Aspirin acts in esophageal cancer: a brief review, J Thorac Dis 2018;10(4):2490–7. 29850157.

[57] Algra AM, Rothwell PM. Effects of regular aspirin on long-term cancer incidence and metastasis: a systematic comparison of evidence from observational studies versus randomised trials, Lancet Oncol 2012;13(5):518–27. 22440112.

[58] Uddin S, Ahmed M, Hussain A, Assad L, Al-Dayel F, Bavi P, et al. Cyclooxygenase-2 inhibition inhibits PI3K/AKT kinase activity in epithelial ovarian cancer, Int J Cancer 2010;126 (2):382–94. 19621391.

[59] Ai G, Dachineni R, Muley P, Tummala H, Jayarama Bhat G. Aspirin and salicylic acid decrease c-Myc expression in cancer cells: a potential role in chemoprevention. Tumor Biol 2016;37(2):1727–38.

[60] Mitrugno A, Sylman JL, Ngo AT, Pang J, Sears RC, Williams CD, et al. Aspirin therapy reduces the ability of platelets to promote colon and pancreatic cancer cell proliferation: implications for the oncoprotein c-MYC, Am J Physiol Cell Physiol 2017;312(2):C176–89. 27903583.

[61] Whitfield JR, Soucek L. Tumor microenvironment: becoming sick of Myc, Cell Mol Life Sci 2012;69(6):931–4. 22033838.

[62] Oronsky B, Reid TR, Oronsky A, Caroen S, Carter CA, Cabrales P. Brief report: RRx-001 is a c-Myc inhibitor that targets cancer stem cells, Oncotarget 2018;9(34):23439–42. 29805745.

[63] Fiorucci S, Antonelli E. NO-NSAIDs: from inflammatory mediators to clinical readouts, Inflamm Allergy Drug Targets 2006;5(2):121–31. 16613571.

[64] Rainsford KD. Anti-inflammatory drugs in the 21st century, Subcell Biochem 2007;42:3–27. 17612044.

[65] Katsuyama K, Shichiri M, Marumo F, Hirata Y. NO inhibits cytokine-induced iNOS expression and NF-kappaB activation by interfering with phosphorylation and degradation of IkappaB-alpha, Arterioscler Thromb Vasc Biol 1998;18(11):1796–802. 9812920.

[66] Hogaboam CM, Befus AD, Wallace JL. Modulation of rat mast cell reactivity by IL-1 beta. Divergent effects on nitric oxide and platelet-activating factor release, J Immunol 1993;151(7):3767–74. 8397260.

[67] Naseem KM. The role of nitric oxide in cardiovascular diseases, Mol Aspects Med 2005;26(1–2):33–65. 15722114.

[68] Huang JQ, Sridhar S, Chen Y, Hunt RH. Meta-analysis of the relationship between Helicobacter pylori seropositivity and gastric cancer, Gastroenterology 1998;114(6):1169–79. 9609753.

[69] Fiorucci S, Antonelli E, Burgaud JL, Morelli A. Nitric oxide-releasing NSAIDs: a review of their current status, Drug Saf 2001;24(11):801–11. 11665868.

[70] Kashfi K. Anti-inflammatory agents as cancer therapeutics, Adv Pharmacol 2009;57:31–89. 20230759.

[71] Hawkey CJ, Jones JI, Atherton CT, Skelly MM, Bebb JR, Fagerholm U, et al. Gastrointestinal safety of AZD3582, a cyclooxygenase inhibiting nitric oxide donator: proof of concept study in humans, Gut 2003;52(11):1537–42. 14570719.

[72] Wallace JL, Del Soldato P. The therapeutic potential of NO-NSAIDs, Fundam Clin Pharmacol 2003;17(1):11–20. 12588626.

[73] Wallace JL, Miller MJ. Nitric oxide in mucosal defense: a little goes a long way, Gastroenterology 2000;119(2):512–20. 10930387.

[74] Dijkstra G, van Goor H, Jansen PL, Moshage H. Targeting nitric oxide in the gastrointestinal tract, Curr Opin Investig Drugs 2004;5(5):529–36. 15202727.

[75] Kashfi K, Rigas B. Molecular targets of nitric-oxide-donating aspirin in cancer, Biochem Soc Trans 2005;33(Pt 4):701–4. 16042578.

[76] Tesei A, Ricotti L, Ulivi P, Medri L, Amadori D, Zoli W. NCX 4016, a nitric oxide-releasing aspirin derivative, exhibits a significant antiproliferative effect and alters cell cycle progression in human colon adenocarcinoma cell lines, Int J Oncol 2003;22(6):1297–302. 12738997.

[77] Paul BD, Snyder SH. H2S: a novel gasotransmitter that signals by sulfhydration, Trends Biochem Sci 2015;40(11):687–700. 26439534.

[78] Gheibi S, Jeddi S, Kashfi K, Ghasemi A. Regulation of vascular tone homeostasis by NO and H2S: Implications in hypertension, Biochem Pharmacol 2018;149:42–59. 29330066.

[79] Bucci M, Papapetropoulos A, Vellecco V, Zhou Z, Pyriochou A, Roussos C, et al. Hydrogen sulfide is an endogenous inhibitor of phosphodiesterase activity, Arterioscler Thromb Vasc Biol 2010;30(10):1998–2004. 20634473.

[80] Kashfi K, Esmaili M. NO-H2S-releasing chimeras as a multifaceted approach to cancer therapy. Nitric oxide (donor/induced) in chemosensitization. 1st ed. Elsevier; 2017105–42.

[81] Cao X, Ding L, Xie ZZ, Yang Y, Whiteman M, Moore PK, et al. A review of hydrogen sulfide synthesis, metabolism, and measurement: is modulation of hydrogen sulfide a novel therapeutic for cancer?, Antioxid Redox Signal 2018;29790379.

[82] Bhatia M, Wong FL, Fu D, Lau HY, Moochhala SM, Moore PK. Role of hydrogen sulfide in acute pancreatitis and associated lung injury, FASEB J 2005;19(6):623–5. 15671155.

[83] Zanardo RC, Brancaleone V, Distrutti E, Fiorucci S, Cirino G, Wallace JL. Hydrogen sulfide is an endogenous modulator of leukocyte-mediated inflammation, FASEB J 2006;20 (12):2118–20. 16912151.

[84] Kashfi K. The dichotomous role of H2S in cancer cell biology? Deja vu all over again, Biochem Pharmacol 2018;149:205–23. 29397935.

[85] Kashfi K. The role of hydrogen sulfide in health and disease, Biochem Pharmacol 2018;149:1–4. 29501583.

[86] Kodela R, Nath N, Chattopadhyay M, Nesbitt DE, Velazquez-Martinez CA, Kashfi K. Hydrogen sulfide-releasing naproxen suppresses colon cancer cell growth and inhibits NF-kappaB signaling, Drug Des Devel Ther 2015;9:4873–82. 26346117.

[87] Chattopadhyay M, Kodela R, Nath N, Barsegian A, Boring D, Kashfi K. Hydrogen sulfide-releasing aspirin suppresses NF-kappaB signaling in estrogen receptor negative breast cancer cells in vitro and in vivo, Biochem Pharmacol 2012;83(6):723–32. 22209867.

[88] Lee ZW, Teo XY, Tay EY, Tan CH, Hagen T, Moore PK, et al. Utilizing hydrogen sulfide as a novel anti-cancer agent by targeting cancer glycolysis and pH imbalance, Br J Pharmacol 2014;171(18):4322–36. 24827113.

[89] Elsheikh W, Blackler RW, Flannigan KL, Wallace JL. Enhanced chemopreventive effects of a hydrogen sulfide-releasing anti-inflammatory drug (ATB-346) in experimental colorectal cancer, Nitric Oxide 2014;41:131–7. 24747869.

[90] Magierowski M, Magierowska K, Surmiak M, Hubalewska-Mazgaj M, Kwiecien S, Wallace JL, et al. The effect of hydrogen sulfide-releasing naproxen (ATB-346) versus naproxen on formation of stress-induced gastric lesions, the regulation of systemic inflammation, hypoxia and alterations in gastric microcirculation, J Physiol Pharmacol 2017;68(5):749–56. 29375050.

[91] http://www.antibethera.com/pipeline/atb-346/. Accessed Sep 9, 2018.

[92] Chattopadhyay M, Kodela R, Olson KR, Kashfi K. NOSH-aspirin (NBS-1120), a novel nitric oxide- and hydrogen sulfide-releasing hybrid is a potent inhibitor of colon cancer cell growth in vitro and in a xenograft mouse model, Biochem Biophys Res Commun 2012;419 (3):523–8. 22366248.

[93] Kodela R, Chattopadhyay M, Kashfi K. NOSH-aspirin: a novel nitric oxide-hydrogen sulfide-releasing hybrid: a new class of anti-inflammatory pharmaceuticals, ACS Med Chem Lett 2012;3(3):257–62. 22916316.

[94] Chattopadhyay M, Kodela R, Duvalsaint PL, Kashfi K. Gastrointestinal safety, chemotherapeutic potential, and classic pharmacological profile of NOSH-naproxen (AVT-219) a dual NO- and H2S-releasing hybrid, Pharmacol Res Perspect 2016;4(2):e00224, 27069635.

[95] Kashfi K, Chattopadhyay M, Kodela R. NOSH-sulindac (AVT-18A) is a novel nitric oxide- and hydrogen sulfide-releasing hybrid that is gastrointestinal safe and has potent anti-inflammatory, analgesic, antipyretic, anti-platelet, and anti-cancer properties, Redox Biol 2015;6:287–96. 26298203.

[96] Kodela R, Chattopadhyay M, Kashfi K. Synthesis and biological activity of NOSH-naproxen (AVT-219) and NOSH-sulindac (AVT-18A) as potent anti-inflammatory agents with chemotherapeutic potential, Medchemcomm 2013;4(11), 24273639.

[97] Ansari MI, Khan MM, Saquib M, Khatoon S, Hussain MK. Dithiolethiones: a privileged pharmacophore for anticancer therapy and chemoprevention, Future Med Chem 2018;10(10):1241–60. 29749746.

[98] Kodela R, Chattopadhyay M, Velazquez-Martinez CA, Kashfi K. NOSH-aspirin (NBS-1120), a novel nitric oxide- and hydrogen sulfide-releasing hybrid has enhanced chemo-preventive properties compared to aspirin, is gastrointestinal safe with all the classic therapeutic indications, Biochem Pharmacol 2015;98(4):564–72. 26394025.

[99] Kashfi K. Unpublished results.

[100] Chattopadhyay M, Velazquez CA, Pruski A, Nia KV, Abdellatif KR, Keefer LK, et al. Comparison between 3-Nitrooxyphenyl acetylsalicylate (NO-ASA) and O2-(acetylsalicyloxymethyl)-1-(pyrrolidin-1-yl)diazen-1-ium-1,2-diolate (NONO-ASA) as safe anti-inflammatory, analgesic, antipyretic, antioxidant prodrugs, J Pharmacol Exp Ther 2010;335(2):443–50. 20679133.

[101] Cortese-Krott MM, Kuhnle GG, Dyson A, Fernandez BO, Grman M, DuMond JF, et al. Key bioactive reaction products of the NO/H2S interaction are S/N-hybrid species, polysulfides, and nitroxyl, Proc Natl Acad Sci U S A 2015;112(34):E4651–60. 26224837.

[102] Cortese-Krott MM, Fernandez BO, Kelm M, Butler AR, Feelisch M. On the chemical biology of the nitrite/sulfide interaction, Nitric Oxide 2015;46:14–24. 25541073.

[103] Nava M, Martin-Drumel MA, Lopez CA, Crabtree KN, Womack CC, Nguyen TL, et al. Spontaneous and selective formation of HSNO, a crucial intermediate linking H2S and nitroso chemistries, J Am Chem Soc 2016;138(36):11441–4. 27540860.

[104] Vannini F, Chattopadhyay M, Kodela R, Rao PP, Kashfi K. Positional isomerism markedly affects the growth inhibition of colon cancer cells by NOSH-aspirin: COX inhibition and modeling, Redox Biol 2015;6:318–25. 26319435.

[105] Nath N, Chattopadhyay M, Rodes DB, Nazarenko A, Kodela R, Kashfi K. Nitric oxide-releasing aspirin suppresses NF-kappaB signaling in estrogen receptor negative breast cancer cells in vitro and in vivo, Molecules 2015;20(7):12481–99. 26184135.

[106] Chattopadhyay M, Nath N, Kodela R, Sobocki T, Metkar S, Gan ZY, et al. Hydrogen sulfide-releasing aspirin inhibits the growth of leukemic Jurkat cells and modulates beta-catenin expression, Leuk Res 2013;37(10):1302–8. 23896061.

[107] Kashfi K. Nitric oxide-releasing hybrid drugs target cellular processes through S-nitrosylation, For Immunopathol Dis Therap 2012;3(2):97–108. 23730526.

[108] Myatt SS, Lam EW. The emerging roles of forkhead box (Fox) proteins in cancer, Nat Rev Cancer 2007;7(11):847–59. 17943136.

[109] Zhang X, Cheng L, Minn K, Madan R, Godwin AK, Shridhar V, et al. Targeting of mutant p53-induced FoxM1 with thiostrepton induces cytotoxicity and enhances carboplatin sensitivity in cancer cells, Oncotarget 2014;5(22):11365–80. 25426548.

[110] Kalinichenko VV, Major ML, Wang X, Petrovic V, Kuechle J, Yoder HM, et al. Foxm1b transcription factor is essential for development of hepatocellular carcinomas and is negatively regulated by the p19ARF tumor suppressor, Genes Dev 2004;18(7):830–50. 15082532.

[111] Wonsey DR, Follettie MT. Loss of the forkhead transcription factor FoxM1 causes centrosome amplification and mitotic catastrophe, Cancer Res 2005;65(12):5181–9. 15958562.

[112] Kim IM, Ackerson T, Ramakrishna S, Tretiakova M, Wang IC, Kalin TV, et al. The Forkhead Box m1 transcription factor stimulates the proliferation of tumor cells during development of lung cancer, Cancer Res 2006;66(4):2153–61. 16489016.

[113] Kalin TV, Wang IC, Ackerson TJ, Major ML, Detrisac CJ, Kalinichenko VV, et al. Increased levels of the FoxM1 transcription factor accelerate development and progression of prostate carcinomas in both TRAMP and LADY transgenic mice, Cancer Res 2006;66(3):1712–20. 16452231.

[114] Chan DW, Yu SY, Chiu PM, Yao KM, Liu VW, Cheung AN, et al. Over-expression of FOXM1 transcription factor is associated with cervical cancer progression and pathogenesis, J Pathol 2008;215(3):245–52. 18464245.

[115] Douard R, Moutereau S, Pernet P, Chimingqi M, Allory Y, Manivet P, et al. Sonic Hedgehog-dependent proliferation in a series of patients with colorectal cancer, Surgery 2006;139 (5):665–70. 16701100.

[116] Wang Z, Banerjee S, Kong D, Li Y, Sarkar FH. Down-regulation of Forkhead Box M1 transcription factor leads to the inhibition of invasion and angiogenesis of pancreatic cancer cells, Cancer Res 2007;67(17):8293–300. 17804744.

[117] Liu M, Dai B, Kang SH, Ban K, Huang FJ, Lang FF, et al. FoxM1B is overexpressed in human glioblastomas and critically regulates the tumorigenicity of glioma cells, Cancer Res 2006;66 (7):3593–602. 16585184.

[118] Wang K, Zhu X, Zhang K, Zhu L, Zhou F. FoxM1 inhibition enhances chemosensitivity of docetaxel-resistant A549 cells to docetaxel via activation of JNK/mitochondrial pathway, Acta Biochim Biophys Sin (Shanghai) 2016;48(9):804–9. 27521795.

[119] Li J, Chen Y, Jin M, Wang J, Li S, Chen Z, et al. MicroRNA-134 reverses multidrug resistance in human lung adenocarcinoma cells by targeting FOXM1, Oncol Lett 2017;13(3):1451–5. 28454276.

[120] Li X, Liang J, Liu YX, Wang Y, Yang XH, Luan BH, et al. miR-149 reverses cisplatin resistance of gastric cancer SGC7901/DDP cells by targeting FoxM1, Pharmazie 2016;71(11):640–3. 29441968.

[121] Zhang N, Wu X, Yang L, Xiao F, Zhang H, Zhou A, et al. FoxM1 inhibition sensitizes resistant glioblastoma cells to temozolomide by downregulating the expression of DNA-repair gene Rad51, Clin Cancer Res 2012;18(21):5961–71. 22977194.

[122] Oronsky BT, Knox SJ, Scicinski JJ. Is nitric oxide (NO) the last word in radiosensitization? A review, Transl Oncol 2012;5(2):66–71. 22496921.

[123] Oronsky B, Scicinski J, Ning S, Peehl D, Oronsky A, Cabrales P, et al. Rockets, radiosensitizers, and RRx-001: an origin story part I, Discov Med 2016;21(115):173–80. 27115167.

[124] Oronsky BT, Knox SJ, Scicinski J. Six degrees of separation: the oxygen effect in the development of radiosensitizers, Transl Oncol 2011;4(4):189–98. 21804913.

[125] Watt D, Cliff M. TNAZ based melt-cast explosives: technology review and AMRL research directions. Melbourne, Australia: DSTO-TR-0702, DSTO Aeronautical and Maritime Research Laboratory.

[126] Hiskey M, Johnson M, Chavez D. Preparation of 1-substituted-3, 3-dinitroazetidines. J Energetic Mater 1999;17:233–52.

[127] Ning S, Bednarski M, Oronsky B, Scicinski J, Saul G, Knox SJ. Dinitroazetidines are a novel class of anticancer agents and hypoxia-activated radiation sensitizers developed from highly energetic materials, Cancer Res 2012;72(10):2600–8. 22589277. Epub 2012/05/17.

[128] Scicinski J, Oronsky B, Ning S, Fanger G, Knox S, Bednarski M. Discovery and development of RRx-001, a novel nitric oxide and ROS mediated epigenetic modulator. In: Bonavida B, editor. Nitric oxide and cancer: pathogenesis and therapy. Switzerland: Springer International Publishing; 2015. p. 259–77.

[129] Raghunand N, Scicinski J, Guntle GP, Jagadish B, Mash EA, Bruckheimer E, et al. Magnetic resonance imaging of RRx-001 pharmacodynamics in preclinical tumors, Oncotarget 2017;8 (60):102511–20. 29254266.

[130] Das DS, Ray A, Das A, Song Y, Tian Z, Oronsky B, et al. A novel hypoxia-selective epigenetic agent RRx-001 triggers apoptosis and overcomes drug resistance in multiple myeloma cells, Leukemia 2016;30(11):2187–97. 27118403.

[131] Vitturi DA, Sun CW, Harper VM, Thrash-Williams B, Cantu-Medellin N, Chacko BK, et al. Antioxidant functions for the hemoglobin beta93 cysteine residue in erythrocytes and in the vascular compartment in vivo, Free Radic Biol Med 2013;55:119–29. 23159546.

[132] Scicinski J, Oronsky B, Taylor M, Luo G, Musick T, Marini J, et al. Preclinical evaluation of the metabolism and disposition of RRx-001, a novel investigative anticancer agent, Drug Metab Dispos 2012;40(9):1810–6. 22699395.

[133] Scicinski J, Oronsky B, Cooper V, Taylor M, Alexander M, Hadar R, et al. Development of methods for the bioanalysis of RRx-001 and metabolites, Bioanalysis 2014;6(7):947–56. 24806903.

[134] Fens M, Larkin S, Morris C, Fitch WL, Scicinski J, Oronsky B, et al. NO or no NO, increased reduction of nitrite to nitric oxide by modified red blood cells. Blood 2011;118: Abstract 2125.

[135] D'Agnillo F, Alayash AI. Interactions of hemoglobin with hydrogen peroxide alters thiol levels and course of endothelial cell death, Am J Physiol Heart Circ Physiol 2000;279(4): H1880–9. 11009476.

[136] Setty BN, Betal SG. Microvascular endothelial cells express a phosphatidylserine receptor: a functionally active receptor for phosphatidylserine-positive erythrocytes, Blood 2008;111 (2):905–14. 17911385.

[137] Cabrales P, Scicinski J, Reid T, Kuypers F, Larkin S, Fens M, et al. A look inside the mechanistic black box: are red blood cells the critical effectors of RRx-001 cytotoxicity?, Med Oncol 2016;33(7):63. 27229330.

[138] Sindrilaru A, Peters T, Wieschalka S, Baican C, Baican A, Peter H, et al. An unrestrained proinflammatory M1 macrophage population induced by iron impairs wound healing in humans and mice, J Clin Invest 2011;121(3):985–97. 21317534.

[139] Ning S, Sekar TV, Scicinski J, Oronsky B, Peehl DM, Knox SJ, et al. Nrf2 activity as a potential biomarker for the pan-epigenetic anticancer agent, RRx-001, Oncotarget 2015;6(25): 21547–56. 26280276.

[140] Zhao H, Ning S, Scicinski J, Oronsky B, Knox SJ, Peehl DM. Epigenetic effects of RRx-001: a possible unifying mechanism of anticancer activity, Oncotarget 2015;6(41):43172–81. 26657731.

[141] Zhao H, Ning S, Nolley R, Scicinski J, Oronsky B, Knox SJ, et al. The immunomodulatory anticancer agent, RRx-001, induces an interferon response through epigenetic induction of viral mimicry, Clin Epigenetics 2017;9:4. 28149332.

[142] Oronsky B, Oronsky N, Cabrales P. Platelet inhibitory effects of the Phase 3 anticancer and normal tissue cytoprotective agent, RRx-001, J Cell Mol Med 2018;16: 30010241. Epub 2018/07/17.

[143] Strauss J, Figg WD. Epigenetic approaches to overcoming chemotherapy resistance, Lancet Oncol 2015;16(9):1013–5. 26296953.

[144] Reid T, Oronsky B, Scicinski J, Scribner CL, Knox SJ, Ning S, et al. Safety and activity of RRx-001 in patients with advanced cancer: a first-in-human, open-label, dose-escalation phase 1 study, Lancet Oncol 2015;16(9):1133–42. 26296952.

[145] Carter CA, Oronsky B, Caroen S, et al. Partial response to carboplatin in an RRx-001 pretreated patient with EGFR-inhibitor-resistance and T790M-negative NSCLC. Respir Med Case Rep. 2016;18:62–5.

[146] Reid T, Dad S, Korn R, Oronsky B, Knox S, Scicinski J. Two case reports of resensitization to previous chemotherapy with the novel hypoxia-activated hypomethylating anticancer agent RRx-001 in metastatic colorectal cancer patients, Case Rep Oncol 2014;7(1):79–85. 24575021.

[147] Carter CA, Oronsky BT, Caroen SZ, Scicinski JJ, Degesys A, Kim MM, et al. RRx-001 in refractory small-cell lung carcinoma: a case report of a partial response after a third reintroduction of platinum doublets, Case Rep Oncol 2016;9(1):171–6. 27065849.

[148] Carter CA, Oronsky B, Caroen S, Scicinski J, Degesys A, Cabrales P, et al. Partial response in an RRx-001-primed patient with refractory small-cell lung cancer after a third introduction of platinum doublets, Case Rep Oncol 2016;9(2):285–9. 27403127.

[149] Brzezniak C, Oronsky B, Scicinski J, Caroen S, Cabrales P, Dean Abrouk N, et al. Conversion of platinum-etoposide-resistant to sensitive SCLC after treatment with the epi-immunotherapeutic RRx-001: a case report, Oncol Res Treat 2016;39(11):720–3. 27855386.

[150] Oronsky B, Caroen S, Zeman K, Quinn M, Brzezniak C, Scicinski J, et al. A partial response to reintroduced chemotherapy in a resistant small cell lung cancer patient after priming with RRx-001, Clin Med Insights Oncol 2016;10:105–8. 27840583.

[151] Brzezniak C, Schmitz BA, Peterson PG, Degesys A, Oronsky BT, Scicinski JJ, et al. RRx-001-induced tumor necrosis and immune cell infiltration in an EGFR mutation-positive NSCLC with resistance to EGFR tyrosine kinase inhibitors: a case report, Case Rep Oncol 2016;9(1):45–50. 26933418.

[152] Carter CA, Oronsky BT, Caroen SZ, Scicinski JJ, Cabrales P, Reid T, et al. Partial response to platinum doublets in refractory EGFR-positive non-small cell lung cancer patients after RRx-001: evidence of episensitization, Case Rep Oncol 2016;9(1):62–7. 26933421.

[153] Carter CA, Oronsky B, Caroen S, Scicinski J, Cabrales P, Degesys A, et al. Partial response to carboplatin in an RRx-001 pretreated patient with EGFR-inhibitor-resistance and T790M-negative NSCLC, Respir Med Case Rep 2016;18:62–5. 27330954.

[154] Brzezniak C, Oronsky B, Trepel J, Summers Jr. TA, Cabrales P, Lee MJ, et al. RRx-001 priming of PD-1 inhibition in the treatment of small cell carcinoma of the vagina: a rare gynecological tumor, Case Rep Oncol 2017;10(1):276–80. 28512410.

[155] Brzezniak C, Oronsky B, Aggarwal R. A complete metabolic response of metastatic castration-resistant neuroendocrine carcinoma of the prostate after treatment with RRx-001 and reintroduced platinum doublets, Eur Urol 2018;73(2):306–7. 28943185.

[156] Cottrill H, Cason S, Caroen S, Oronsky B, Donaldson E. A recurrent platinum refractory ovarian cancer patient with a partial response after RRx-001 resensitization to platinum doublet, J Investig Med High Impact Case Rep 2018;6: 232470961876008029552569.

[157] Carter CA, Schmitz B, Peterson PG, Quinn M, Degesys A, Jenkins J, et al. Immune reactivity and pseudoprogression or tumor flare in a serially biopsied neuroendocrine patient treated with the epigenetic agent RRx-001, Case Rep Oncol 2016;9(1):164–70. 27065848.

[158] Carter CA, Degesys A, Oronsky B, Scicinski J, Caroen SZ, Oronsky AL, et al. Flushing out carcinoid syndrome: beneficial effect of the anticancer epigenetic agent RRx-001 in a patient with a treatment-refractory neuroendocrine tumor, Case Rep Oncol 2015;8(3):461–5. 26600780.

[159] Oronsky B, Reid TR, Larson C, Carter CA, Brzezniak CE, Oronsky A, et al. RRx-001 protects against cisplatin-induced toxicities, J Cancer Res Clin Oncol 2017;143(9):1671–7. 28417195.

[160] Fonseca MD, Cunha FQ, Kashfi K, Cunha TM. NOSH-aspirin (NBS-1120), a dual nitric oxide and hydrogen sulfide-releasing hybrid, reduces inflammatory pain, Pharmacol Res Perspect 2015;3(3):e00133, 26236481.

[161] Rozemuller AJ, van Gool WA, Eikelenboom P. The neuroinflammatory response in plaques and amyloid angiopathy in Alzheimer's disease: therapeutic implications, Curr Drug Targets CNS Neurol Disord 2005;4(3):223–33. 15975026.

[162] Hoozemans JJ, Veerhuis R, Rozemuller AJ, Eikelenboom P. The pathological cascade of Alzheimer's disease: the role of inflammation and its therapeutic implications, Drugs Today (Barc) 2002;38(6):429–43. 12532179.

[163] Drochioiu G, Tudorachi L, Murariu M. NOSH aspirin may have a protective role in Alzheimer's disease, Med Hypotheses 2015;84(3):262–7. 25636606.

[164] Lee M, McGeer E, Kodela R, Kashfi K, McGeer PL. NOSH-aspirin (NBS-1120), a novel nitric oxide and hydrogen sulfide releasing hybrid, attenuates neuroinflammation induced by microglial and astrocytic activation: a new candidate for treatment of neurodegenerative disorders, Glia 2013;61(10):1724–34. 23918470.

[165] Ji J, Xiang P, Li T, Lan L, Xu X, Lu G, et al. NOSH-NBP, a novel nitric oxide and hydrogen sulfide-releasing hybrid, attenuates ischemic stroke-induced neuroinflammatory injury by modulating microglia polarization, Front Cell Neurosci 2017;11:154. 28603491.

[166] Wang X, Wang L, Sheng X, Huang Z, Li T, Zhang M, et al. Design, synthesis and biological evaluation of hydrogen sulfide releasing derivatives of 3-n-butylphthalide as potential antiplatelet and antithrombotic agents, Org Biomol Chem 2014;12(31):5995–6004. 24988475.

[167] Brouse C, Ortiz D, Su Y, Oronsky B, Scicinski J, Cabrales P. Impact of hemoglobin nitrite to nitric oxide reductase on blood transfusion for resuscitation from hemorrhagic shock, Asian J Transfus Sci 2015;9(1):55–60. 25722574.

[168] Cabrales P, Caroen S, Oronsky A, Carter C, Trepel J, Summers T, et al. The macrophage stimulating anti-cancer agent, RRx-001, protects against ischemia-reperfusion injury, Expert Rev Hematol 2017;10(6):575–82. 28448172.

[169] Metcalf B, Chuang C, Dufu K, Patel MP, Silva-Garcia A, Johnson C, et al. Discovery of GBT440, an orally bioavailable R-state stabilizer of sickle cell hemoglobin. ACS Med Chem Lett 2017;8(3):321–6.

[170] Mack AK, Kato GJ. Sickle cell disease and nitric oxide: a paradigm shift?, Int J Biochem Cell Biol 2006;38(8):1237–43. 16517208.

[171] Gladwin MT, Kato GJ, Weiner D, Onyekwere OC, Dampier C, Hsu L, et al. Nitric oxide for inhalation in the acute treatment of sickle cell pain crisis: a randomized controlled trial, JAMA 2011;305(9):893–902. 21364138.

[172] Fens MH, Cabrales P, Scicinski J, Larkin SK, Suh JH, Kuypers FA, et al. Targeting tumor hypoxia with the epigenetic anticancer agent, RRx-001: a superagonist of nitric oxide generation, Med Oncol 2016;33(8):85. 27377482.

[173] Fens MHAM, Larkin SK, Fitch WL, Scicinski J, Oronsky B, Kuypers FA. Treatment with a novel dinitroazetidine, abdnaz, improves nitrite reductase activity of sickle red blood cells. Blood 2012;120(21):3246.

[174] Yalcin O, Oronsky B, Carvalho LJ, Kuypers FA, Scicinski J, Cabrales P. From METS to malaria: RRx-001, a multi-faceted anticancer agent with activity in cerebral malaria, Malar J 2015;14:218. 26017006.

[175] Mehta M, Sonawat HM, Sharma S. Glycolysis in Plasmodium falciparum results in modulation of host enzyme activities, J Vector Borne Dis 2006;43(3):95–103. 17024857.

[176] Oronsky B, Scicinski J, Reid T, Oronsky A, Carter C, Oronsky N, et al. RRx-001, a novel clinical-stage chemosensitizer, radiosensitizer, and immunosensitizer, inhibits glucose 6-phosphate dehydrogenase in human tumor cells, Discov Med 2016;21(116):251–65. 27232511.

[177] Esu E, Effa EE, Opie ON, Uwaoma A, Meremikwu MM. Artemether for severe malaria, Cochrane Database Syst Rev 2014;11(9):CD010678, 25209020.

CHAPTER 8

Nitric Oxide Scavenging-Based Therapies for Targeting Colorectal Cancer

Jon Peñarando*, Kevin Myant†, Enrique Aranda*,‡,§, Antonio Rodriguez-Ariza*,‡,§

**Instituto Maimónides de Investigación Biomédica de Córdoba (IMIBIC), Córdoba, Spain*

†The Institute of Genetics and Molecular Medicine (IGMM), University of Edinburgh, Western General Hospital, Edinburgh, Scotland

‡Centro de Investigación Biomédica en Red de Cáncer (CIBERONC), Madrid, Spain

§Unidad de Gestión Clínica de Oncología Médica, Hospital Universitario Reina Sofía, Córdoba, Spain

ABSTRACT

Initially discovered as a powerful regulator of vascular tone, nitric oxide (NO) has also been highlighted as a key factor in different diseases, including cancer. This molecule is aberrantly produced during tumor progression mainly through the enzymatic catabolism of arginine by three different proteins belonging to the nitric oxide synthase (NOS) family. The relevance of each isoenzyme as the main source of NO in carcinogenesis may be critically different depending on the tumor context. Thus, the endothelial NOS (eNOS) protein is the most relevant source of NO in different scenarios where the cancer stem cell (CSC) phenotype is enhanced, such as the mesenchymal poor-prognosis subtype colorectal tumors. Besides, eNOS upregulation has been shown in different *Apc* conditional mouse models where *Apc*-deficient cells maintain a crypt progenitor-like phenotype associated with early colorectal lesions. Since NO signaling is independent of how it was produced, NO scavengers represent the best approach to target this molecule produced from any source. In this regard, the NO scavenger carboxy-2-phenyl-4,4,5,5-tetramethylimidazoline-1-oxyl 3-oxide (c-PTIO) has been shown to improve the efficacy of immunotherapy, to impair the CSC phenotype, and to reduce tumor growth in vivo. Therefore, the control of the aberrant production of NO may represent a key upstream hub for targeting different cancer-related events and different NO-related pathologies.

Therapeutic Application of Nitric Oxide in Cancer and Inflammatory Disorders. https://doi.org/10.1016/B978-0-12-816545-4.00008-6

ABBREVIATIONS

CRC	colorectal cancer
CSC	cancer stem cell
CTLs	cytotoxic T lymphocytes
eNOS	endothelial nitric oxide synthase
iNOS	inducible nitric oxide synthase
L-nil	L-N^6-(1-iminoethyl)lysine
L-NMMA	N^G-methyl-L-arginine
L-NNA	N^G-nitro-L-arginine
nNOS	neuronal nitric oxide synthase
NO	nitric oxide
NOS	nitric oxide synthase
PTIO	2-phenyl-4,4,5,5-tetramethylimidazoline-1-oxyl 3-oxide

Conflict of Interest

No potential conflicts of interest were disclosed.

INTRODUCTION

It has been a long road since nitric oxide (NO) was considered as a mere environmental pollutant derived from industry, automobile exhausts, and electrical storms [1]. Because of its exponential scientific interest, NO was proclaimed "molecule of the year" by the journal *Science* in 1992 [2], and 6 years later, Furchgott, Murad, and Ignarro were awarded with the Physiology and Medicine Nobel Prize for their discoveries concerning NO as a signaling molecule in the cardiovascular system. Since then, the role of NO in the cardiovascular system has been the focus of an intensive research. However, over the last decades, it has emerged an increasing interest about the role of NO in carcinogenesis [3] where it has been highlighted as an important agent, including colon cancer [4]. It has been shown that NO has a dual role in tumoral processes, being cytotoxic or cytostatic at high levels, whereas low levels can have the opposite effect and promote tumor growth [5]. This molecule is endogenously synthesized by three different NO synthase (NOS) isoforms using L-arginine and molecular oxygen as substrates: neuronal NOS (nNOS), endothelial NOS (eNOS), and inducible NOS (iNOS) [6]. The first two isoforms are incorrectly named as constitutive and require an increase of intracellular calcium and calmodulin binding to synthesize NO. The induction of iNOS is mediated by cytokines, and it can produce greater amounts of NO than the other isoforms [7]. There are also NOS-independent NO sources such as dietary nitrates that are firstly reduced to nitrite by bacterial nitrate reductases to eventually produce NO in acidic or hypoxic environments or enzymatically [8]. Because of its biochemical properties, NO is a powerful signaling molecule that regulates a great variety of biological responses [9] and can exert its effects through a cGMP-dependent pathway, acting on cGMP-dependent protein kinases, cGMP-gated ion channels, and phosphodiesterases [10]. Besides, there is a cGMP-independent pathway where NO reacts with molecular oxygen, superoxide thiols, and transition metals to exert its activity. Finally, nitration and *S*-nitrosylation are post-translational modifications by which NO modifies proteins resulting in the regulation of their function and different cell signaling pathways [5, 11].

The biological effect of NO is independent of how it is produced [12], and therefore, it is important to consider all the potential sources when designing any strategy implying this molecule. In this chapter, we will review several NO-based strategies, with special emphasis on those aimed at blocking the action of this molecule with prominent functions in the processes of tumor formation and progression.

TARGETING NITRIC OXIDE PRODUCTION

The manipulation of NO levels has been an encouraging strategy to obtain a better therapeutic gain and to enhance the efficacy of both chemotherapy and radiotherapy [13]. In this regard, the inhibition of the three NOS isoforms has been one of the main approaches addressed (Fig. 1). One feature that makes the NOS family as an optimal therapeutic target is that there are different small molecules that inhibit these enzymes. Historically, the first compounds considered as NOS inhibitors were arginine derivatives and similar analogues [14]. Thus, with advances in computer modeling and once structural data of NOS were available, many inhibitors with the minimal structure to get the optimal binding were designed [14]. N^G-Methyl-L-arginine (L-NMMA), N^G-nitro-L-arginine (L-NNA), and its methyl ester prodrug (L-NAME) have been the most widely used NOS inhibitors [15]. This latter compound is a pan-NOS inhibitor that has been clinically evaluated for the treatment of septic and cardiogenic shock [16, 17] and has also shown an inhibitory effect in several in vivo cancer models [18–21]. Studies support the clinical evaluation of L-NAME for the treatment of KRAS mutation-positive non-small cell lung cancer considering that it was able to decrease tumor growth and provide a survival advantage in a lung cancer mouse model [22]. This inhibitor has been shown effective with other types of therapies such as radiotherapy, where NOS inhibition improved the radiation tumor response through Th1 immune polarization within the tumor microenvironment [23]. This result was previously found by using other inhibitors such as L-NNA, where NOS inhibition increased the antitumor effect of radiation in squamous carcinoma xenografts [24]. Compounds with a higher affinity for a particular isoform have also shown to be a plausible therapeutic option in cancer. Thus, L-N^6-(1-iminoethyl)lysine (L-nil), which is an

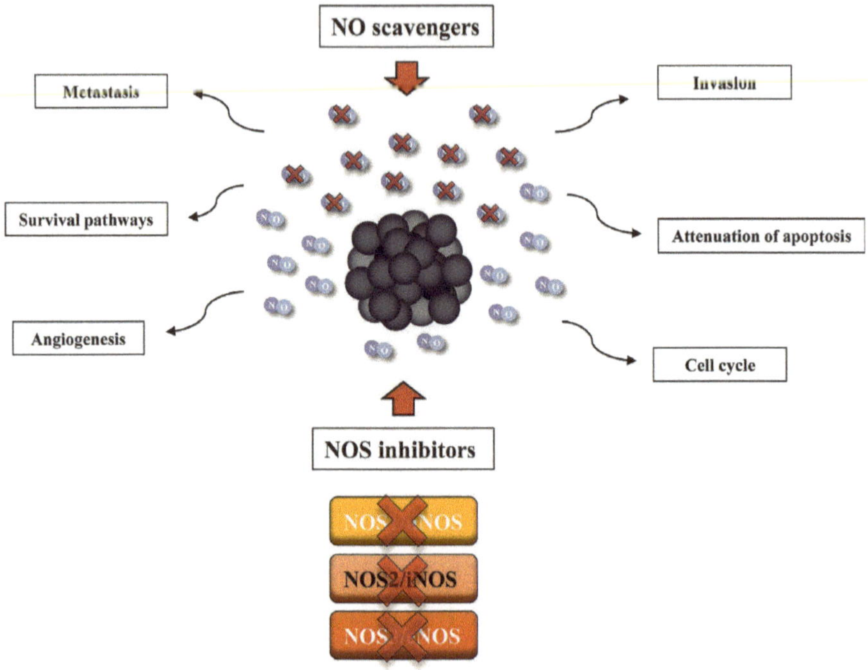

FIG. 1
Different strategies for targeting the aberrant NO production.

iNOS-selective small-molecule inhibitor, showed antitumor effects in a preclinical human melanoma model and synergized with chemotherapeutic agents in vivo without apparent associated toxicity [25]. Other iNOS inhibitors were found effective in reducing tumor growth and enhancing the survival rate in triple-negative aggressive form of breast cancer [26]. The inhibition of the endothelial NOS isoform may also contribute to decrease other important characteristics in tumorigenesis such as the tumor cell adhesion to intact microvessels under physiological flows [27]. All these evidences make NOS inhibitors relevant pharmacological tools to potentially increase the current clinical results. Nevertheless, although selective NOS inhibitors have been shown to improve the unspecific effects of nonselective inhibitors in animal studies [28], there are issues that seem difficult to address, such as the cardioprotective role of iNOS in sepsis [29]. In fact, a phase III clinical trial conducted with the nonselective NOS inhibitor L-NMMA had to be prematurely ended since it resulted in depressed cardiac function and increased mortality [30].

NITRIC OXIDE SCAVENGING-BASED APPROACHES

There is a huge controversy about the role of iNOS in colon cancer [31–35], and targeting this isoenzyme offers a limited therapeutic benefit in contexts where traditionally considered constitutive isoforms are the main or even the only source of NO [36]. In fact, we have recently shown that the mesenchymal molecular subtype of colorectal cancer (CRC) shows an upregulation of eNOS comparing with the rest of the CRC subtypes, whereas iNOS was found absent [37]. Besides, when considering NO as a potential target, it must be taken into account that there are also NOS-independent NO sources that are not blocked by NOS inhibitors, such as dietary nitrates [8]. Actually, it has been recently suggested that nitrate-derived NO in the gut may play an important role in the metabolism of NO in mammals, linking diet and health problems [38]. The serum albumin protein can also be a storage and transport NO increasing its half-life from a few seconds to several hours, having a fundamental role in maintaining the NO pool, particularly in pathological settings [39]. It is important to consider all the potential sources of NO when designing any NO-based therapeutic strategy since the biological effect of this molecule is independent of how it was produced [12]. In this regard, NO scavengers emerge as an alternative approach to target directly this molecule before it can exert its effects, trapping NO from all sources (Fig. 1). Besides, NO-trapping compounds show a second-order reaction kinetic, decreasing NO preferentially in regions of aberrant overproduction and having a light impact on the essential basal NO levels [40]. However, despite having been shown promising as an alternative to NOS inhibitors to remove the excess of NO in different experimental models, NO scavengers have received considerably less attention as a therapeutic option [41]. Compounds such as curcumins and ruthenium have demonstrated pharmacological activity and efficiently targeted NO [42, 43]. The reduction of NO levels with curcumin has been shown to be a potential adjuvant treatment to imatinib in patients with chronic myeloid leukemia [44]. Since NO constitutes a ligand for transition metals, different ruthenium complexes have been shown to scavenge this molecule in several biological systems [45]. Scavenging NO with Ru(III) polycarboxylates resulted in a reduced capacity of tumor growth in a rat carcinosarcoma model, whereas iNOS-specific inhibitors had no effect on tumor development, highlighting the need of a full blockade in NO production [40]. Similarly, since the metalloprotein hemoglobin contains the iron transition metal, NO can react with both the Fe(II) and Fe(III) heme, and this protein can act both as an NO scavenger and an NO donor [46]. Cobalamins have been traditionally recommended as an effective antidote against cyanide poisoning [47]; however, it has been also reported their ability to bind NO and their application in preventing abnormal vascular responses in vivo [48]. Thus, most of NO scavengers have been shown to be effective targeting this molecule in NO-mediated diseases both in vitro and in vivo. However, there are a limited number of studies about the potential therapeutic benefit of NO scavenging-based therapies in cancer. Nitronyl nitroxide compounds, such as 2-phenyl-4,4,5,5-tetramethylimidazoline-1-oxyl 3-oxide (PTIO), and its derivatives, carboxy-PTIO (c-PTIO) and carboxymethoxy-PTIO, scavenge NO in a radical-radical reaction [46]. The NO-suppressing activity of PTIO resulted in a decreased enhanced vascular permeability in the murine sarcoma S-180 [49, 50]. The NO scavenger c-PTIO has been tested in several in vivo models such as dogs, rats, and mice for different pathologies [51–53]. In a zebra fish rat glioma xenograft model, c-PTIO treatment resulted in a decreased tumor vascularization and tumor growth, downregulating the expression of cyclin D1. In this study, authors supported again the need of total NO blockade to efficiently target this molecule [54]. NO scavenging-based therapies also have shown their potential benefit in the encouraging field of immunotherapy. Hirano and colleagues showed that targeting NO with c-PTIO restores the function and proliferative capacity of cytotoxic T lymphocytes (CTLs) from the tumor, improving the efficacy of immunotherapy in a melanoma murine model [55]. Although the NOS inhibitor L-NMMA restored CTL proliferation in vitro, it failed in increasing the antitumor activity of this cell population in vivo, suggesting that c-PTIO might be a more suitable option for targeting NO from all available sources [55].

We have recently shown that NO constitutes a key regulator in the maintenance of the cancer stem cell (CSC) phenotype in CRC [37]. Scavenging NO with c-PTIO resulted in an impairment of important molecular signaling pathways in CSC biology both in vitro and in vivo and reduced the tumor growth in a mesenchymal cell line-derived xenograft model [37]. To explore the role of NO in the intestinal stem cell population, we developed $VilCre^{ERT2}$ $Apc^{fl/fl}$ mouse organoids that maintain a "crypt progenitor-like" phenotype associated with the expansion of a stem Lgr5-positive cell population [56]. When intestinal epithelial cells from wt or $Apc^{fl/fl}$ mice were cultured using the organoid-forming assay, wt organoids grew forming crypt-like structures, whereas $Apc^{fl/fl}$ organoids completely lost their epithelial intestinal structure (Fig. 2A). Notably, analysis of NOS isoform expression showed that eNOS was highly expressed in *Apc*-deficient organoids compared with their normal counterparts, whereas iNOS was downregulated (Fig. 2B). This result supports that the eNOS isoform could have an important role during the expansion of the Lgr5-positive stem cell subpopulation after *Apc* loss. Importantly, scavenging of NO with c-PTIO caused a significant decrease of $Apc^{fl/fl}$ organoid proliferation, whereas wt organoids grew normally (Fig. 2C and D). Besides, NO depletion also caused a marked decrease in the $Lgr5^+$ cell population both in wt and $Apc^{fl/fl}$ organoids (Fig. 2E). Furthermore, other stem cell markers, including Troy, Vav3, and Slc14a1, were downregulated after NO scavenging (Fig. 2E). Thus, our results indicated an important regulatory role of NO in intestinal CSC biology. Furthermore, NO depletion abolished the capacity of human CRC cells to form tumorspheres and organoids in vitro. Importantly, the effect was molecule-specific since the use of NO donors allowed tumor cells to recover their initial phenotype and abrogated the anti-CSC effect of c-PTIO.

Therefore, to further explore the antitumor capacity of NO depletion, two different xenograft experiments were performed using the mesenchymal HCT116

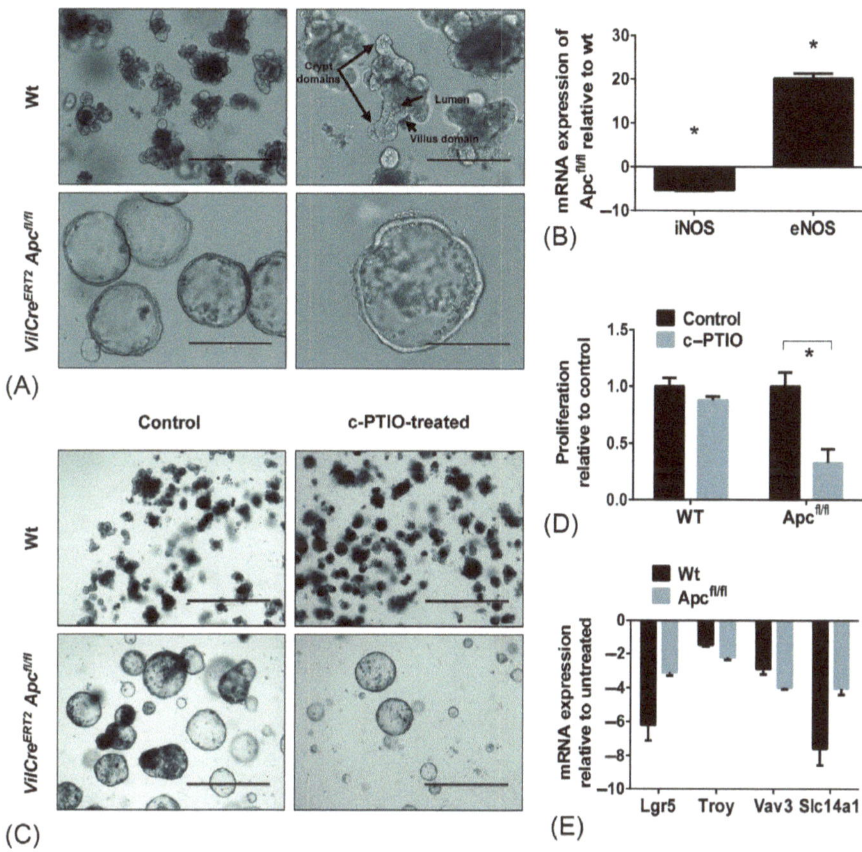

FIG. 2

eNOS is upregulated in organoids derived from the *Apc$^{fl/fl}$* mouse intestine, and NO trapping decreases their proliferation and downregulates the expression of stem cell markers. (A) Representative images from typical wt and *Apc$^{fl/fl}$* organoids. Scale bars: 400 μm (*left*) and 200 μm (*right*). (B) qRT-PCR analysis of iNOS and eNOS expression from wt and *Apc$^{fl/fl}$* organoids. $*P < 0.05$ compared with the wt organoids. (C and D) Proliferation assay of control or 48 h c-PTIO (500 μM)-treated wt and *Apc$^{fl/fl}$* organoids. Proliferation was determined by image analysis with ImageJ software. $*P < 0.05$ compared with the control. Scale bars: 1000 μm. (E) qRT-PCR analysis of Lgr5, Troy, Vav3, and Slc14a1 expression from control or c-PTIO (500 μM)-treated wt and *Apc$^{fl/fl}$* organoids.

CRC cancer cell line [57]. Thus, control and c-PTIO-treated HCT116 cells were subcutaneously injected into immunocompromised mice, and then, tumor progression was evaluated. As shown in Fig. 3A, NO-depleted cells generated tumors half the size of those generated from untreated control cells. The decreased capacity of pretreated cancer cells to form larger tumors suggests a long-term effect of scavenging NO and an important role of this molecule in the regulation of CSCs. On the other hand, to elucidate the antitumor effectiveness of c-PTIO in vivo, HCT116 tumor cells were subcutaneously injected into

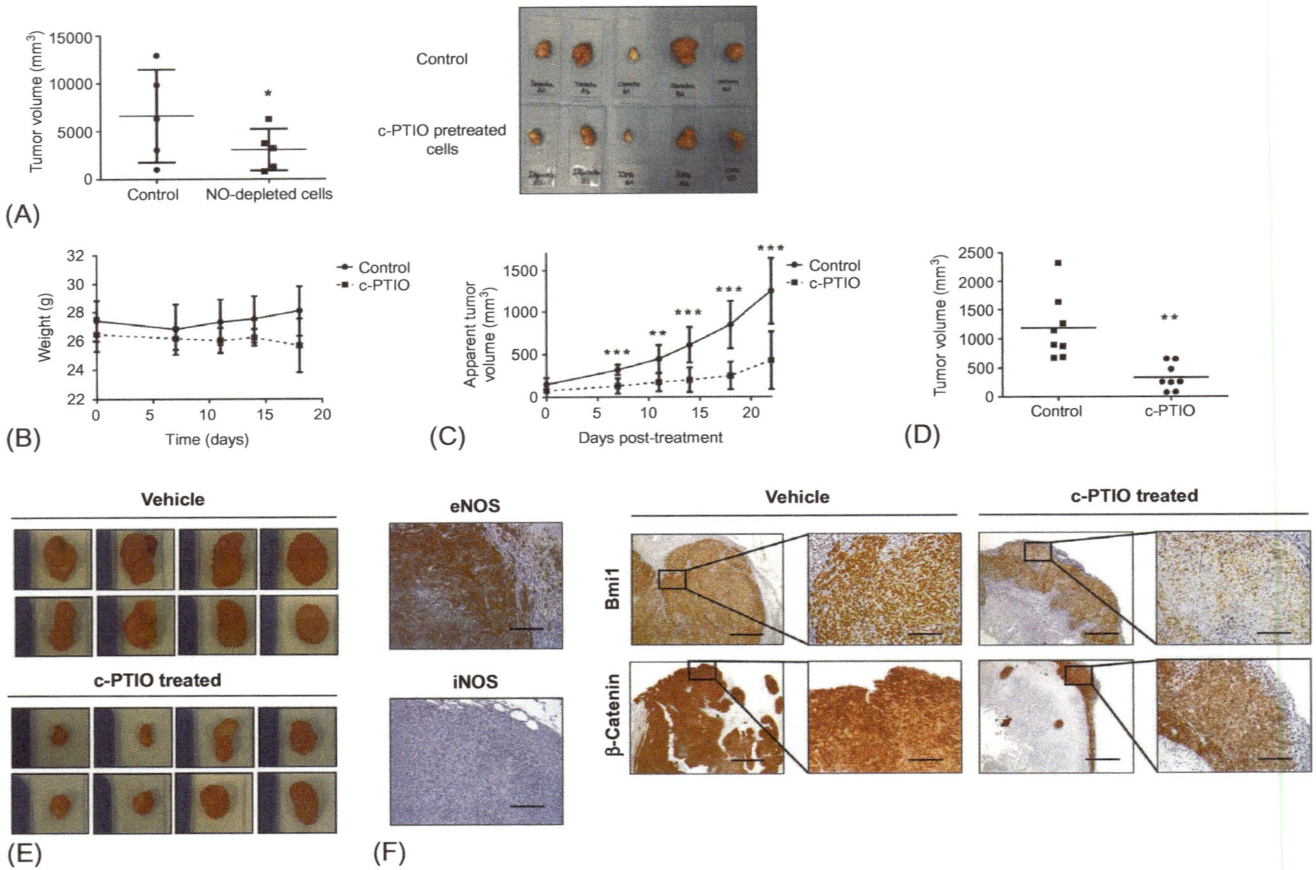

FIG. 3

NO trapping with c-PTIO reduces the capacity of mesenchymal HCT116 cells to form tumors in xenografted mice and decreases the expression in vivo of β-catenin and Bmi1. (A) Control or 24 h c-PTIO (100 μM)-pretreated HCT116 cells were subcutaneously injected into the flanks of immunocompromised mice. Final tumor volumes were analyzed after 37 days of inoculation. On the other hand, mice xenografted with HCT116 cells were treated with vehicle or c-PTIO (320 mg/kg) once a week. (B) Mouse weight was monitored over time. (C) Tumor size was measured during the course of treatment. (D and E) Final tumor volume was analyzed after harvested. (F) Representative images from immunohistochemistry of eNOS and iNOS from HCT116-derived tumors and β-catenin and Bmi1 from vehicle or c-PTIO-treated mouse tumors. $*P<0.05$, $**P<0.01$, and $***P<0.001$ compared with control.

immunocompromised mice, and animals were then treated with 320 mg/kg of c-PTIO three times a week. The scavenger had no toxic effect or caused significant weight loss during the experiment (Fig. 3B). Besides, in vivo c-PTIO treatment impaired tumor growth (Fig. 3C), and final tumor volume was significantly decreased compared with that of untreated control mice, even using a threefold lower dose (Fig. 3D and E). Interestingly, tumor xenografts did not show iNOS expression, whereas eNOS was found highly expressed in both epithelial and endothelial cells (Fig. 3F). Thus, the effect of c-PTIO was the result of eNOS-produced NO scavenging, the main source of this molecule in the mesenchymal tumor. Furthermore, the expression of β-catenin and Bmi1, which is a relevant target for cancer therapeutics [58], was found decreased in tumors after c-PTIO treatment (Fig. 3F). Again, these results further support a key role of eNOS-derived NO in CSC maintenance and show NO scavenging as an attractive alternative to target this tumor subpopulation.

CONCLUSIONS

The huge pleiotropic effect of NO in carcinogenesis and the fact that NO scavengers have been widely and safely used in different in vivo models support NO scavenging-based therapies as an attractive approach to target the tumor biology and improve current clinical outcomes. We have shown that the NO scavenging with c-PTIO impairs stem-related signaling pathways essential for colorectal CSCs in vitro and to have an antiproliferative effect in vivo. The development of strategies that allow to increase the effective dose of the scavenger in the tumor microenvironment might result in encouraging antitumor effects that could be soon successfully translated to the bedside.

Acknowledgments

This work was supported by the Plan Estatal de I + D + I, ISCIII-Subdirección General de Evaluación y Fomento de la Investigación, Spanish Government, and European Regional Development Fund (ERDF) (grant PI16-01508).

References

[1] Carpenter AW, Schoenfisch MH. Nitric oxide release: part II. Therapeutic applications. Chem Soc Rev 2012;41(10):3742–52.

[2] Koshland Jr. DE. The molecule of the year. Science 1992;258(5090):1861.

[3] Huerta S. Nitric oxide for cancer therapy. Future Sci OA 2015;1(1):FSO44.

[4] Chowdhury R, Godoy LC, Thiantanawat A, Trudel LJ, Deen WM, Wogan GN. Nitric oxide produced endogenously is responsible for hypoxia-induced HIF-1alpha stabilization in colon carcinoma cells. Chem Res Toxicol 2012;25(10):2194–202.

[5] Aranda E, Lopez-Pedrera C, De La Haba-Rodriguez JR, Rodriguez-Ariza A. Nitric oxide and cancer: the emerging role of S-nitrosylation. Curr Mol Med 2012;12(1):50–67.

[6] Forstermann U, Sessa WC. Nitric oxide synthases: regulation and function. Eur Heart J 2012;33 (7):829–37. 37a–37d.

[7] Piazza M, Guillemette JG, Dieckmann T. Dynamics of nitric oxide synthase-calmodulin interactions at physiological calcium concentrations. Biochemistry 2015;54(11):1989–2000.

[8] Webb AJ, Patel N, Loukogeorgakis S, Okorie M, Aboud Z, Misra S, Rashid R, Miall P, Deanfield J, Benjamin N, MacAllister R, Hobbs AJ, Ahluwalia A. Acute blood pressure lowering, vasoprotective, and antiplatelet properties of dietary nitrate via bioconversion to nitrite. Hypertension 2008;51(3):784–90.

[9] Vasudevan D, Bovee RC, Thomas DD. Nitric oxide, the new architect of epigenetic landscapes. Nitric Oxide Biol Chem 2016;59:54–62.

[10] Francis SH, Busch JL, Corbin JD, Sibley D. cGMP-dependent protein kinases and cGMP phosphodiesterases in nitric oxide and cGMP action. Pharmacol Rev 2010;62(3):525–63.

[11] Huerta S, Chilka S, Bonavida B. Nitric oxide donors: novel cancer therapeutics (review). Int J Oncol 2008;33(5):909–27.

[12] Socco S, Bovee RC, Palczewski MB, Hickok JR, Thomas DD. Epigenetics: the third pillar of nitric oxide signaling. Pharmacol Res 2017;121:52–8.

[13] Singh S, Gupta AK. Nitric oxide: role in tumour biology and iNOS/NO-based anticancer therapies. Cancer Chemother Pharmacol 2011;67(6):1211–24.

[14] Vitecek J, Lojek A, Valacchi G, Kubala L. Arginine-based inhibitors of nitric oxide synthase: therapeutic potential and challenges. Mediators Inflamm 2012;2012:318087.

[15] Alderton WK, Cooper CE, Knowles RG. Nitric oxide synthases: structure, function and inhibition. Biochem J 2001;357:593–615. Pt 3.

[16] Cotter G, Kaluski E, Milo O, Blatt A, Salah A, Hendler A, Krakover R, Golick A, Vered Z. LINCS: L-NAME (a NO synthase inhibitor) in the treatment of refractory cardiogenic shock: a prospective randomized study. Eur Heart J 2003;24(14):1287–95.

[17] Avontuur JA, Tutein Nolthenius RP, van Bodegom JW, Bruining HA. Prolonged inhibition of nitric oxide synthesis in severe septic shock: a clinical study. Crit Care Med 1998;26(4):660–7.

[18] Edwards P, Cendan JC, Topping DB, Moldawer LL, MacKay S, Copeland E, Lind DS. Tumor cell nitric oxide inhibits cell growth in vitro, but stimulates tumorigenesis and experimental lung metastasis in vivo. J Surg Res 1996;63(1):49–52.

[19] Lampson BL, Kendall SD, Ancrile BB, Morrison MM, Shealy MJ, Barrientos KS, Crowe MS, Kashatus DF, White RR, Gurley SB, Cardona DM, Counter CM. Targeting eNOS in pancreatic cancer. Cancer Res 2012;72(17):4472–82.

[20] Gallo O, Masini E, Morbidelli L, Franchi A, Fini-Storchi I, Vergari WA, Ziche M. Role of nitric oxide in angiogenesis and tumor progression in head and neck cancer. J Natl Cancer Inst 1998;90(8):587–96.

[21] Weiss JM, Ridnour LA, Back T, Hussain SP, He P, Maciag AE, Keefer LK, Murphy WJ, Harris CC, Wink DA, Wiltrout RH. Macrophage-dependent nitric oxide expression regulates tumor cell detachment and metastasis after IL-2/anti-CD40 immunotherapy. J Exp Med 2010;207 (11):2455–67.

[22] Pershing NL, Yang CJ, Xu M, Counter CM. Treatment with the nitric oxide synthase inhibitor L-NAME provides a survival advantage in a mouse model of Kras mutation-positive, non-small cell lung cancer. Oncotarget 2016;27:42835–92.

[23] Ridnour LA, Cheng RY, Weiss JM, Kaur S, Soto-Pantoja DR, Basudhar D, Heinecke JL, Stewart CA, DeGraff W, Sowers AL, Thetford A, Kesarwala AH, Roberts DD, Young HA, Mitchell JB, Trinchieri G, Wiltrout RH, Wink DA. NOS inhibition modulates immune polarization and improves radiation-induced tumor growth delay. Cancer Res 2015;75(14):2788–99.

[24] Cardnell RJ, Mikkelsen RB. Nitric oxide synthase inhibition enhances the antitumor effect of radiation in the treatment of squamous carcinoma xenografts. PLoS One 2011; 6(5)e20147.

[25] Sikora AG, Gelbard A, Davies MA, Sano D, Ekmekcioglu S, Kwon J, Hailemichael Y, Jayaraman P, Myers JN, Grimm EA, Overwijk WW. Targeted inhibition of inducible nitric oxide synthase inhibits growth of human melanoma in vivo and synergizes with chemotherapy. Clin Cancer Res 2010;16(6):1834–44.

[26] Granados-Principal S, Liu Y, Guevara ML, Blanco E, Choi DS, Qian W, Patel T, Rodriguez AA, Cusimano J, Weiss HL, Zhao H, Landis MD, Dave B, Gross SS, Chang JC. Inhibition of iNOS as a novel effective targeted therapy against triple-negative breast cancer. Breast Cancer Res 2015;17:25.

[27] Zhang L, Zeng M, Fu BM. Inhibition of endothelial nitric oxide synthase decreases breast cancer cell MDA-MB-231 adhesion to intact microvessels under physiological flows. Am J Physiol Heart Circ Physiol 2016;310(11):H1735–47.

[28] Ichinose F, Hataishi R, Wu JC, Kawai N, Rodrigues AC, Mallari C, Post JM, Parkinson JF, Picard MH, Bloch KD, Zapol WM. A selective inducible NOS dimerization inhibitor prevents systemic, cardiac, and pulmonary hemodynamic dysfunction in endotoxemic mice. Am J Physiol Heart Circ Physiol 2003;285(6):H2524–30.

[29] Price S, Mitchell JA, Anning PB, Evans TW. Type II nitric oxide synthase activity is cardioprotective in experimental sepsis. Eur J Pharmacol 2003;472(1–2):111–8.

[30] Lopez A, Lorente JA, Steingrub J, Bakker J, McLuckie A, Willatts S, Brockway M, Anzueto A, Holzapfel L, Breen D, Silverman MS, Takala J, Donaldson J, Arneson C, Grove G, Grossman S, Grover R. Multiple-center, randomized, placebo-controlled, double-blind study of the nitric oxide synthase inhibitor 546C88: effect on survival in patients with septic shock. Crit Care Med 2004;32(1):21–30.

[31] Ambs S, Merriam WG, Bennett WP, Felley-Bosco E, Ogunfusika MO, Oser SM, Klein S, Shields PG, Billiar TR, Harris CC. Frequent nitric oxide synthase-2 expression in human colon adenomas: implication for tumor angiogenesis and colon cancer progression. Cancer Res 1998;58(2):334–41.

[32] Moochhala S, Chhatwal VJ, Chan ST, Ngoi SS, Chia YW, Rauff A. Nitric oxide synthase activity and expression in human colorectal cancer. Carcinogenesis 1996;17(5):1171–4.

[33] Chhatwal VJ, Ngoi SS, Chan ST, Chia YW, Moochhala SM. Aberrant expression of nitric oxide synthase in human polyps, neoplastic colonic mucosa and surrounding peritumoral normal mucosa. Carcinogenesis 1994;15(10):2081–5.

[34] Ahn B, Ohshima H. Suppression of intestinal polyposis in Apc(Min/+) mice by inhibiting nitric oxide production. Cancer Res 2001;61(23):8357–60.

[35] Scott DJ, Hull MA, Cartwright EJ, Lam WK, Tisbury A, Poulsom R, Markham AF, Bonifer C, Coletta PL. Lack of inducible nitric oxide synthase promotes intestinal tumorigenesis in the Apc(Min/+) mouse. Gastroenterology 2001;121(4):889–99.

[36] Jadeski LC, Hum KO, Chakraborty C, Lala PK. Nitric oxide promotes murine mammary tumour growth and metastasis by stimulating tumour cell migration, invasiveness and angiogenesis. Int J Cancer 2000;86(1):30–9.

[37] Penarando J, Lopez-Sanchez LM, Mena R, Guil-Luna S, Conde F, Hernandez V, Toledano M, Gudino V, Raponi M, Billard C, Villar C, Diaz C, Gomez-Barbadillo J, De la Haba-Rodriguez J, Myant K, Aranda E, Rodriguez-Ariza A. A role for endothelial nitric oxide synthase in intestinal stem cell proliferation and mesenchymal colorectal cancer. BMC Biol 2018;16(1):3.

[38] Tiso M, Schechter AN. Correction: nitrate reduction to nitrite, nitric oxide and ammonia by gut bacteria under physiological conditions. PLoS One 2015;10(5)e0127490.

[39] Luo M, Boudier A, Pallotta A, Maincent P, Vincourt JB, Leroy P. Albumin as a carrier for NO delivery: preparation, physicochemical characterization, and interaction with gold nanoparticles. Drug Dev Ind Pharm 2016;42(12):1928–37.

[40] Flitney FW, Pritchard RJ, Kennovin GD, Bisland SK, Hirst DG, Fricker SP. Antitumor actions of ruthenium(III)-based nitric oxide scavengers and nitric oxide synthase inhibitors. Mol Cancer Ther 2011;10(9):1571–80.

[41] Harbrecht BG. Therapeutic use of nitric oxide scavengers in shock and sepsis. Curr Pharm Des 2006;12(27):3543–9.

[42] Sreejayan RMN. Nitric oxide scavenging by curcuminoids. J Pharm Pharmacol 1997;49(1):105–7.

[43] Marmion CJ, Cameron B, Mulcahy C, Fricker SP. Ruthenium as an effective nitric oxide scavenger. Curr Top Med Chem 2004;4(15):1585–603.

[44] Ghalaut VS, Sangwan L, Dahiya K, Ghalaut PS, Dhankhar R, Saharan R. Effect of imatinib therapy with and without turmeric powder on nitric oxide levels in chronic myeloid leukemia. J Oncol Pharm Pract 2012;18(2):186–90.

[45] Fricker SP, Slade E, Powell NA, Vaughan OJ, Henderson GR, Murrer BA, Megson IL, Bisland SK, Flitney FW. Ruthenium complexes as nitric oxide scavengers: a potential therapeutic approach to nitric oxide-mediated diseases. Br J Pharmacol 1997;122(7):1441–9.

[46] Fricker SP. Nitric oxide scavengers as a therapeutic approach to nitric oxide mediated disease. Expert Opin Investig Drugs 1999;8(8):1209–22.

[47] Mushett CW, Kelley KL, Boxer GE, Richards JC. Antidotal efficacy of vitamin B12a (hydroxocobalamin) in experimental cyanide poisoning. Proc Soc Exp Biol Med Soc 1952;81(1):234–7.

[48] Greenberg SS, Xie J, Zatarain JM, Kapusta DR, Miller MJ. Hydroxocobalamin (vitamin B12a) prevents and reverses endotoxin-induced hypotension and mortality in rodents: role of nitric oxide. J Pharmacol Exp Ther 1995;273(1):257–65.

[49] Maeda H, Noguchi Y, Sato K, Akaike T. Enhanced vascular permeability in solid tumor is mediated by nitric oxide and inhibited by both new nitric oxide scavenger and nitric oxide synthase inhibitor. Jap J Cancer Res 1994;85(4):331–4.

[50] Wu J, Akaike T, Maeda H. Modulation of enhanced vascular permeability in tumors by a bradykinin antagonist, a cyclooxygenase inhibitor and a nitric oxide scavenger. Cancer Res 1998;58(1):159–65.

[51] Mitaka C, Hirata Y, Yokoyama K, Nagura T, Tsunoda Y, Amaha K. Beneficial effect of carboxy-PTIO on hemodynamic and blood gas changes in septic shock dogs. Crit Care 1997;1(1):45–50.

[52] Yoshida M, Akaike T, Wada Y, Sato K, Ikeda K, Ueda S, Maeda H. Therapeutic effects of imidazolineoxyl N-oxide against endotoxin shock through its direct nitric oxide-scavenging activity. Biochem Biophys Res Commun 1994;202(2):923–30.

[53] Abu-Amara M, Yang SY, Quaglia A, Rowley P, de Mel A, Tapuria N, Seifalian A, Davidson B, Fuller B. Nitric oxide is an essential mediator of the protective effects of remote ischaemic preconditioning in a mouse model of liver ischaemia/reperfusion injury. Clin Sci 2011;121(6):257–66.

[54] Yousfi N, Pruvot B, Lopez T, Magadoux L, Franche N, Pichon L, Salvadori F, Solary E, Garrido C, Laurens V, Chluba J. The impact of tumor nitric oxide production on VEGFA expression and tumor growth in a zebrafish rat glioma xenograft model. PLoS One 2015;10(3)e0120435.

[55] Hirano K, Hosoi A, Matsushita H, Iino T, Ueha S, Matsushima K, Seto Y, Kakimi K. The nitric oxide radical scavenger carboxy-PTIO reduces the immunosuppressive activity of myeloid-derived suppressor cells and potentiates the antitumor activity of adoptive cytotoxic T lymphocyte immunotherapy. Oncoimmunology 2015;4(8)e1019195.

[56] Barker N, van Es JH, Kuipers J, Kujala P, van den Born M, Cozijnsen M, Haegebarth A, Korving J, Begthel H, Peters PJ, Clevers H. Identification of stem cells in small intestine and colon by marker gene Lgr5. Nature 2007;449(7165):1003–7.

[57] Sadanandam A, Lyssiotis CA, Homicsko K, Collisson EA, Gibb WJ, Wullschleger S, Ostos LC, Lannon WA, Grotzinger C, Del Rio M, Lhermitte B, Olshen AB, Wiedenmann B, Cantley LC, Gray JW, Hanahan D. A colorectal cancer classification system that associates cellular phenotype and responses to therapy. Nat Med 2013;19(5):619–25.

[58] Yu T, Chen X, Zhang W, Colon D, Shi J, Napier D, Rychahou P, Lu W, Lee EY, Weiss HL, Evers BM, Liu C. Regulation of the potential marker for intestinal cells, Bmi1, by beta-catenin and the zinc finger protein KLF4: implications for colon cancer. J Biol Chem 2012;287(6):3760–8.

CHAPTER 9

Therapeutic Aspects of Protein Denitrosylation

Salvatore Rizza*, Paola Giglio†, Fiorella Faienza†, Giuseppe Filomeni*,†

*Redox Signaling and Oxidative Stress Research Group, Cell Stress and Survival Unit, Center for Autophagy, Recycling and Disease (CARD), Danish Cancer Society Research Center, Copenhagen, Denmark

†Department of Biology, University of Rome Tor Vergata, Rome, Italy

ABSTRACT

Nitric oxide is a fundamental second messenger regulating cell physiology through a variety of different signaling mechanisms. The majority of NO bioactivity is exerted via binding to cysteine residues of proteins, a reaction called *S*-nitrosylation that affects protein function and stability. Among NO-mediated signaling, *S*-nitrosylation is emerging as a central mechanism in cellular homeostasis by controlling processes that are often deregulated in pathological conditions. For these reasons, it is a constantly growing interest in targeting *S*-nitrosylation for therapeutic purposes.

To overcome the difficulties in manipulating *S*-nitrosylation without interfering with other NO-mediated signaling pathways, researchers moved their attention to the molecular systems assigned to remove NO from proteins (denitrosylases). Here, we review the most recent therapeutic approaches targeting protein denitrosylation, focusing on the two main systems: GSNOR and thioredoxin.

ABBREVIATIONS

AGT	O^6-alkylguanine-DNA alkyltransferase
CBR1	carbonyl reductase 1
CF	cystic fibrosis
CFTR	cystic fibrosis transmembrane conductance regulator
CoA	coenzyme A
GSH	reduced glutathione
GSNO	*S*-nitrosoglutathione

Therapeutic Application of Nitric Oxide in Cancer and Inflammatory Disorders. https://doi.org/10.1016/B978-0-12-816545-4.00009-8

GSNOR	*S*-nitrosoglutathione reductase
GSSG	glutathione disulfide
HCC	hepatocellular carcinoma
HIF-1α	ηypoxia-inducible factor-1α
NF-κB	nuclear factor kappa-light-chain-enhancer of activated B cells
NO	nitric oxide
NOS	nitric oxide synthase
Nrf2	nuclear factor erythroid 2-related factor 2
PSNO	*S*-nitrosylated protein
PTM	posttranslational modification
SDH	succinate dehydrogenase
SNO	*S*-nitrosothiol
SNO-CoA	*S*-nitroso-coenzyme A
TR	thioredoxin reductase
TRAP1	TNF receptor-associated protein 1
Trx	thioredoxin
VEGF	vascular endothelial growth factor

Conflict of Interest

No potential conflicts of interest were disclosed.

INTRODUCTION

NO-Mediated Signaling

Nitric oxide (NO) is a gaseous free radical produced by nitric oxide synthases (NOSs) in a NADPH-dependent reaction combining arginine and molecular oxygen [1]. Nitric oxide acts as a fast and highly diffusible signaling molecule that can target cellular biomolecules directly or by the activity of its metabolites generated upon reaction with transition metals, oxygen, or superoxide. The main effects of an NO burst are time-, organ-, and concentration-dependent. NO is a well-known activator of the soluble guanylate cyclase (sGC) [2], shaping, in such a way, the production of cyclic GMP (cGMP) with the subsequent modulation of calcium channels and protein phosphorylation (Fig. 1A). Fluctuations of NO levels are proficiently sensed by mitochondria that adapt, in such a way, the metabolic rate of ATP production. Nearly all the complexes of the respiratory chain can be modulated by direct NO attachment to *Fe—S* centers or by the covalent modification of specific residues, e.g., tyrosines and cysteines [3–5] (Fig. 1B). In particular, upon reaction with superoxide, NO can generate peroxynitrite ($ONOO^-$), which is an extremely reactive

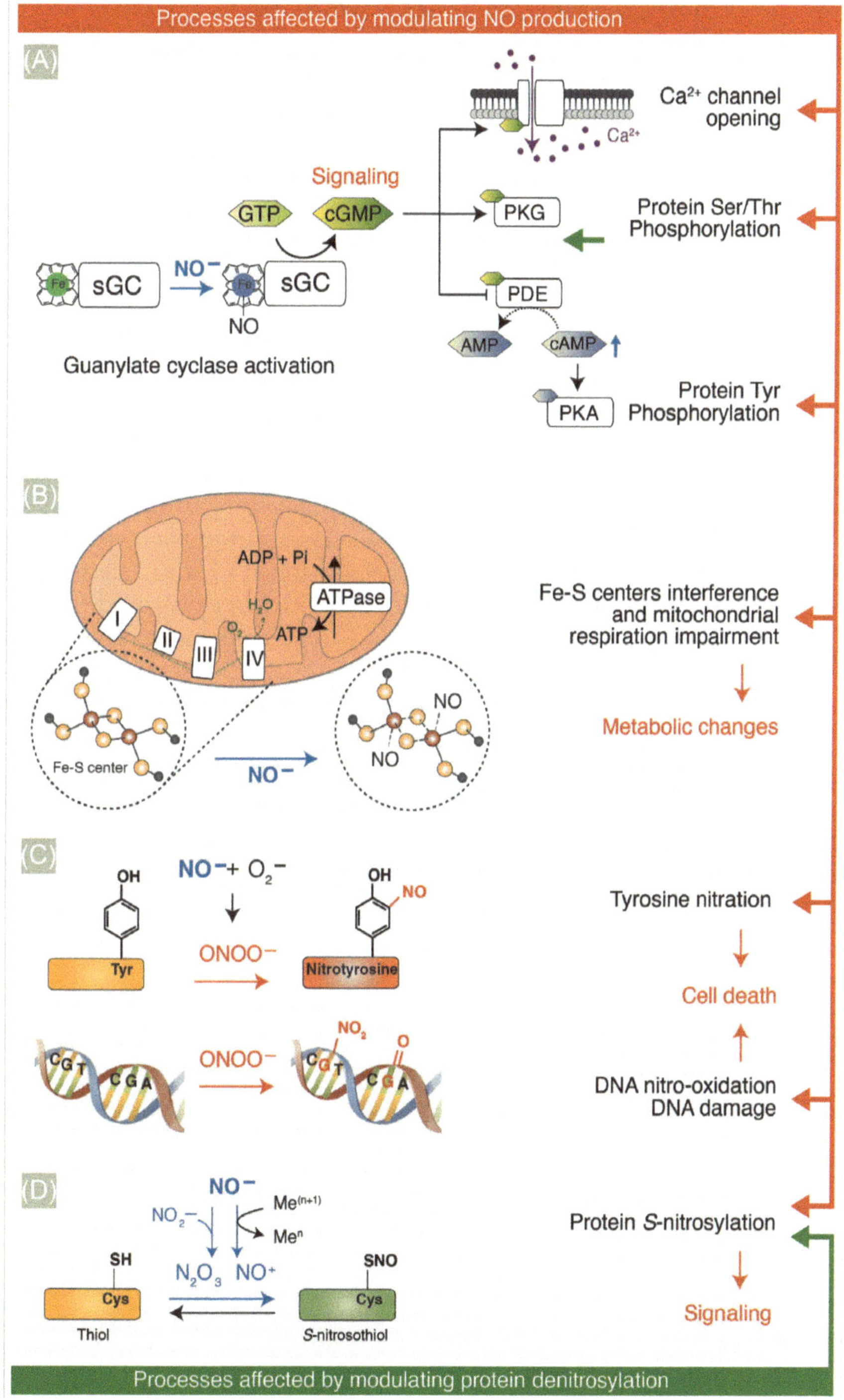

FIG. 1 Targets and processes modulated by nitric oxide.

nitrogen species responsible for protein tyrosine nitration and DNA nitrooxidation (Fig. 1C). This irreversible reaction has been associated with irreversible protein and DNA damage. It leads to cell death, and the mechanism is widely used by immune cells as a lethal weapon against pathogens and cancer cells. Nitric oxide can also easily react with nitrogen dioxide (NO_2) or transition metals (e.g., Fe^{3+} and Cu^{2+}), giving rise to dinitrogen trioxide (N_2O_3) or nitrosonium ion (NO^+), respectively. In the case that both of these species come into contact with cysteine residues of proteins and small peptides, an *S*-nitrosothiol (SNO) is generated (Fig. 1D) by a reaction termed *S*-nitrosation, more commonly known as *S*-nitrosylation. Protein *S*-nitrosylation is emerging as the main redox-based thiol-dependent posttranslational modification (PTM) and accounts for the majority of NO bioactivity [6, 7]. This covalent reaction retains all the features of a classical PTM: It is regulated by enzymatic activities (i.e., induced by nitrosylases and shut down by denitrosylases) [8, 9]; it depends on the presence of consensus motifs surrounding the reactive cysteine [8]; it takes place in specific subcellular compartments [10]. *S*-Nitrosylation modulates protein-protein interactions, enzymatic activities, turnover, and localization, and it has been demonstrated to cross talk with all other PTMs [6, 7, 11–13]. *S*-Nitrosylation of specific proteins could be also achieved by means of *trans*nitrosylation, which represents the reaction of exchange of an NO moiety between proteins and (usually) small nonprotein thiols [7, 14]. The same NOSs exist in complexes with other accessory proteins, whose main role is to mediate *trans*nitrosylation of a given target [8, 9]. To date, about 3000 proteins have been identified as targets of *S*-nitrosylation, and due to its role of protein activity modulator, *S*-nitrosylation might represent a new therapeutic target for the treatment of several pathological states.

FIG. 1—CONT'D (A) NO activates the soluble guanylate cyclase (sGC) by binding to its heme group and stimulates the production of cyclic GMP (cGMP). cGMP accumulation (i) induces the opening of calcium channels; (ii) activates protein kinase G (PKG) and the downstream phosphorylation cascade; and (iii) stimulates phosphodiesterase (PDE) to produce cAMP, a second messenger that, in turn, activates protein kinase A (PKA). (B) NO modulates mitochondrial physiology and metabolism by direct attachment to *Fe—S* centers. (C) Upon reaction with superoxide (O_2^-), NO generates peroxynitrite ($ONOO^-$), which induces protein tyrosine nitration and DNA nitrooxidation and leads to cell death. (D) The reaction between NO and nitrogen dioxide (NO_2) or redox metals (e.g., Fe^{3+} and Cu^{2+}) gives rise to dinitrogen trioxide (N_2O_3) or nitrosonium ion (NO^+), respectively. Both these species react with cysteine residues (—SH) of proteins and small peptides to generate *S*-nitrosothiols (—SNO). The reaction is termed *S*-nitrosylation, and it acts as a posttranslational modification affecting protein function, stability, localization, and signaling. The pharmacological modulation of NO levels affects all the aforementioned processes (*red arrows*), whereas protein *S*-nitrosylation is selectively affected by targeting denitrosylating systems (*green arrow*).

Protein Denitrosylation as a Therapeutic Target

In the last decades, the list of processes affected by protein *S*-nitrosylation is exponentially increasing and includes cell proliferation [15], cell death [16–19], gene transcription [20], hypoxia response [21], mitochondrial homeostasis and metabolism [19, 22–25], autophagy/mitophagy [22, 26–28], and differentiation [15, 29, 30]. As many of the aforementioned processes are deregulated in numerous human diseases, developing new tools to selectively manipulate *S*-nitrosylation without interfering with other NO-mediated signaling pathways is attracting the interests of many researchers. Actually, the great therapeutic potential of the modulation of NO signaling through the pharmacological inhibition of NOSs or, by contrast, the use of NO donors has been already tested in several pathological settings, such as cancer, neurodegeneration and cardiovascular and inflammatory diseases. However, the response to compounds that affect the levels of free NO is often complex and unpredictable. Indeed, due to the plethora of effects that NO exerts on human physiology and, mostly, on the different molecular mechanisms in which it is involved, it is virtually impossible to control protein *S*-nitrosylation without affecting sGC activity or mitochondrial respiration (Fig. 1). Conversely, the increasing knowledge on denitrosylating systems—enzymatic reactions able to metabolize NO and restore the reduced form of cysteine—suggests a new perspective for therapeutic approaches aiming at selectively modulating protein *S*-nitrosylation.

THERAPEUTIC ASPECTS OF PROTEIN DENITROSYLATION

GSNOR

GSNOR Functions and Activity

The spontaneous reaction between NO and glutathione (GSH) gives rise to *S*-nitrosoglutathione (GSNO). GSNO represents a bioavailable pool of NO because of (i) the relatively high content of GSH in the cell; (ii) its long half-life; and (iii) its marked propensity to *trans*nitrosylation [1, 7], the last feature making it the main determinant of the levels of *S*-nitrosylated proteins (PSNOs). GSNO levels are kept low as GSNO is rapidly and irreversibly decomposed by the ubiquitously expressed and evolutionary conserved enzyme *S*-nitrosoglutathione reductase (GSNOR). GSNOR is a member of the alcohol dehydrogenase family and is encoded by the *ADH5* gene. However, its activity toward alcohols is limited, whereas its selective substrate is GSNO [31–33]. GSNO degradation is a two-step reaction that requires NADH/NADPH as cofactor and releases glutathione disulfide (GSSG) and NO moiety as ammonia (NH_3)—or in some cases, hydroxylamine (NH_2OH) [33–35] (Fig. 2A). Since GSNO and PSNO concentrations are maintained in equilibrium through *trans*-nitrosylation reactions, by reducing GSNO, GSNOR indirectly regulates PSNOs

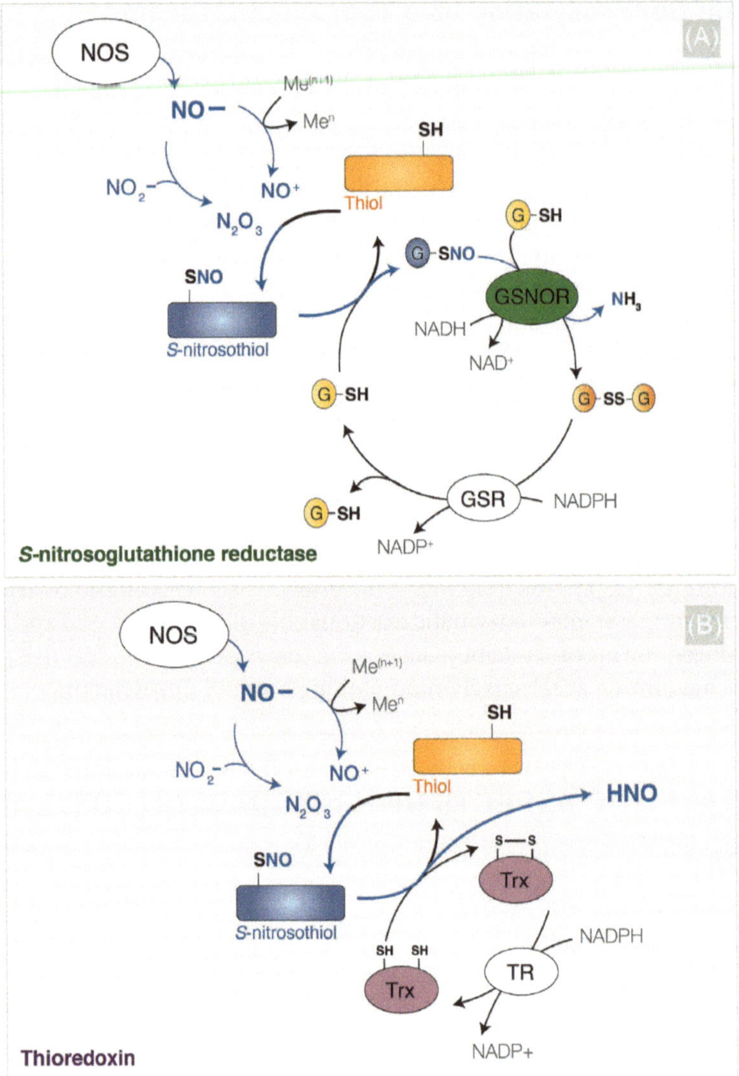

FIG. 2

Primary denitrosylating systems. (A) *S*-nitrosylated proteins can exchange NO moiety with other thiol-containing proteins or with low-molecular-weight thiols, the most important of which is glutathione (GSH). By charging the NO moiety, GSH turns into *S*-nitrosoglutathione (GSNO), which is actively metabolized by the GSNO reductase (GSNOR) at the expenses of NADH in a reductive process that generates ammonia (NH_3) or hydroxylamine (not shown here), and releases glutathione disulfide (GSSG). The NADPH-dependent activity of glutathione disulfide reductase (GSR) reduces GSSG back to GSH, which becomes again available for *trans*nitrosylation reactions. (B) Reduced Trx $(\text{—SH})_2$ mediates denitrosylation of target proteins by releasing the NO moiety as nitroxyl (HNO). Trx reductase (TR) reduces back the disulfide form of Trx (-S-S-) at the expenses of NADPH.

and acts as the main cellular denitrosylase. Such a fundamental role implies that GSNOR is tightly regulated at multiple (transcriptional, translational, and posttranslational) levels [32, 34, 36]. Studies of the last two decades have highlighted the relevance of GSNOR in pathophysiology [34]. In particular, it has been found out that GSNOR activity plays a role in smooth muscle relaxation [37–39], neuronal signaling [29, 40], cardiovascular system homeostasis [39, 41], and inflammatory processes [30, 42]. As a consequence, conditions in which GSNOR is lacking—and, thus, PSNOs are extremely increased—have been associated with a plethora of pathological states, such as increased susceptibility to bacterial infections [35], hypervascularization and resistance to stroke [43], spontaneous hepatocellular carcinoma (HCC) development [44, 45], neuromuscular atrophy [46], neurodegenerative diseases [47], and aging [22].

GSNOR Direct Inhibition as Therapeutic Tool

Given the central role of protein *S*-nitrosylation in cellular homeostasis, GSNOR might represent an attractive therapeutic target. Along this line of reasoning, it is worthwhile to mention that GSNOR-null ($GSNOR^{-/-}$) mice show increased vasculogenesis upon myocardial ischemic injuries, which is fundamental to preserve the heart from possible damages [43, 48]. It has been demonstrated that this phenomenon relies upon *S*-nitrosylation of the hypoxia-inducible factor-1α (HIF-1α). This modification leads to HIF-1α stabilization, which, in turn, activates the transcription of the vascular endothelial growth factor (VEGF) in order to induce vasculogenesis [43]. The therapeutic potential of GSNOR inhibition inspired the synthesis of selective pharmacological inhibitors that are currently under evaluation [49, 50]. Among this class of compounds, the small molecule N6338 has been successfully tested in in vivo models of hypertension [51].

Similarly, to the other members of the alcohol dehydrogenase family, GSNOR is highly expressed in the liver, where—acting as formaldehyde dehydrogenase—it also contributes to xenobiotic detoxification. It has been demonstrated that SNOs promote proliferation and protection of hepatocytes upon liver resection or injury in zebra fish and mice by inducing the antioxidant response mediated by the nuclear factor erythroid 2-related factor 2 (Nrf2) [52]. Based on this evidence, GSNOR inhibitors, by maintaining high levels of SNOs, have been proposed to stimulate liver regeneration [52], being, thus, potentially beneficial as coadjuvants in the treatment of liver diseases (e.g., cirrhosis), infections, or postoperative phases that could severely compromise liver structure and function.

It is well documented that GSNO, by releasing NO, acts as endogenous bronchodilator. As a consequence, $GSNOR^{-/-}$ mice, by retaining elevate levels of SNOs in the lungs, display a low airway susceptibility to allergens [38]. Based on this evidence, many studies have been performed with the aim to validate the use of GSNOR inhibitors in the treatment of asthma. Specifically, it has been shown that administrations of the GSNOR inhibitor N6022 restore SNO levels in the airways and, eventually, result in decreased bronchoconstriction and eosinophil infiltrations [50, 53, 54]. Asthma is a multifactorial disease characterized, inter alia, by the presence of cytokines from different cell types in the airways, such as epithelial cells, smooth muscle cell, and $CD4^+$ T cells. Cytokine production is induced by few regulators, including the nuclear factor kappa-light-chain-enhancer of activated B cells (NF-κB), and strongly participates to the inflammatory process. Using the specific GSNOR inhibitor SPL-334, Ferrini and colleagues [55] observed a decrease in proinflammatory cytokines in asthma mouse models, arguing that this is due to GSNOR inhibition-dependent suppression of NF-κB. Similarly, the GSNOR inhibitor N6022 has been used with success in mouse models of multiple sclerosis, an autoimmune disease characterized by extensive and progressive demyelination of brain and spinal cord neurons. In particular, N6022 has been demonstrated to downregulate the inflammatory response driven by $CD4^+$ T cells [56]. This result takes on an even greater importance as it is supported by the evidence that $GSNOR^{-/-}$ mice suffer from lymphopenia due to an increased rate of apoptosis of CD4-single-positive thymocytes [30]—from which originate the naive form of helper T lymphocytes—indicating that GSNOR is implicated in the correct lymphocyte development and maturation.

It has been demonstrated that $GSNOR^{-/-}$ mice are protected from airway hyperreactivity. Moreover, patients with cystic fibrosis (CF) or bronchial epithelial cells expressing the mutant form of the cystic fibrosis transmembrane conductance regulator (CFTR) show low SNO levels due to an upregulation of GSNOR activity [57]. Cavosonstat (N91115), a novel compound acting as a potent and selective GSNOR inhibitor, has been recently synthesized and proposed to increase CFTR stability and to be beneficial in experimental models of CF [58]. Among the mechanisms proposed, it is worthwhile to note that, by inhibiting GSNOR activity, it preserves GSNO concentrations that have been previously shown being important to stabilize CFTR and prevent its degradation by the proteasome [59].

Targeting GSNOR Dysfunction in Chemotherapy

GSNOR expression has been recently linked to cancer, and the impairment of its denitrosylase activity seems to be involved in the activation of crucial pathways associated with tumor induction, survival, and progression. GSNOR acts, indeed, as a prototypical tumor suppressor [60] by modulating DNA repair [45, 61], cancer metabolism and autophagy [19, 28], gene expression [6, 62], cell proliferation [15, 63], and apoptosis resistance [19, 64]. The first correlation between GSNOR and cancer dates back to the evidence that *Gsnor* genetic ablation in mice induced spontaneous formation of hepatocellular carcinoma (HCC) due to the aberrant *S*-nitrosylation of cys^{145} of the DNA repair enzyme O^6-alkylguanine-DNA alkyltransferase (AGT) [44, 45]. In particular, it was demonstrated that *S*-nitrosylation affects the stability of AGT, thereby promoting its degradation via the proteasome [44]. In addition, analyses performed on human HCC patients showed a significant decrease of GSNOR protein levels and activity in the half of the cases examined, arguing for a functional link between GSNOR-dependent *S*-nitrosylation and hepatocarcinogenesis. This further evidence suggests that an accumulation of PSNOs might, in general, affect DNA repair efficiency and promote spontaneous mutagenesis [45]. Along this line of reasoning, it has been proposed that GSNO accumulation induces the excessive *S*-nitrosylation and consequent impairment of other enzymes involved in DNA damage repair [60]. Therefore, it is plausible that GSNOR, by controlling GSNO levels, preserves genomic stability by underpinning the DNA quality control [60].

So far, the only pharmacological tool proposed to face up the tumorigenic effects of excessive *S*-nitrosylation is the inhibition of NOS activity. [65]. However, such a nonspecific approach could impact on the whole NO-mediated signaling, allowing to question on the quantity (and severity) of side effects that would arise from that. Therefore, the strategies aiming at targeting the oncogenic effects of any decreased expressions of GSNOR in cancer are mostly indirect. We recently reported that, in HCC cells, the excessive *S*-nitrosylation deriving from GSNOR downregulation results in mitochondrial abnormalities, a characteristic that could be exploited for therapeutic purposes [19]. In particular, we provided evidence that GSNOR deficiency results in a general impairment of the mitochondrial complexes, except from the mitochondrial complex II (also known as succinate dehydrogenase, SDH), which, by contrast, is induced reasonably as an adaptive response for sustaining ATP production. As a consequence, GSNOR-deficient HCC cells are extremely sensitive to SDH-directed mitochondrial toxins [19].

Given the role of GSNOR as oncosuppressor, the discovery of the signal transduction pathways modulated by GSNOR deficiency/GSNO accumulation becomes crucial in order to characterize defective conditions/proteins resulting from excessive nitrosylation that, as previously described for SDH, can act as new targets in cancer therapy for synthetic lethal approaches [66, 67].

Thioredoxin

Thioredoxin Function and Activity

Together with GSNOR, thioredoxin (Trx) represents a critical player in the regulation of protein *S*-nitrosylation [68–70]. The capability of Trx dithiol active site to undergo oxidation by concomitantly reducing low-molecular-weight and protein disulfides or SNOs contributes to detach NO moiety from

S-nitrosoglutathione and *S*-nitrosylated proteins. In this reaction, NO moiety is released as nitroxyl (HNO) concomitantly with the reduction of the sulfhydryl group [70] (Fig. 2B). In the same process, Trx active cysteines are oxidized to form a disulfide that can be promptly reduced back by the NADPH-dependent flavoprotein thioredoxin reductase (TR). During the denitrosylation process, the intermediate *S*-nitrosothioredoxin is stable enough to *trans*nitrosylate proteins, such as caspase-3, which is now well established to undergo *S*-nitrosylation in a Trx-mediated manner [71]. Due to its clear capability to *trans*nitrosylate and/or release NO^-, Trx acts essentially as a *S*-nitrosylation tuner, more than a denitrosylase.

Thioredoxin Inhibition as Therapeutic Tool

The Trx/TR system has been found to be frequently deregulated in diverse pathological contexts. However, at variance with GSNOR whose main task is largely related to the control of PSNO levels, the Trx/TR system mainly oversees the reduction/oxidation balance in the cell by cooperating with other antioxidant systems (i.e., glutathione/glutathione reductase) and enzymes (i.e., protein disulfides isomerases and peroxiredoxins) [68]. Based on this high degree of versatility, it is nearly impossible to discern the denitrosylating from the antioxidant capabilities of the Trx/TR system and, thus, to selectively target only one of these cellular functions. Nonetheless, the Trx/TR system exhibits a clear clinical potential for the treatment of several pathological conditions, e.g., thrombosis [72] or different types of cancer [73–75], with several drugs targeting Trx or TR being already validated and currently under clinical trials [73]. If these effects are related to the denitrosylating capability, however, this has been elucidated thus far.

Thioredoxin Mimetics

Trx/TR dysfunctions and the consequent increase of nitrooxidative stress are among the shared mechanisms of a number of pathological states and drug toxicity. For this reasons, few years ago, Bachnoff and colleagues developed potent and selective thioredoxin-mimetic peptides capable of counteracting the oxidative stress resulting from the Trx/TR inhibition or malfunctioning [76]. By emulating the structure of the active site of Trx, they synthetized and tested a number of small peptides bearing two cysteine residues flanking two amino acids (CxxC motif) or a single amino acid (CxC motif). The peptides were characterized by the ability to overcome the inadequacy of a deficient Trx system [76] and were successfully tested in pathological conditions, such as high-glucose-dependent oxidative stress [77] or traumatic brain injury [78]. Remarkably, Trx mimetic peptides show efficient denitrosylating capabilities, and they have been suggested as potential therapeutic tools in treating pathological conditions related to nitrosative stress [79].

Additional Denitrosylating Systems

Additional minor denitrosylating systems have been identified; however, their role in the maintenance of the nitrooxidative balance is still not clear yet. It has been reported that the human carbonyl reductase 1 (CBR1), but not the isoform 3 (CBR3), shows a NADPH-dependent GSNO reductase activity [80, 81]. This cytosolic enzyme is assigned to the reduction of quinones, prostaglandins, and other carbonyl-containing molecules. However, CBR1 efficiency toward GSNO is much lower than GSNOR [34]. TR has been also proposed to catalyze the homolytic cleavage of GSNO in GSH and NO, acting, in such a way, as a denitrosylase [82]. However, due to the kinetic properties of this reaction, the contribution of TR to GSNO reduction in physiological conditions is marginal. Coenzyme A (CoA), the central molecule on which thiol-based acyl-group transfer relies, has been recently reported to be involved in the thiol-based transfer of NO groups as well. As a reservoir of NO, *S*-nitroso-CoA (SNO-CoA) seems contributing to *S*-nitrosylation-mediated signaling in cell metabolism [83]. The downstream signaling network can be controlled by SNO-CoA reductases, a class of enzymes conserved from bacteria to humans that has been recently discovered. To date, the role in pathophysiology of these denitrosylating systems is largely unknown; however, they can constitute a new class of enzymes that, similarly to GSNOR, regulate SNO levels.

CONCLUSIONS

Nitric oxide is a fundamental second messenger regulating cell physiology and several human diseases that derive from an incorrect functioning of its signaling. Therapies based on NO modulation are still considered with caution, and few molecules are currently in clinical trials due to the high complexity of the signaling and the wide variety of pathways activated upon direct modulation of NO levels. For these reasons, the interest in targeting *S*-nitrosylation by modulating denitrosylases is becoming broader. Research of new compounds is getting more appealing and competitive as it poses the challenge (advantage) of being extremely selective: it should not interfere with the whole spectrum of NO signaling, but rather tune only a specific subgroup of pathways.

In some pathological settings (e.g., ischemic injuries, hypertension, liver toxicity, or inflammatory disease such as asthma), a number of synthetic GSNOR inhibitors have been already developed and successfully tested. On the contrary, conditions of GSNOR deficiency, such as those reported in cancer, are still far from being druggable directly with denitrosylase mimics. However, recent lines of evidence have shown that the resulting accumulation of SNOs could be indirectly exploited to design anticancer agents for synthetic lethal approaches.

Acknowledgments

This work has been supported by Kræftens Bekæmpelse grant KBVU R146-A9414 and Associazione Italiana per la Ricerca sul Cancro (AIRC) grant IG20719.

References

[1] Nakamura T, Lipton SA. Protein S-nitrosylation as a therapeutic target for neurodegenerative diseases. Trends Pharmacol Sci 2016;37:73–84. https://doi.org/10.1016/j.tips.2015.10.002.

[2] Stone JR, Marletta MA. Soluble guanylate cyclase from bovine lung: activation with nitric oxide and carbon monoxide and spectral characterization of the ferrous and ferric states. Biochemistry 1994;33:5636–40. https://doi.org/10.1021/bi00184a036.

[3] Brown GC, Borutaite V. Nitric oxide inhibition of mitochondrial respiration and its role in cell death. Free Radic Biol Med 2002;33:1440–50. https://doi.org/10.1016/S0891-5849(02)01112-7.

[4] Dahm CC, Moore K, Murphy MP. Persistent S-nitrosation of complex I and other mitochondrial membrane proteins by S-nitrosothiols but not nitric oxide or peroxynitrite: implications for the interaction of nitric oxide with mitochondria. J Biol Chem 2006;281:10056–65. https://doi.org/10.1074/jbc.M512203200.

[5] Giulivi C, Kato K, Cooper CE. Nitric oxide regulation of mitochondrial oxygen consumption I: cellular physiology. Am J Physiol Cell Physiol 2006;291:C1225–31. https://doi.org/10.1152/ajpcell.00307.2006.

[6] Hess DT, Matsumoto A, Kim S-OO, Marshall HE, Stamler JS. Protein S-nitrosylation: purview and parameters. NatRevMolCell Biol 2005;6:150–66. https://doi.org/10.1038/nrm1569.

[7] Nakamura T, Lipton SA. Emerging role of protein-protein transnitrosylation in cell signaling pathways. Antioxid Redox Signal 2012;18:120709084345003. https://doi.org/10.1089/ars.2012.4703.

[8] Jia J, Arif A, Terenzi F, Willard B, Plow EF, Hazen SL, Fox PL. Target-selective protein S-nitrosylation by sequence motif recognition. Cell 2014;159:623–34. https://doi.org/10.1016/j.cell.2014.09.032.

[9] Seth D, Hess DT, Hausladen A, Wang L, Wang Y, Stamler JS. A multiplex enzymatic machinery for cellular protein S-nitrosylation. Mol Cell 2018;69:451–464.e6. https://doi.org/10.1016/j.molcel.2017.12.025.

[10] Wolhuter K, Eaton P. How widespread is stable protein S-nitrosylation as an end-effector of protein regulation? Free Radic Biol Med 2017;109:156–66. https://doi.org/10.1016/j.freeradbiomed.2017.02.013.

[11] Hess DT, Stamler JS. Regulation by S-nitrosylation of protein post-translational modification. J Biol Chem 2012;287:4411–8. https://doi.org/10.1074/jbc.R111.285742.

[12] Rizza S, Montagna C, Di Giacomo G, Cirotti C, Filomeni G. S-Nitrosation and ubiquitin-proteasome system interplay in neuromuscular disorders. Int J Cell Biol 2014;2014:428764 https://doi.org/10.1155/2014/428764.

[13] Zhou H, Stomberski CT, Stamler JS. Cross talk between S-nitrosylation and phosphorylation involving kinases and nitrosylases. Circ Res 2018;122:1485–7. https://doi.org/10.1161/CIRCRESAHA.118.313109.

[14] Nakamura T, Wang L, Wong CCL, Scott FL, Eckelman BP, Han X, Tzitzilonis C, Meng F, Gu Z, Holland EA, Clemente AT, Okamoto SI, Salvesen GS, Riek R, Yates JR, Lipton SA. Transnitrosylation of XIAP regulates caspase-dependent neuronal cell death. Mol Cell 2010;39:184–95. https://doi.org/10.1016/j.molcel.2010.07.002.

[15] Hatzistergos KE, Paulino EC, Dulce RA, Takeuchi LM, Bellio MA, Kulandavelu S, Cao Y, Balkan W, Kanashiro-Takeuchi RM, Hare JM. *S*-Nitrosoglutathione reductase deficiency enhances the proliferative expansion of adult heart progenitors and myocytes post myocardial infarction. J Am Heart Assoc 2015;4:e001974. https://doi.org/10.1161/JAHA.115.001974.

[16] Mannick JB, Schonhoff C, Papeta N, Ghafourifar P, Szibor M, Fang K, Gaston B. S-Nitrosylation of mitochondrial caspases. J Cell Biol 2001;154:1111 LP–1116.

[17] Talbott SJ, Luanpitpong S, Stehlik C, Azad N, Iyer AKV, Wang L, Rojanasakul Y. S-Nitrosylation of FLICE inhibitory protein determines its interaction with RIP1 and activation of NF-κB. Cell Cycle 2014;13:1948–57. https://doi.org/10.4161/cc.28898.

[18] Tsang AHK, Lee Y-I, Ko HS, Savitt JM, Pletnikova O, Troncoso JC, Dawson VL, Dawson TM, Chung KKK. S-Nitrosylation of XIAP compromises neuronal survival in Parkinson's disease. Proc Natl Acad Sci U S A 2009;106:4900–5. https://doi.org/10.1073/pnas.0810595106.

[19] Rizza S, Montagna C, Cardaci S, Maiani E, Di GG, Sanchez-Quiles V, Blagoev B, Rasola A, De Zio D, Stamler JS, Cecconi F, Filomeni G. S-Nitrosylation of the mitochondrial chaperone TRAP1 sensitizes hepatocellular carcinoma cells to inhibitors of succinate dehydrogenase. Cancer Res 2016;76:4170–82. https://doi.org/10.1158/0008-5472.CAN-15-2637.

[20] Sha Y, Marshall HE. S-Nitrosylation in the regulation of gene transcription. Biochim Biophys Acta 2012;1820:701–11. https://doi.org/10.1016/j.bbagen.2011.05.008.

[21] Li F, Sonveaux P, Rabbani ZN, Liu S, Yan B, Huang Q, Vujaskovic Z, Dewhirst M, Li CY. Regulation of HIF-1α stability through S-nitrosylation. Mol Cell 2007;26:63–74. https://doi.org/10.1016/j.molcel.2007.02.024.

[22] Rizza S, Cardaci S, Montagna C, Di Giacomo G, De Zio D, Bordi M, Maiani E, Campello S, Borreca A, Puca AA, Stamler JS, Cecconi F, Filomeni G. S-Nitrosylation drives cell senescence and aging in mammals by controlling mitochondrial dynamics and mitophagy. Proc Natl Acad Sci U S A 2018;201722452. https://doi.org/10.1073/pnas.1722452115.

[23] Clementi E, Brown GC, Feelisch M, Moncada S. Persistent inhibition of cell respiration by nitric oxide: crucial role of S-nitrosylation of mitochondrial complex I and protective action of glutathione. Proc Natl Acad Sci U S A 1998;95:7631–6.

[24] Piantadosi CA. Regulation of mitochondrial processes by protein S-nitrosylation. Biochim Biophys Acta 2012;1820:712–21. doi:http://doi.org/10.1016/j.bbagen.2011.03.008.

[25] Ozawa K, Komatsubara AT, Nishimura Y, Sawada T, Kawafune H, Tsumoto H, Tsuji Y, Zhao J, Kyotani Y, Tanaka T, Takahashi R, Yoshizumi M. S-Nitrosylation regulates mitochondrial quality control via activation of parkin. Sci Rep 2013;3:1–10. https://doi.org/10.1038/srep02202.

[26] Rizza S, Filomeni G. Denitrosylate and live longer: how ADH5/GSNOR links mitophagy to aging. Autophagy 2018; [doi:IN PRESS].

[27] Sarkar S, Korolchuk VI, Renna M, Imarisio S, Fleming A, Williams A, Garcia-Arencibia M, Rose C, Luo S, Underwood BR, Kroemer G, O'Kane CJ, Rubinsztein DC. Complex inhibitory effects of nitric oxide on autophagy. Mol Cell 2011;43:19–32. https://doi.org/10.1016/j.molcel.2011.04.029.

[28] Montagna C, Rizza S, Maiani E, Piredda L, Filomeni G, Cecconi F. To eat, or NOt to eat: S-nitrosylation signaling in autophagy. FEBS J 2016;283:3857–69. https://doi.org/10.1111/febs.13736.

[29] Wu K, Zhang Y, Wang P, Zhang L, Wang T, Chen C. Activation of GSNOR transcription by NF-κB negatively regulates NGF-induced PC12 differentiation. Free Radic Res 2014;48:1011–7. https://doi.org/10.3109/10715762.2014.906743.

[30] Yang Z, Wang Z-E, Doulias P-T, Wei W, Ischiropoulos H, Locksley RM, Liu L. Lymphocyte development requires S-nitrosoglutathione reductase. J Immunol 2010;185:6664–9. https://doi.org/10.4049/jimmunol.1000080.

[31] Liu L, Hausladen A, Zeng M, Que L, Heitman J, Stamler JS. A metabolic enzyme for S-nitrosothiol conserved from bacteria to humans. Nature 2001;410:490–4.

[32] Barnett SD, Buxton ILO. The role of S-nitrosoglutathione reductase (GSNOR) in human disease and therapy. Crit Rev Biochem Mol Biol 2017;52:1–15. https://doi.org/10.1080/10409238.2017.1304353.

[33] Jensen DE, Belka GK, Du Bois GC. S-Nitrosoglutathione is a substrate for rat alcohol dehydrogenase class III isoenzyme. Biochem J 1998;331(Pt 2):659–68. https://doi.org/10.1042/bj3310659.

[34] Rizza S, Filomeni G. Chronicles of a reductase: biochemistry, genetics and physio-pathological role of GSNOR. Free Radic Biol Med 2017;110:19–30. https://doi.org/10.1016/j.freeradbiomed.2017.05.014.

[35] Liu L, Yan Y, Zeng M, Zhang J, Hanes MA, Ahearn G, McMahon TJ, Dickfeld T, Marshall HE, Que LG, Stamler JS. Essential roles of S-nitrosothiols in vascular homeostasis and endotoxic shock. Cell 2004;116:617–28. doi:http://doi.org/10.1016/S0092-8674(04)00131-X.

[36] Guerra D, Ballard K, Truebridge I, Vierling E. S-Nitrosation of conserved cysteines modulates activity and stability of S-nitrosoglutathione reductase (GSNOR). Biochemistry 2016;https://doi.org/10.1021/acs.biochem.5b01373.

[37] Henderson EM, Gaston B. SNOR and wheeze: the asthma enzyme? Trends Mol Med 2005;11:481–4. https://doi.org/10.1016/j.molmed.2005.09.009.

[38] Que LG, Liu L, Yan Y, Whitehead GS, Gavett SH, Schwartz DA, Stamler JS. Protection from experimental asthma by an endogenous bronchodilator. Science 2005;308:1618–21. https://doi.org/10.1126/science.1108228.

[39] Sips PY, Irie T, Zou L, Shinozaki S, Sakai M, Shimizu N, Nguyen R, Stamler JS, Chao W, Kaneki M, Ichinose F. Reduction of cardiomyocyte S-nitrosylation by S-nitrosoglutathione reductase protects against sepsis-induced myocardial depression. Am J Physiol Heart Circ Physiol 2013;304:H1134–46. https://doi.org/10.1152/ajpheart.00887.2012.

[40] Wu K, Ren R, Su W, Wen B, Zhang Y, Yi F, Qiao X, Yuan T, Wang J, Liu L, Izpisua Belmonte JC, Liu GH, Chen C. A novel suppressive effect of alcohol dehydrogenase 5 in neuronal differentiation. J Biol Chem 2014;289:20193–9. https://doi.org/10.1074/jbc.C114.561860.

[41] Beigi F, Gonzalez DR, Minhas KM, Sun Q-A, Foster MW, Khan SA, Treuer AV, Dulce RA, Harrison RW, Saraiva RM, Premer C, Schulman IH, Stamler JS, Hare JM. Dynamic denitrosylation via S-nitrosoglutathione reductase regulates cardiovascular function. Proc Natl Acad Sci U S A 2012;109:4314–9. https://doi.org/10.1073/pnas.1113319109.

[42] Tang CH, Seeley EJ, Huang X, Wolters PJ, Liu L. Increased susceptibility to Klebsiella pneumonia and mortality in GSNOR-deficient mice. Biochem Biophys Res Commun 2013;442:122–6. https://doi.org/10.1016/j.bbrc.2013.11.028.

[43] Lima B, Lam GKW, Xie L, Diesen DL, Villamizar N, Nienaber J, Messina E, Bowles D, Kontos CD, Hare JM, Stamler JS, Rockman HA. Endogenous S-nitrosothiols protect against myocardial injury. Proc Natl Acad Sci U S A 2009;106:6297–302. https://doi.org/10.1073/pnas.0901043106.

[44] Wei W, Li B, Hanes MA, Kakar S, Chen X, Liu L, Parkin DM, Bray F, Ferlay J, Pisani P, El-Serag HB, Davila JA, Petersen NJ, McGlynn KA, Perz JF, Armstrong GL, Farrington LA, Hutin YJ, Bell BP, et al. S-Nitrosylation from GSNOR deficiency impairs DNA repair and promotes hepatocarcinogenesis. Sci Transl Med 2010;2:19ra13. https://doi.org/10.1126/scitranslmed.3000328.

[45] Wei W, Yang Z, Tang CH, Liu L. Targeted deletion of GSNOR in hepatocytes of mice causes nitrosative inactivation of O6-alkylguanine-dna alkyltransferase and increased sensitivity to genotoxic diethylnitrosamine. Carcinogenesis 2011;32:973–7. https://doi.org/10.1093/carcin/bgr041.

[46] Montagna C, Di Giacomo G, Rizza S, Cardaci S, Ferraro E, Grumati P, De Zio D, Maiani E, Muscoli C, Lauro F, Ilari S, Bernardini S, Cannata S, Gargioli C, Ciriolo MR, Cecconi F, Bonaldo P, Filomeni G. *S*-Nitrosoglutathione reductase deficiency-induced *S*-nitrosylation results in neuromuscular dysfunction. Antioxid Redox Signal 2014;21:570–87. https://doi.org/10.1089/ars.2013.5696.

[47] Rizza S, Cirotti C, Montagna C, Cardaci S, Consales C, Cozzolino M, Carrì MT, Cecconi F, Filomeni G. S-Nitrosoglutathione reductase plays opposite roles in SH-SY5Y models of Parkinson's disease and amyotrophic lateral sclerosis. Mediators Inflamm 2015;2015: https://doi.org/10.1155/2015/536238.

[48] Lima B, Forrester MT, Hess DT, Stamler JS. S-Nitrosylation in cardiovascular signaling. Circ Res 2010;106:633–46. https://doi.org/10.1161/CIRCRESAHA.109.207381.

[49] Everett TR, Wilkinson IB, Lees CC. Pre-eclampsia: the potential of GSNO reductase inhibitors. Curr Hypertens Rep 2017;19:1–7. https://doi.org/10.1007/s11906-017-0717-2.

[50] Green LS, Chun LE, Patton AK, Sun X, Rosenthal GJ, Richards JP. Mechanism of inhibition for N6022, a first-in-class drug targeting S-nitrosoglutathione reductase. Biochemistry 2012;51:2157–68. https://doi.org/10.1021/bi201785u.

[51] Chen Q, Sievers RE, Varga M, Kharait S, Haddad DJ, Patton AK, Delany CS, Mutka SC, Blonder JP, Dube GP, Rosenthal GJ, Springer ML. Pharmacological inhibition of S-nitrosoglutathione reductase improves endothelial vasodilatory function in rats in vivo. J Appl Physiol 2013;114:752–60. https://doi.org/10.1152/japplphysiol.01302.2012.

[52] Cox AG, Saunders DC, Kelsey PB, Conway AA, Tesmenitsky Y, Marchini JF, Brown KK, Stamler JS, Colagiovanni DB, Rosenthal GJ, Croce KJ, North TE, Goessling W. S-Nitrosothiol signaling regulates liver development and improves outcome following toxic liver injury. Cell Rep 2014;6:56–69. https://doi.org/10.1016/j.celrep.2013.12.007.

[53] Blonder JP, Mutka SC, Sun X, Qiu J, Green LH, Mehra NK, Boyanapalli R, Suniga M, Look K, Delany C, Richards JP, Looker D, Scoggin C, Rosenthal GJ. Pharmacologic inhibition of S-nitrosoglutathione reductase protects against experimental asthma in BALB/c mice through attenuation of both bronchoconstriction and inflammation. BMC Pulm Med 2014; 14:3https://doi.org/10.1186/1471-2466-14-3.

[54] Sun X, Wasley JWF, Qiu J, Blonder JP, Stout AM, Green LS, Strong SA, Colagiovanni DB, Richards JP, Mutka SC, Chun L, Rosenthal GJ. Discovery of S-nitrosoglutathione reductase inhibitors: potential agents for the treatment of asthma and other inflammatory diseases. ACS Med Chem Lett 2011;2:402–6. https://doi.org/10.1021/ml200045s.

[55] Ferrini ME, Simons BJ, Bassett DJP, Bradley MO, Roberts K, Jaffar Z. S-Nitrosoglutathione reductase inhibition regulates allergen-induced lung inflammation and airway hyperreactivity. PLoS One 2013;8:https://doi.org/10.1371/journal.pone.0070351.

[56] Saxena N, Won J, Choi S, Singh AK, Singh I. S-Nitrosoglutathione reductase (GSNOR) inhibitor as an immune modulator in experimental autoimmune encephalomyelitis. Free Radic Biol Med 2018;121:57–68. https://doi.org/10.1016/j.freeradbiomed.2018.04.558.

[57] Zaman K, Sawczak V, Zaidi A, Butler M, Bennett D, Getsy P, Zeinomar M, Greenberg Z, Forbes M, Rehman S, Jyothikumar V, DeRonde K, Sattar A, Smith L, Corey D, Straub A, Sun F, Palmer L, Periasamy A, et al. Augmentation of CFTR maturation by S-nitrosoglutathione reductase. Am J Physiol Cell Mol Physiol 2015;310:L263–70. https://doi.org/10.1152/ajplung.00269.2014.

[58] Donaldson SH, Solomon GM, Zeitlin PL, Flume PA, Casey A, McCoy K, Zemanick ET, Mandagere A, Troha JM, Shoemaker SA, Chmiel JF, Taylor-Cousar JL. Pharmacokinetics and safety of cavosonstat (N91115) in healthy and cystic fibrosis adults homozygous for F508DEL-CFTR. J Cyst Fibros 2017;16:371–9. https://doi.org/10.1016/j.jcf.2017.01.009.

[59] Chen L, Patel RP, Teng X, Bosworth CA, Lancaster JR, Matalon S. Mechanisms of cystic fibrosis transmembrane conductance regulator activation by S-Nitrosoglutathione. J Biol Chem 2006;281:9190–9. https://doi.org/10.1074/jbc.M513231200.

[60] Rizza S, Filomeni G. Tumor suppressor roles of the Denitrosylase GSNOR. Crit Rev Oncog 2016;21:433–45. https://doi.org/10.1615/CritRevOncog.2017021074.

[61] Pontel LB, Rosado IV, Burgos-Barragan G, Garaycoechea JI, Yu R, Arends MJ, Chandrasekaran G, Broecker V, Wei W, Liu L, Swenberg JA, Crossan GP, Patel KJ. Endogenous formaldehyde is a hematopoietic stem cell genotoxin and metabolic carcinogen. Mol Cell 2015;60:177–88. https://doi.org/10.1016/j.molcel.2015.08.020.

[62] Gomes SA, Rangel EB, Premer C, Dulce RA, Cao Y, Florea V, Balkan W. S-Nitrosoglutathione reductase (GSNOR) enhances vasculogenesis by mesenchymal stem cells. Proc Natl Acad Sci USA 2013;110(8):2834–9. https://doi.org/10.1073/pnas.1220185110/-/DCSupplemental.www.pnas.org/cgi/doi/10.1073/pnas.1220185110.

[63] Yamashita AMS, Ancillotti MTC, Rangel LP, Fontenele M, Figueiredo-Freitas C, Possidonio AC, Soares CP, Sorenson MM, Mermelstein C, Nogueira L. Balance between S-nitrosylation and denitrosylation modulates myoblast proliferation independently of soluble guanylyl cyclase activation. Am J Physiol 2017;313:C11–26. https://doi.org/10.1152/ajpcell.00140.2016.

[64] Cañas A, López-Sánchez LM, Peñarando J, Valverde A, Conde F, Hernández V, Fuentes E, López-Pedrera C, de la Haba-Rodríguez JR, Aranda E, Rodríguez-Ariza A. Altered S-nitrosothiol homeostasis provides a survival advantage to breast cancer cells in HER2 tumors and reduces their sensitivity to trastuzumab. Biochim Biophys Acta 2016;1862:601–10. doi: https://doi.org/10.1016/j.bbadis.2016.02.005.

[65] Tang CH, Wei W, Hanes MA, Liu L. Hepatocarcinogenesis driven by GSNOR deficiency is prevented by iNOS inhibition. Cancer Res 2013;73:2897–904. https://doi.org/10.1158/0008-5472.CAN-12-3980.

[66] Fang B. Development of synthetic lethality anticancer therapeutics. J Med Chem 2014; 57:7859–73. https://doi.org/10.1021/jm500415t.

[67] Lord CJ, Tutt ANJ, Ashworth A. Synthetic lethality and cancer therapy: lessons learned from the development of PARP inhibitors. Annu Rev Med 2015;66:455–70. https://doi.org/10.1146/annurev-med-050913-022545.

[68] Sengupta R, Holmgren A. Thioredoxin and thioredoxin reductase in relation to reversibleS-nitrosylation. Antioxid Redox Signal 2013;18:259–69. https://doi.org/10.1089/ars.2012.4716.

[69] Benhar M, Forrester MT, Hess DT, Stamler JS. Regulated protein denitrosylation by cytosolic and mitochondrial thioredoxins. Science 2008;320:1050LP–1054LP.

[70] Sengupta R, Ryter SW, Zuckerbraun BS, Tzeng E, Billiar TR, Stoyanovsky DA. Thioredoxin catalyzes the denitrosation of low-molecular mass and protein S-nitrosothiols. Biochemistry 2007;46:8472–83. https://doi.org/10.1021/bi700449x.

[71] Mitchell DA, Marletta MA. Thioredoxin catalyzes the S-nitrosation of the caspase-3 active site cysteine. Nat Chem Biol 2005;1:154.

[72] Metcalfe C, Ramasubramoni A, Pula G, Harper MT, Mundell SJ, Coxon CH. Thioredoxin inhibitors attenuate platelet function and thrombus formation. PLoS One 2016;11:1–18. https://doi.org/10.1371/journal.pone.0163006.

[73] Zhang B, Zhang J, Peng S, Liu R, Li X, Hou Y, Han X, Fang J. Thioredoxin reductase inhibitors: a patent review. Expert Opin Ther Pat 2017;27:547–56. https://doi.org/10.1080/13543776.2017.1272576.

[74] Stafford WC, Peng X, Olofsson MH, Zhang X, Luci DK, Lu L, Cheng Q, Trésaugues L, Dexheimer TS, Coussens NP, Augsten M, Ahlzén HSM, Orwar O, Östman A, Stone-Elander S, Maloney DJ, Jadhav A, Simeonov A, Linder S, et al. Irreversible inhibition of cytosolic thioredoxin reductase 1 as a mechanistic basis for anticancer therapy. Sci Transl Med 2018;10:1–14. https://doi.org/10.1126/scitranslmed.aaf7444.

[75] Schroeder A, Warnken U, Röth D, Klika KD, Vobis D, Barnert A, Bujupi F, Oberacker T, Schnölzer M, Nicolay JP, Krammer PH, Gülow K. Targeting Thioredoxin-1 by dimethyl fumarate induces ripoptosome-mediated cell death. Sci Rep 2017;7:1–12. https://doi.org/10.1038/srep43168.

[76] Bachnoff N, Trus M, Atlas D. Alleviation of oxidative stress by potent and selective thioredoxin-mimetic peptides. Free Radic Biol Med 2011;50:1355–67. https://doi.org/10.1016/j.freeradbiomed.2011.02.026.

[77] Lejnev K, Khomsky L, Bokvist K, Mistriel-Zarbib S, Naveh T, Farb TB, Alsina-Fernandez J, Atlas D. Thioredoxin-mimetic peptides (TXM) inhibit inflammatory pathways associated with high-glucose and oxidative stress. Free Radic Biol Med 2016;99:557–71. https://doi.org/10.1016/j.freeradbiomed.2016.09.011.

[78] Baratz-Goldstein R, Deselms H, Heim LR, Khomski L, Hoffer BJ, Atlas D, Pick CG. Thioredoxin-mimetic-peptides protect cognitive function after mild traumatic brain injury (mTBI). PLoS One 2016;11:1–19. https://doi.org/10.1371/journal.pone.0157064.

[79] Kronenfeld G, Engelman R, Weisman-Shomer P, Atlas D, Benhar M. Thioredoxin-mimetic peptides as catalysts of S-denitrosylation and anti-nitrosative stress agents. Free Radic Biol Med 2015;79:138–46. https://doi.org/10.1016/j.freeradbiomed.2014.11.021.

[80] Staab CA, Hartmanová T, El-Hawari Y, Ebert B, Kisiela M, Wsol V, Martin HJ, Maser E. Studies on reduction of S-nitrosoglutathione by human carbonyl reductases 1 and 3. Chem Biol Interact 2011;191:95–103. https://doi.org/10.1016/j.cbi.2011.01.016.

[81] Bateman RL, Rauh D, Tavshanjian B, Shokat KM. Human carbonyl reductase 1 is an S-nitrosoglutathione reductase. J Biol Chem 2008;283:35756–62. https://doi.org/10.1074/jbc.M807125200.

[82] Nikitovic D, Holmgren A. Enzymology: S-Nitrosoglutathione is cleaved by the thioredoxin system with liberation of glutathione and redox regulating nitric oxide. J Biol Chem 1996;271:19180–5. doi: https://doi.org/10.1074/jbc.271.32.19180.

[83] Anand P, Hausladen A, Wang Y-J, Zhang G-F, Stomberski C, Brunengraber H, Hess DT, Stamler JS. Identification of S-nitroso-CoA reductases that regulate protein S-nitrosylation. Proc Natl Acad Sci U S A 2014;111:201417816 https://doi.org/10.1073/pnas.1417816112.

CHAPTER 10

Nitric Oxide-Releasing Functional Nanomaterials for Anticancer Therapy

Dongsik Park, Gurusamy Saravanakumar, Won Jong Kim
Department of Chemistry, Pohang University of Science and Technology (POSTECH), Pohang, Republic of Korea

ABSTRACT

Nitric oxide (NO) is a signaling gas molecule that plays important roles in various physiological processes and also in cancer biology. Many of its functions are greatly dependent on its concentration and duration in vivo. However, the clinical applications of most of the available low-molecular-weight NO donors are limited by lower aqueous stability and poor bioavailability. Therefore, in recent years, a number of NO-delivering nanoplatforms have been designed by integrating NO donors with functional nanomaterials to improve the local concentration of NO at the targeted disease site, thereby leading to enhanced therapeutic effects. This chapter highlights the recent progress of NO-releasing functional nanoplatforms with a special emphasis on chemo- and radiotherapy.

ABBREVIATIONS

BNN6	*N,N′*-di-*sec*-butyl-*N,N′*-dinitroso-1,4-phenylenediamine
BNNs	*N,N′*-bis(carboxymethyl)-*N,N′*-dinitroso-*p*-phenylenediamines
CDI	coefficient of drug interaction
CysNO	*S*-nitroso-L-cysteine
DEA/NO	diethylamine NONOates
DOX	doxorubicin
GSH	glutathione
GSNO	*S*-nitroso-L-glutathione
GST	glutathione *S*-transferase
GTN	glyceryl trinitrate
HIF-1	hypoxia-inducible factor 1
ISDN	isosorbide dinitrate

Therapeutic Application of Nitric Oxide in Cancer and Inflammatory Disorders. https://doi.org/10.1016/B978-0-12-816545-4.00010-4

ISMN	isosorbide mononitrate
MDR	multidrug resistance
NO	nitric oxide
NOS	nitric oxide synthase
OEGMA	oligoethylene glycol methacrylate
PAPA/NO	propylamine propylamine NONOates
PDA	polydopamine
PETN	pentaerythritol tetranitrate
Pgp	*P*-glycoprotein
PLGA	poly(DL-lactic-*co*-glycolic acid)
PROLI/NO	proline NONOates
RAFT	reversible addition-fragmentation chain transfer
RBS	Roussin's black salt
RNS	reactive nitrogen species
RSNO	*S*-nitrosothiol
SNAP	*S*-nitroso-*N*-acetylpenicillamine
SNP	sodium nitroprusside
SPER/NO	spermine NONOates
Tf	transferrin
UCNP	upconversion nanoparticle
VDM	2-vinyl-4,4-dimethyl-5-oxazolone monomer

Conflict of Interest

No potential conflicts of interest were disclosed.

INTRODUCTION

Ever since the discovery of nitric oxide (NO) as a signaling molecule in the cardiovascular system in the 1980s, there has been immense scientific research focused on this small-molecule free radical, and this continues to grow exponentially [1, 2]. NO has been implicated in various biological activities such as angiogenesis, immune response, cardiovascular homeostasis, and apoptosis [2]. Recently, several roles of NO in cancer biology have been extensively explored [3]. At low concentrations (about picomolar), NO promotes angiogenesis and antiapoptotic effects, resulting in enhanced nutrient delivery and improved cancer cell growth [3, 4]. However, when the concentration of NO is about micromolar to millimolar, NO could lead to cell-cycle arrest and apoptosis through the processes, such as DNA deamination, nitrosylation of enzymes, and inhibition of mitochondrial respiration [3, 5]. In addition, a high concentration of NO improves the efficacy of chemotherapy and radiotherapy [6]. Considering the concentration-dependent activity of NO, controlling the amount of NO at the tumor tissue is crucial for cancer therapy. To increase the local concentration of NO, numerous exogenous low-molecular-weight NO donors have been developed. Although these NO donors increased antitumor effect when combined with other therapeutic methods, the use of low-molecular-weight NO donors suffers from several issues when employed in vivo. For example, low-molecular-weight NO donors such as *N*-diazeniumdiolates (NONOates) exhibit poor aqueous stability and thus spontaneous release of NO during circulation, causing undesired side effects and a low concentration of NO at the tumor tissues. Therefore, many researchers have engaged with the development of carrier systems that can enhance the stability and release of a therapeutic concentration of NO at the targeted site of action in a spatiotemporally controlled manner. In recent years, nanosized delivery platforms have received enormous attention owing to the opportunity to target tumor tissue passively via enhanced permeation and retention (EPR) effects [7]. Encouraged by this tumor-targeting strategy, several NO-delivering nanoplatforms have been developed to improve therapeutic efficacy.

In this chapter, we have highlighted the recent progress of NO-releasing nanoplatforms, with a specific emphasis on the chemo- and radiotherapy. In the first section, we briefly introduced an overview of diverse NO donors and their preparation conditions and NO-releasing mechanisms. In the second section, we discussed the potential role of NO in anticancer therapy and other physiological and pathophysiological processes. In the final section, we comprehensively discussed several NO-releasing functional nanoplatforms for anticancer therapy.

NITRIC OXIDE (NO) DONORS

In general, the term NO donors represent functional moieties or pharmacologically active agents that can generate NO upon being administered exogenously. Considering the availability of numerous NO donors for potential biomedical applications, this section provides a brief overview of the following few representative NO donors: *N*-diazeniumdiolates (NONOates), *S*-nitrosothiol (RSNO), organic nitrates, metal-nitrosyl complex, and photo-triggered NO donors. We hope that a brief discussion on the NO donors could allow the readers to have a clear insight into their chemical reactivities and how these NO donors have been efficiently integrated with diverse carrier systems and subsequent NO-releasing mechanisms for successful therapeutic applications.

N-Diazeniumdiolates (NONOates)

NONOates are one of the most widely investigated NO donors because of their ability to release two molar equivalents of NO per mole of donor under physiological conditions [8]. In their seminal paper of 1960, Drago and Paulik first described this class of compounds in the form of an adduct of diethylamine and

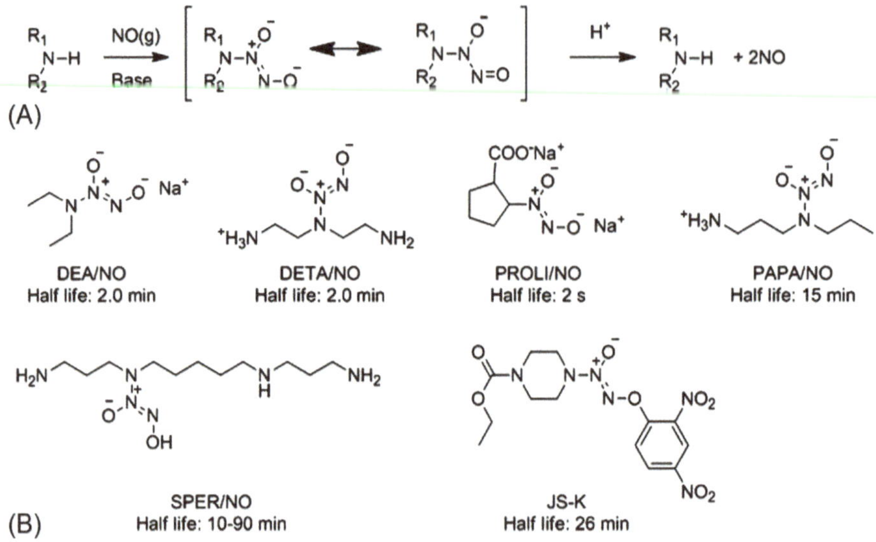

FIG. 1

Chemical structure of NONOates. A) Mechanism of formation of NONOates group and NO generation from this moiety. B) Chemical structure of representative NONOates compounds and its half-life. *Reproduced with permission from Sergio H. Nitric oxide for cancer therapy. Future Sci OA 2015;1:FSO44.*

NO [9]. The chemical structure of NONOates consists of an N=N (diazen) group, a formal positive charge (ium), and two negatively charged oxygen atoms (diolates) (Fig. 1A). In general, NONOates can be equally synthesized by reacting secondary amines with gaseous NO under high pressure. This procedure has allowed to synthesize diverse NONOates ranging from small molecules to polymers, such as diethylamine NONOate (DEA/NO), spermine NONOate (SPER/NO), proline NONOate (PROLI/NO), propylamine propylamine NONOates (PAPA/NO), and polyethylenimine NONOate [9–12] (Fig. 1B). The release of NO from these compounds simply follows a first-order kinetics. Furthermore, the half-lives of NONOates, ranging from seconds to hours, were greatly dependent on the chemical structure, molecular size, and inter-/intramolecular hydrogen bonding of the amine precursors. In contrast to other NO donors, the NONOates do not require any activation mechanism, and their NO release can spontaneously proceed via hydrolysis at physiological pH. Although this fast NO release is considered to be beneficial for some therapeutic applications, their selectivity to a given organ, tissue, or cell type is rather limited. For example, NONOates in free form or integrated with drug delivery systems may release most of their NO during systemic circulation before reaching the desired site of action. Therefore, to improve the stability and biodistribution of NONOates, the protection of terminal oxygen of the diolate moiety with stimuli-responsive moieties has received considerable attention because it allows to facilitate target-specific NO release. One of the widely used protected NONOate molecules was O^2-arylated diazeniumdiolate JS-K (Fig. 1B), which releases NO by the interaction with glutathione (GSH) and glutathione S-transferase (GST) [13].

S-Nitrosothiols (RSNOs)

RSNOs are endogenous metabolites of NO and have shown many important biological activities including as mediators in numerous cell signaling pathways. In fact, the endogenous RSNOs, *S*-nitroso-L-glutathione (GSNO), *S*-nitroso-L-cysteine (CysNO), and *S*-nitrosoalbumin act as an important reservoir of NO in vivo [14]. Unlike NONOates, RSNOs are synthesized by reacting thiols (RSH) with NO derivatives such as NO_2, N_2O_4, and N_2O_3. The direct reaction between NO and thiols results in disulfides instead of RSNOs. However, if NO is used along with O_2 or other oxidants, *S*-nitrosation may be induced via oxidation of NO to the aforementioned NO derivatives. In general, nitrosation reactions can be carried out in aqueous or nonaqueous methods, while the latter one may avoid undesired side reactions such as hydrolysis of the starting materials. The release of NO from RSNOs can be induced by various triggers such as heat, UV light, ascorbate, and Cu ions (Fig. 2A). A large number of studies have utilized two relatively stable RNSOs, namely,

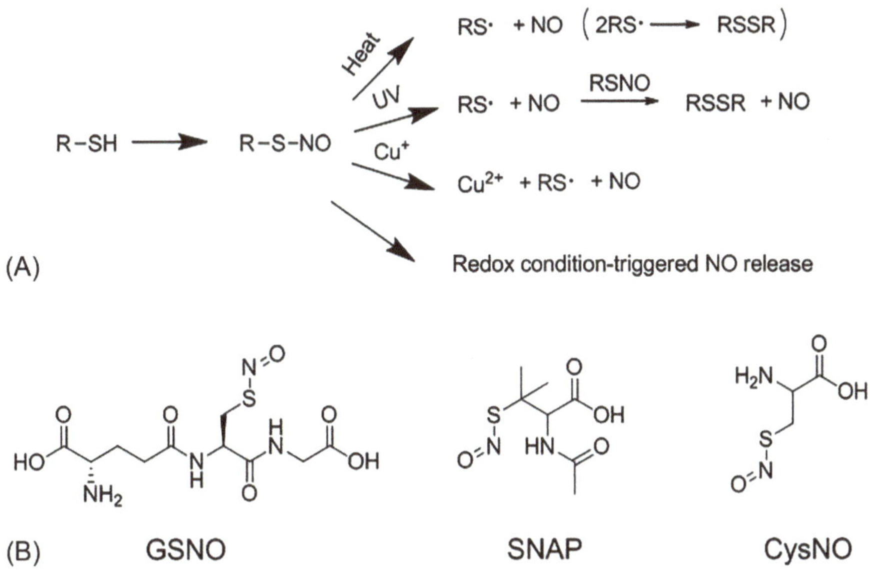

FIG. 2

Chemical structure of RSNO. (A) Mechanism of the formation of RSNO and its triggered NO release. (B) Chemical structure of representative RSNO compounds.

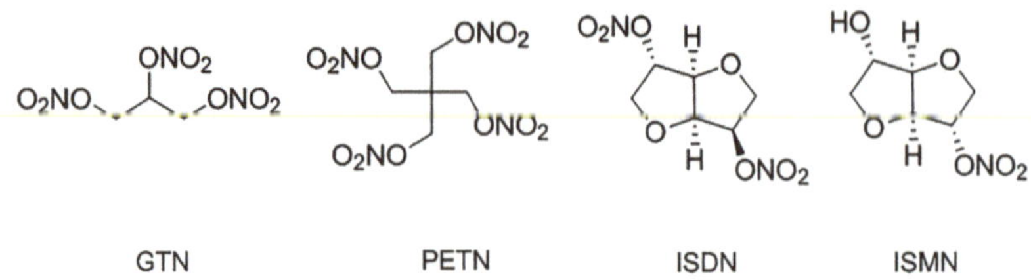

FIG. 3
Chemical structure of representative organic nitrate compounds.

S-nitroso-*N*-acetylpenicillamine (SNAP) and GSNO as NO donors for in vivo preclinical studies (Fig. 2B). In spite of the promising potential of RSNOs, the clinical development of RSNO therapeutics is hindered by its poor storage stability and the lack of specificity.

Organic Nitrates

Organic nitrates, the oldest class of NO donors, are nitric acid esters of alcohol groups (R-ONO_2) [15]. Commonly used organic nitrates include glyceryl trinitrate (GTN), pentaerythritol tetranitrate (PETN), isosorbide dinitrate (ISDN), and isosorbide mononitrate (ISMN) and have been used as vasodilators in clinical trials [15, 16] (Fig. 3). Especially, GTN has been routinely used for a long time to the treatment of cardiovascular diseases such as angina pectoris and chronic angina [16, 17]. NO release of GTN does not induce itself, but it commonly produces NO by the enzyme-mediated process in vivo [16, 17].

It is proposed that several enzymes such as GST, cytochrome P450 reductase, xanthine oxidoreductase, and aldehyde dehydrogenase-2 could be involved in the generation of NO [17]. In other cases, organic nitrates oxidize the ferrous heme of hemoglobin and myoglobin to release NO. Similarly, it is observed that the ferrous active site with NADPH reduces the organic nitrates and releases NO [15]. However, the mechanism of NO release from organic nitrates has not yet been fully identified [17].

Metal-Nitrosyl Complexes

Metal-nitrosyl complexes are another interesting class of NO donors. The reaction of NO with various transition metals has been studied for many years. Early examples of these types of NO donors include iron-nitrosyl complexes such as Roussin's black salt (RBS) anion $[Fe_4S_3(NO)_7]^-$ and red salt dianinion $[Fe_2S_2(NO)_4]^{2-}$, which have been investigated as photochemical NO delivery agents to release NO at the physiological targets on demand [18, 19] (Fig. 4). In another case, sodium nitroprusside (SNP) ($Na_2[Fe(CN)_5NO]$), which can generate NO by the interaction with sulfhydryl-containing molecules, also has been clinically studied [19].

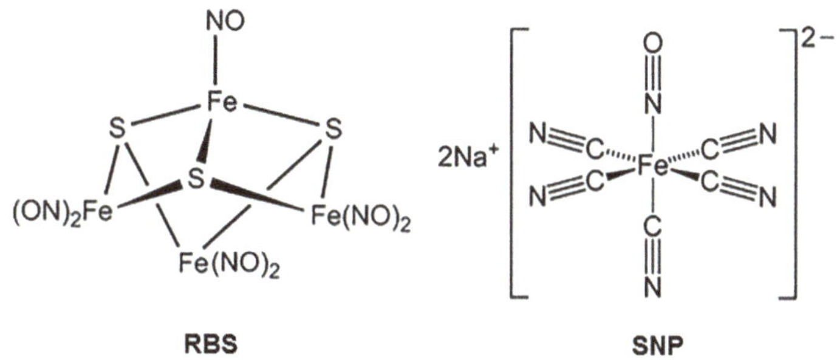

FIG. 4
Representative NO-releasing metal-nitrosyl complexes.

The stability and the structure of the metal-nitrosyl complexes are largely dependent on the oxidation state of the metal center. Among various metal-nitrosyl complexes that have been reported, ruthenium-nitrosyl (Ru-NO) complexes are reported to exhibit high stability compared with other metal-nitrosyl complexes. However, most of the Ru-NO complexes are activated to release NO under UV light (<400 nm), and this harmful high-energy radiation not only damaged the skin but also exhibited poor deep tissue penetration. Near-infrared (NIR) light (with wavelengths of about 700–1100 nm) is less damaging and has a deeper penetration depth owing to reduced light absorption and scattering by biological tissues. Therefore, designing of Ru-NO complexes, which undergo facile NO photorelease upon NIR irradiation, is gaining more attention [20]. Interestingly, the properties of Ru-NOs can be fine-tuned by varying the nature of the ligands around the metal. Although notable progress has been made in this area, extensive toxicity studies on metal-nitrosyl complexes are required before applying to the practical purposes.

Photo-Triggered Organic NO Donors

Besides metal-nitrosyl-based photoactive NO donors, several types of photochemically triggered organic NO donors have also been developed. A representative example of this class includes nitrobenzene derivatives [21, 22]. As shown in Fig. 5A, the ortho positons of the nitrobenzene derivatives are often substituted with bulky substituents such as methyl, trifluoromethyl, or arenes. Owing to the steric hindrance from the bulky substituents, the nitro group in the ring is directed almost perpendicular to the aromatic plane [23]. In such out-of-plane geometry, the p orbital of the $-NO_2$ group oxygen atom is overlapped with the p orbital of the aromatic ring in both the ground and excited states. Thus, the generation of NO is facilitated via nitro-to-nitrite photorearrangement followed by the rupture of the O–N bond. Other photo-triggered organic NO donors are derivatives of *N,N′*-bis(carboxymethyl)-*N,N′*-dinitroso-*p*-phenylenediamines (BNNs) [24, 25], which generate two NO

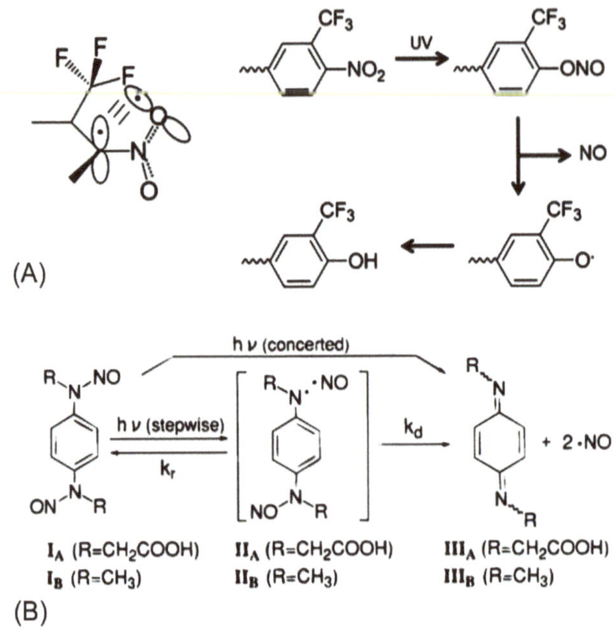

FIG. 5

Mechanism of photo-triggered NO release. (A) NO-generating mechanism of nitrobenzene. (B) Mechanism of the release of NO from BNNs. *(A) Reproduced from Kim J, Saravanakumar G, Choi HW, Park D, Kim WJ. A platform for nitric oxide delivery. J Mater Chem B, 2014;2:341–356 with permission from Royal Society of Chemistry. (B) Reproduced from Namiki S, Arai T, Fujimori K: High-performance caged nitric oxide: a new molecular design, synthesis, and photochemical reaction. J Am Chem Soc 1997;119:3840–3841 with permission from American Chemical Society.*

molecules per BNN through UV-light-mediated photocleavage of the *N–N* bond [24] (Fig. 5B).

However, the NO release from the photo-triggered NO donors is mostly accomplished through UV light irradiation, which may induce deleterious effect to normal tissues. Therefore, the development of organic donors that trigger NO under NIR light is highly desirable.

THERAPEUTIC ROLE OF NO IN THE TREATMENT OF CANCER

NO itself has shown to induce apoptosis or necrosis, thus emerging as a potential agent for antitumor therapy. In addition, NO acts as chemosensitizer and radiosensitizer (Fig. 6).

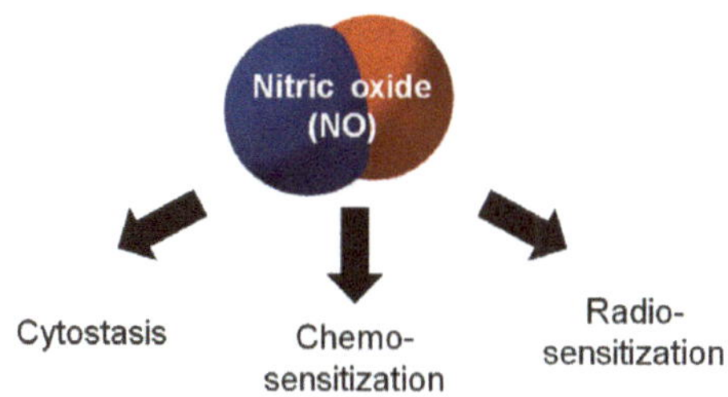

FIG. 6
Major therapeutic role of nitric oxide in the treatment of cancer.

Direct Role of NO in Inducing Cell Cytotoxicity

Endogenous NO plays an important role in cancer biology. In the past few decades, several studies have shown that the induction of apoptosis in tumor cells is mediated through macrophage-derived NO [26, 27]. In particular, activated macrophages produce large amounts of NO that induce both cytostasis and cytotoxicity to tumor cells, and this process is associated with the activation of nitric oxide synthase (NOS), an enzyme that exists in several isoforms [27–29]. The NO generated from the activated macrophage could interact with intracellular iron to form iron-nitrosyl complex, resulting in the loss of activity of enzymes that have catalytically active nonheme-iron coordinated to sulfur. The loss of a significant amount of intracellular iron, inhibition of mitochondrial respiration and aconitase, and DNA synthesis may result in the antitumor effects [29, 30]. Furthermore, NO could also interact with other metallic ions such as zinc, thus inhibiting the binding between cysteine and zinc. This process could alter cysteine-zinc finger-based proteins and transcription factors binding to DNA.

Other possible NO cytotoxicity mechanisms could be achieved through the conversion of NO into higher reactive nitrogen species (RNS) [6]. When NO reacts with superoxide generated during respiration, highly reactive species such as peroxynitrite ($ONOO^-$) and dinitrogen trioxide (N_2O_3) are generated that have a critical impact to the cell cycle and apoptosis of tumor cells [26, 31]. This generated RNS could induce mitochondrial inhibition, suppression of ribonucleotide reductase, deamination of DNA, lipid peroxidation, and protein nitrosylation, resulting in cell growth inhibition. All of the above mechanisms suggest that NO could be potentially used as a therapeutic agent for antitumor therapy [6, 26, 31].

NO as Chemo-Sensitizer by Reducing Drug Resistance

Resistance to chemotherapy is one of the major factors for failure of treatment in progressive and inoperable cancers. Recently, it is widely accepted that hypoxic condition of the tumor microenvironment mediated by the transcription factor hypoxia-inducible factor 1 (HIF-1) has been implicated in chemotherapy failure and drug resistance [32–34]. In hypoxic condition, HIF-1 activates the multidrug resistance 1 (*MDR1*) gene that encodes for the *P*-glycoprotein (Pgp), which functions as an efflux pump and decreases the intracellular concentration of chemotherapeutic drugs in tumor cells [35–39]. In addition, hypoxia induces tumor cells to be less sensitive to p53-mediated apoptosis and downregulates the DNA mismatch repair system, which contributes to cisplatin resistance in human cancer cells [34, 40]. Therefore, targeting hypoxia is considered to be one of the important strategies for the enhancement of antitumor therapy.

NO has a potential to inhibit HIF-1 activation and promotes HIF-1 proteasomal degradation in hypoxic malignant cells [41–43]. In addition to HIF-1 reduction, NO inhibits Pgp to reduce the MDR effect [44]. Besides, NO was also shown to activate p53 in cancer cells and promotes apoptosis of cancer cells by p53-dependent anticancer agents including cisplatin [45, 46]. In this regard, numerous NO donors have been investigated as potent chemosensitizing agents to improve the therapeutic index of the chemotherapy. DETA-NONOate has been shown to reverse hypoxia-mediated drug resistance and enhanced antitumor effects of drugs such as 5-fluorouracil and doxorubicin (DOX) in human breast carcinoma cell (MDA-MB-231) and mouse melanoma cell (B16F10) in vitro [47, 48]. In the head and neck squamous cell carcinoma, DETA/NO significantly enhanced the antitumor effect of cisplatin [49]. Similarly, the combination of DETA/NO and cisplatin increased tumor regression and survival rate in B16F1 melanoma-bearing mice compared with those treated with cisplatin alone [50]. The combination of Taxol with *S*-nitrosocaptopril, an organic-nitrate-based NO donor, enhanced the cytotoxicity effect of Taxol in prostate cancer cells (PC-3 and LNCaP), but no significant effect was observed in neuroblastoma cell lines (SKNDZ and CHP212) [51]. The combination of another organic-nitrate-based NO donor, bis(nitroxymethyl) oxetane, with chemotherapeutics such as cyclophosphamide and DOX induced tumor regression and significantly improved the survival rate of leukemia-bearing mice compared with that of only drugs [52]. Irrespective of the types and chemical structures, these NO donors have shown promising chemosensitizing effects for subsequent chemotherapy.

NO as Radio-Sensitizer for Effective Radio-Therapy

Hypoxic condition in the tumor microenvironment has been known as an important factor that negatively impacts radiotherapy [53]. It has been reported that hypoxic tumor cells were 2.5–3 times less radiosensitive than well-oxygenated cells [54, 55]. The most widely accepted mechanism of radiotherapy involves the direct ionization of DNA or the generation of oxygen-centered radical such as hydroxyl radical or superoxide anions by the ionization of water [56, 57]. Under sufficient oxygen condition, oxygen can react with the damaged DNA, resulting in a stable organic peroxide, which is not easily repaired by a cell [53]. It has been shown that well-oxygenated cells are more radiosensitive than hypoxic cells. Thus, reoxygenation on tumor tissue is one of the effective approaches to increase radiosensitivity [58]. However, this strategy has been limited by several factors such as limited oxygen diffusion inside tumors because of the surrounding actively respiring cells [59]. To overcome these issues, NO has been actively studied as a radiosensitizer [58]. Several studies have demonstrated that the combination of radiotherapy and NO donors could exhibit enhanced radiosensitivity than radiation alone in hypoxic tumor cells [6, 60, 61]. In particular, the NO concentration at a micromolar level was able to enhance 2.1- to 2.5-fold high radiosensitivity, which is significantly higher than any other radiosensitizing drugs [6]. For the treatment of brain malignant tumors, it has been suggested that a combination of radiotherapy, NO donors or inducers (such as L-arginine), and a hyperoxygenation could permit the diffusion of oxygen into the cells, thus increasing sensitivity to radiation. The generation of NO by endogenous NO donors through the activation of the NO synthase or by administrating exogenous NO donors (such as ISDN, SNP, SPER/NO, or SNAP) was shown to enhance radiotherapy in several hypoxic tumor cells [61–65].

Although the exact mechanism of NO radiosensitization is not clearly understood, there have been several mechanisms proposed in the literature. One of these mechanisms suggested that NO could improve tumor oxygenation by increasing the blood flow via vasodilation and decreasing oxygen consumption near blood vessels. The later process could allow deep penetration of oxygen in the target tissues [66–68]. In addition, NO can also decrease tumor oxygen consumption by inhibition of respiration through binding reversibly to cytochrome *c* oxidase [59, 67]. Some studies have indicated NO-mediated tumor reoxygenation may increase tumor radiosensitivity through HIF-1 degradation [59]. It has also been suggested that NO-mediated hypoxic radiosensitization may proceed a process similar to that of oxygen in radiotherapy [60, 69].

DEVELOPMENT OF NO DELIVERY NANOPLATFORMS BY COMBINING FUNCTIONAL NANOMATERIALS AND NO DONORS

Stimuli-Responsive NO Delivery Platforms Using *S*-Nitrosothiol

As discussed earlier, low-molecular-weight RSNOs are known to act as reservoir/transporter of NO in biological systems that are capable of generating NO spontaneously under physiological conditions [70–72]. However, such a

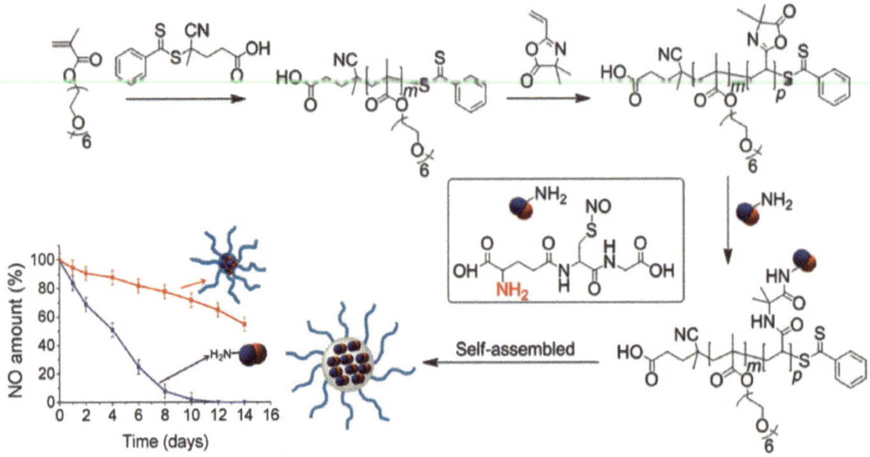

FIG. 7

Preparation of the GSNO-conjugated polymeric nanoparticle and NO-releasing property. *Reprinted with permission from Duong HTT, Kamarudin ZM, Erlich RB, Li Y, Jones MW, Kavallaris M, Boyer C, Davis TP. Intracellular nitric oxide delivery from stable NO-polymeric nanoparticle carriers. Chem Commun 2013;49:4190–4192. Copyright 2013, Royal Society of Chemistry.*

rapid and uncontrolled release of NO from low-molecular-weight S-nitrosothiol may limit its potential therapeutic application in vivo [72]. In addition, some of these NO donors have shown poor aqueous solubility and lack both stability and tumor specificity. Therefore, to improve the bioavailability and target specificity of NO, RSNOs have been conjugated to carrier systems. This strategy not only would allow to load a large amount of NO with improved aqueous stability but also facilitates sustained and controlled release of the loaded NO payload.

Duong et al. developed GSNO-conjugated polymeric nanoparticles for efficient intracellular delivery of NO [73] (Fig. 7). To prepare the nanoparticles, they polymerized oligoethylene glycol methacrylate (OEGMA) and 2-vinyl-4,4-dimethyl-5-oxazolone monomer (VDM) by the reversible addition-fragmentation chain transfer (RAFT) polymerization, and then, the NO donor GSNO was conjugated to the resulting block copolymer via nucleophilic addition at the azlactone group of the VDM moiety [74], thus allowing the NO-releasing moiety to accommodate on the hydrophobic inner core of the polymeric nanoparticles upon self-assembly. Consequently, the stability of GSNO was significantly improved and, thereby, rendered a long half-life time compared with that of free GSNO. The release of NO from the nanoparticles was stimulated by intracellular redox conditions, such as GSH, ascorbic acid, and metallic ions. The combination of these NO-releasing polymeric nanoparticles and chemotherapeutic drug cisplatin demonstrated an enhanced antitumor activity in BE(2)-C neuroblastoma cells than cisplatin alone, whereas no such synergistic effects were observed on the noncancerous fibroblast cell line MRC-5, indicating the potential of the GSNO-conjugated polymeric nanoparticles as a chemosensitizer.

It is widely known that high-energy X-ray radiation can break down certain chemical bonds having low bond energy such as disulfide (S–S, 240 kJ mol^{-1}) and diselenide bonds (Se–Se, 172 kJ mol^{-1}) [75, 76]. Therefore, it was speculated that the S–N bond (with bond energy 150 kJ mol^{-1}) in RSNO can also be easily cleaved by X-ray radiation. Based on this premise, Fan et al. developed an X-ray radiation-controlled NO-releasing system by functionalizing upconversion nanoparticles with RNSO-conjugated mesoporous silica (Fig. 8) for on-demand hypoxic radiosensitization [75]. As expected, the as-constructed system showed accelerated NO release upon X-ray irradiation, likely through cleavage of the S–N bond, and further, the NO release rate was able to control by simply regulating the dose of X-rays. These nanoparticles with X-ray irradiation demonstrated high antitumor effect in 4T1-tumor-bearing mice compared with X-ray irradiation alone, which was primarily ascribed to strong radiosensitizing effects by the X-ray-mediated NO release. Since X-rays have high penetrating power, this system has the potential to treat deep-seated solid tumors.

The photothermally generated heat was also investigated as a potential stimulus to promote NO release from RSNO via scission of S–NO bonds [77]. To accomplish this trait, the carrier should be integrated with photothermal agents. By taking the advantages of the efficient photothermal characteristic of nanoparticles having a spherical magnetite core (Fe_3O_4) and a polydopamine (PDA) shell, Guo et al. developed photothermally triggered NO nanogenerators based on the RSNO for the reversal of MDR in cancer (Fig. 9). The outer mesoporous shell of the nanogenerator was loaded with the drug doxorubicin, while the surface was further functionalized with transferrin (Tf) to endow tumor-targeting characteristics. Upon NIR light irradiation, the nanogenerator efficiently converted the absorbed photon of light into heat that triggered NO release both in MCF-7/ADR cells and MCF-7/ADR tumor-bearing mice. The high local NO concentrations were able to inhibit the Pgp, which is an important cause for MDR. This, consequently, improved the intracellular accumulation of DOX and resulted in high in vitro and in vivo antitumor efficacies. The results from this study clearly suggest that combining the photothermal agent, RSNO, and an anticancer drug into a single nanoplatform may have a high potential to overcome MDR in cancer therapy.

Development of NONOates-Based NO Delivery Nanoplatform

As discussed before, NONOates have high NO loading capacity and can deliver two moles of NO per mole of NONOate-functionalized molecules. However, uncontrolled spontaneous release of NO molecules from NONOates under

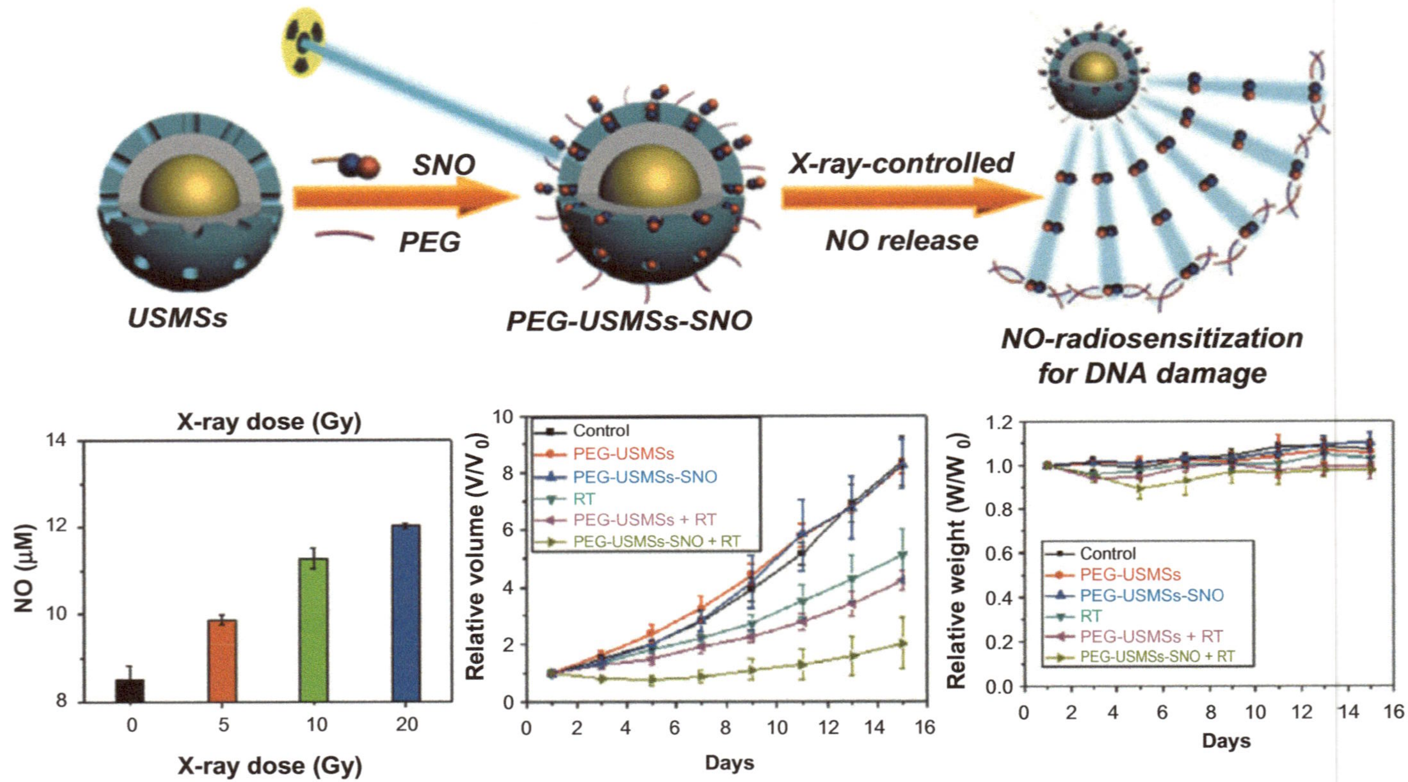

FIG. 8

Overall strategy of X-ray-triggered NO-releasing nanoparticles for antitumor therapy. The nanoparticle showed X-ray dose-dependent NO release and enhancement of radiotherapy without toxicity. *Reprinted with permission from Fan W, Bu W, Zhang Z, Shen B, Zhang H, He Q, Ni D, Cui Z, Zhao K, Bu J, Du J, Liu J, Shi J. X-ray radiation-controlled NO-release for on-demand depth-independent hyxpoxic radiosensitization. Angew Chem Int Ed 2015;54:14026–14030. Copyright 2015, John Wiley & Sons.*

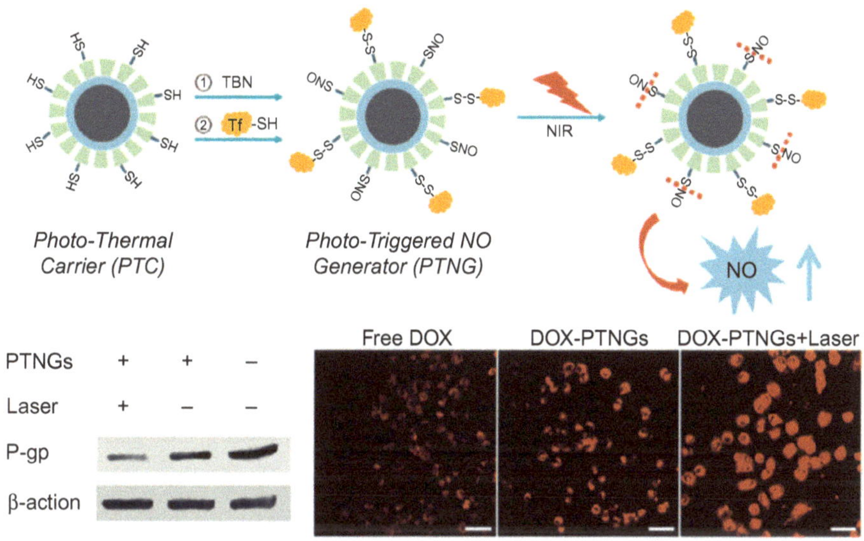

FIG. 9

Schematic diagram of photothermally triggered NO-releasing nanoparticles and their activities. NO release from photothermal activity showed p-glycoprotein inhibition with enhancement of drug accumulation. *Reprinted with permission from Guo R, Tian Y, Wang Y, Yang W. Near-infrared laser-triggered nitric oxide nanogenerators for the reversal of multidrug resistance in cancer. Adv Funct Mater 2017;1606398. Copyright 2017, John Wiley & Sons.*

physiological conditions has caused significant concern in utilizing them for therapeutic applications. To surmount these issues, several strategies have been adopted including conjugating NONOates into macromolecular scaffolds or physically encapsulating NONOate donors into the carrier systems.

Recently, Chung et al. physically loaded DETA/NO and the anticancer agent irinotecan (CPT-11) into hollow microspheres composed of poly(DL-lactic-*co*-glycolic acid) (PLGA) and further investigated its potential for reversing Pgp-mediated MDR [78] (Fig. 10). Once the active-agent-loaded microspheres were extravasated into the acidic tumor microenvironment, the protons were infiltrated into the shell of the hollow microsphere and triggered the release of gaseous NO molecules from DETA/NO that not only inhibited the Pgp but also facilitated the release of CPT-11 by destabilizing the integrity of the carrier structure [78–80], which eventually improved the intracellular accumulation of CPT-11 and demonstrated high antitumor activity. Besides improving the antitumor efficacy by chemosensitization and reversal of MDR in cancer, NO-donor-loaded nanoparticles were also reported to induce tumor-specific vasodilation due to the enhanced accumulation of nanoparticles within the tumor tissues via the EPR effect. In a report, Tahara et al. showed that PEGylated liposomes loaded with DETA/NO were able to accumulate two times more in

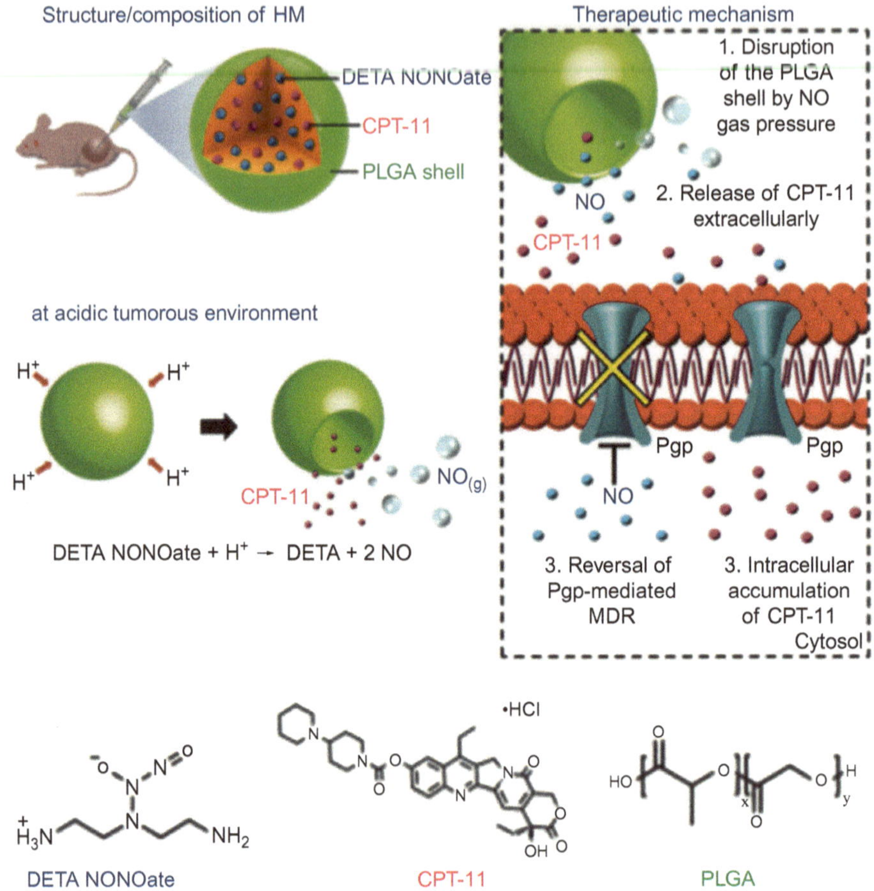

FIG. 10

Illustration of the structure of NO-generating PLGA hollow nanoparticle and its therapeutic mechanism. *Reprinted with permission from Chung MF, Liu HY, Lin KJ, Chia WT, Sung HW. A pH-responsive carrier system that generates NO bubbles to trigger drug release and reverse P-glycoprotein-mediated multidrug resistance. Angew Chem Int Ed 2015;54:9890–9893. Copyright 2015, John Wiley & Sons.*

tumors than empty liposomes, with no appreciable change in systemic blood pressure [81] (Fig. 11). These improved accumulations were attributed to the improved EPR effect, caused by the specific vasodilation induced by NO released from the DETA/NO liposomes. It is also gratifying to note that even a small dose of NO-releasing NONOates was able to induce efficient tumor-specific biodistribution of the NONOate-loaded liposomes.

It should be noted that the physical encapsulation of NONOate donors in their native forms inside nanoparticles always does not necessarily guarantee complete inhibition of unintended NO release during formulation or circulation in the blood. Therefore, Kumar et al. prepared block copolymer micelles loaded

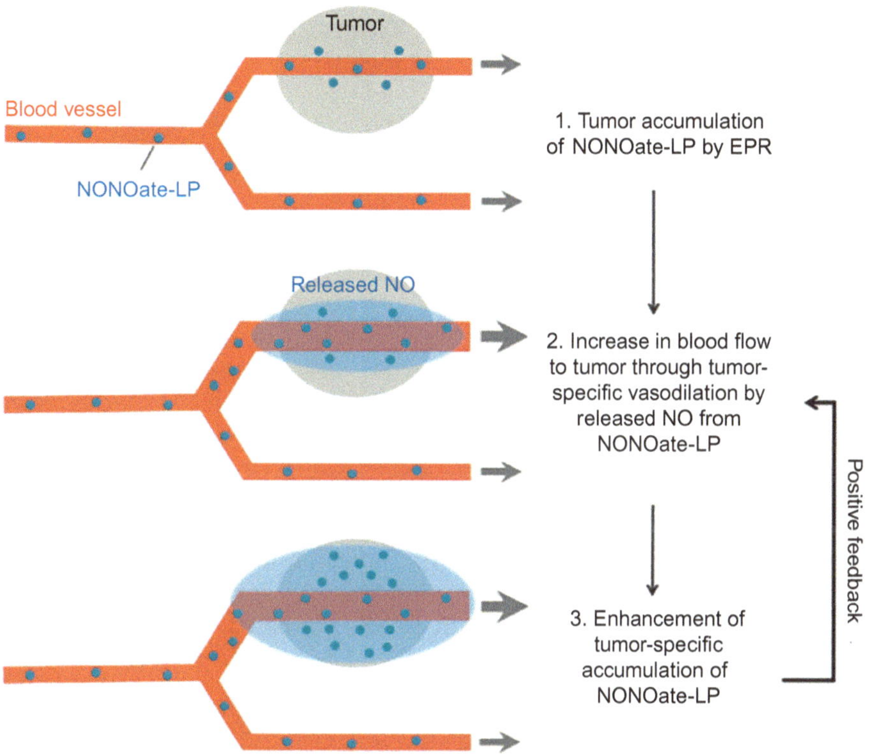

FIG. 11

Overall scheme of the improved accumulation of NO-donor-loaded liposomes. *Reprinted with permission from Tahara Y, Yoshikawa T, Sato H, Mori Y, Zahangir MH, Kishimura A, Mori T, Katayama Y. Encapsulation of a nitric oxide donor into a liposome to boost the enhanced permeation and retention (EPR) effect. Med Chem Communi 2017;8:415–421. Copyright 2016, Royal Society of Chemistry.*

with NO prodrug JS-K [82], which have shown potent antineoplastic activity [13, 83] (Fig. 12). The incorporation of JS-K prodrugs into the micelles prevented its activation by glutathione during circulation, thereby significantly improving the decomposition time from 15 min to 5 h, and this time interval was believed to be optimal for targeting micellar carriers to solid tumors via the EPR effect. These prodrug-loaded micelles showed similar antitumor effects as native prodrugs in vitro. After uptake by the tumor cells, the NO release from the prodrugs was facilitated by the elevated intracellular glutathione present inside the tumor cells.

Organic Nitrate-Based NO Delivery Platforms

Among several NO donors, low-molecular-weight organic-nitrate-based NO donors such as glyceryl trinitrate (GTN) have been widely employed as

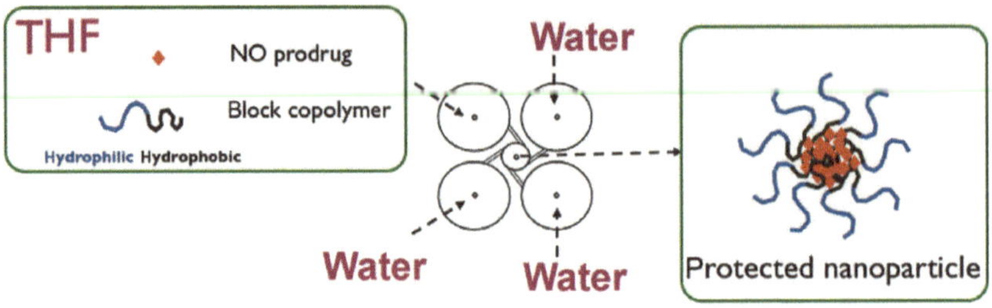

FIG. 12

Illustration of NO-prodrug-loaded nanoparticle. *Reprinted with permission from Kumar V, Hong SY, Maciag AE, Saavedra JE, Adamson DH, Prud'homme RK, Keefer LK, Chakrapani H. Stabilization of the nitric oxide (NO) prodrugs and anticancer leads, PABA/NO and double JS-K, through incorporation into PEG-protected nanoparticles. Mol Pharmaceut 2010;7:291–298. Copyright 2010, American Chemical Society.*

therapeutic agents for cardiovascular diseases. However, these low-molecular-weight NO donors are more susceptible to reticuloendothelial system exposure and easily cleared from the body [84]. Therefore, in recent years, organic-nitrate-based nanoplatforms have been increasingly investigated. Song et al. synthesized a new type of a polymeric NO donor based on organic nitrates, namely, nitrate-functionalized D-α-tocopheryl polyethylene glycol succinate (TNO_3) (Fig. 13) [85]. Owing to its amphiphilic characteristics, the TNO_3 was able to self-assemble into stable micelles in an aqueous condition [86, 87]. These TNO_3 micelles exhibited a negligible NO release in cell growth medium, whereas controlled and much higher NO release was observed in human hepatocellular carcinoma HepG2 cells due to an elevated GSH concentration in the tumor cells. When these micelles were treated along with DOX to HepG2 cells, the NO released from the micelles induced enhanced cellular uptake of DOX, resulting in improved cytotoxicity. The coefficient of drug interaction (CDI) values of TNO_3 and DOX were found to be much lower than 0.7 at various time intervals, indicating synergistic effect of TNO_3 and DOX. In vivo administration of the TNO_3 and DOX combination showed higher antitumor effect compared with TNO_3 or DOX alone. This study also suggests that improving the NO concentration at the tumor may enhance the therapeutic efficacy of chemotherapeutic drugs.

Recently, Deepagan et al. have prepared amphiphilic NO-releasing block copolymer (PEG-*b*-NO-Dex) micelles composed of the hydrophilic PEG shell and the hydrophobic nitrated dextran inner core [86] (Fig. 14). When administered in human colon cancer HT29-bearing mice, the PEG-*b*-NO-Dex micelles of the size of about 130 nm had about 1.75-fold higher accumulation in tumor tissue compared with that of control micelles without NO-generating capabilities. Monitoring the tumor blood flow by the three-dimensional ultrasound power Doppler indicated about 30% increase in the blood flow of the tumor

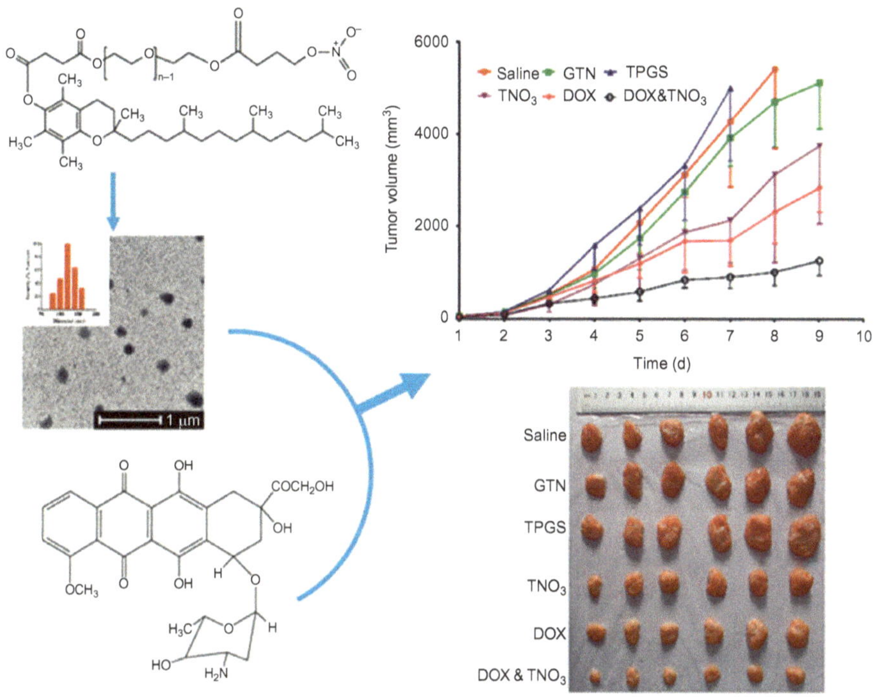

FIG. 13

Schematic illustration of a nitrate-functionalized nanoparticle as chemosensitizer. Combination of DOX and NO-releasing nanoparticles showed high antitumor effect. *Reprinted with permission from Song Q, Tan S, Zhuang X, Guo Y, Zhao Y, Wu T, Ye Q, Si L, Zhang Z. Nitric oxide releasing D-α-tocopheryl polyethylene glycol succinate for enahncing antitumor activity of doxorubicin. Mol Pharmaceut 2014;11:4118–4129. Copyright 2014, American Chemical Society.*

vasculature at 6 h post injection. To investigate the therapeutic potential, DOX was physically loaded into the micelles by solvent exchange method. Because of the NO-mediated vasodilation, the DOX accumulation in the tumors of the mice treated with PEG-*b*-NO-Dex micelles (14.50 μg/g tissue) was significantly higher than in those treated with the control micelles (9.33 μg/g tissue) and free DOX (2.16 μg/g tissue). Consequently, DOX-loaded PEG-*b*-NO-Dex micelles showed higher in vivo antitumor efficacy over the control groups. Thus, this study clearly suggests that NO-releasing micelles could boost the EPR effect by means of dilating tumor blood vessels.

Photo-Induced NO Donors With Functional Nanomaterials

Photo-triggered NO donors are highly promising for in vivo applications because of their ability to release NO remotely in a spatiotemporal controlled

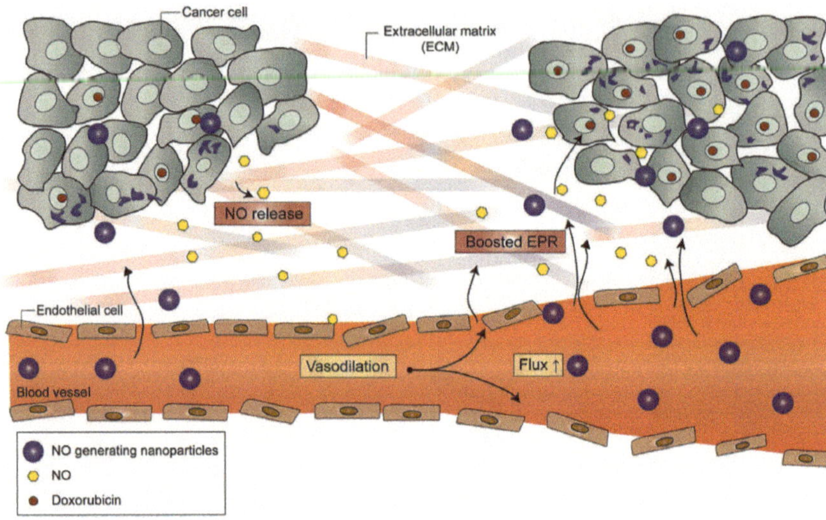

FIG. 14

Schematic illustration of the boosted EPR effect by using NO-generating nanoparticles. *Reprinted with permission from Deepagan VG, Ko H, Kwon S, Rao NV, Kim SK, Um W, Lee S, Min J, Lee J, Choi KY, Shin S, Suh M, Park JH. Intracellularly activatable nanovasodilators to enhance passive cancer targeting regime. Nano Lett 2018;18:2637–2644. Copyright 2018, American Chemical Society.*

manner at the target disease site [87, 88]. However, some of the photo-triggered NO donors suffer from cytotoxic issues, caused by the photodegradation products of the donors [22]. For example, metal-nitrosyl compounds especially sodium nitroprusside can induce severe side effects such as hypotension and cyanide toxicity [89]. To alleviate the undesired side effects, nanoparticle-based photo-triggered NO donors have been developed because this could improve the tumor specificity and reduce the nonspecific activation of NO at the normal tissue. In this section, we describe some of the representative photo-triggered NO delivery nanoplatforms.

Fan et al. developed biodegradable polymeric nanoparticle, coloaded with photo-triggered NO donor (*N,N′*-di-*sec*-butyl-*N,N′*-dinitroso-1,4-phenylenediamine (BNN6)) and DOX, to reverse the MDR of tumor cells and enhance chemosensitization for enhanced anticancer therapy [90] (Fig. 15). The photo-triggered BNN6 donors are sensitive to UV light. In vitro cytotoxicity studies against the MDR ovarian cancer cell line OVCAR-8/ADR showed increased cytotoxicity under UV light irradiation compared with the control micelles loaded with only DOX and free DOX. As anticipated, the NO generated from BNN6 was found to be effective in reversing the MDR and increasing the anticancer potency of DOX.

Unfortunately, most of the photo-triggered NO donors are induced by UV/blue-light irradiation with limited tissue penetration [91–93]. Recently,

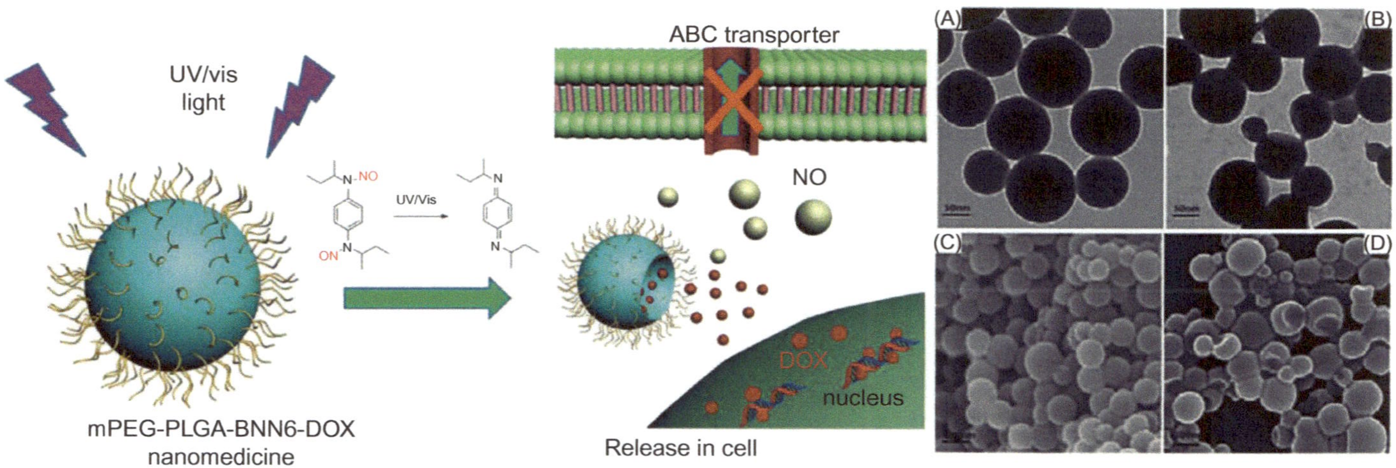

FIG. 15

Schematic illustration of BNN6-loaded nanomedicine and its role as chemosensitizer. UV-triggered NO release inhibited the ABC transporter, which improved chemotherapy. *Reprinted with permission from Fan J, He Q, Zhang F, Yang X, Wang Z, Lu N, Fan W, Lin L, Niu G, He N, Song J, Chen X. Light-responsive biodegradable nanomedicine overcomes multidrug resistance via NO-enahced chemosensitization. ACS Appl Mater Interf 2016;8:13804–13811. Copyright 2016, American Chemical Society.*

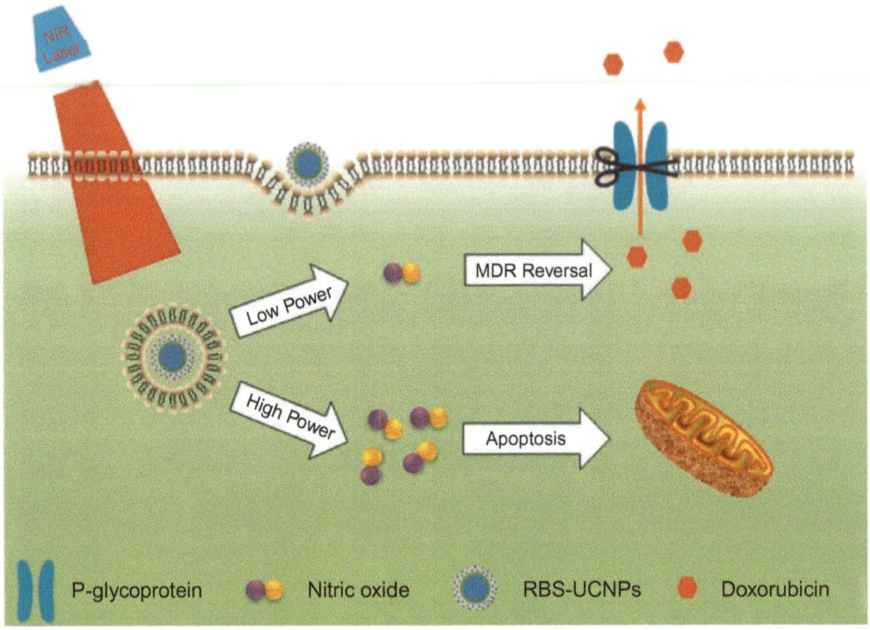

FIG. 16

Schematic illustration of RBS-loaded UCNPs. UCNPs turned NIR photon to UV photons, which generate NO from RBS. The nanoparticle showed enhanced anticancer effect due to NO-mediated MDR reversal and apoptosis. *Reprinted with permission from Zhang X, Tian G, Yin W, Wang L, Zheng X, Yan L, Li J, Su H, Chen C, Gu Z, Zhao Y. Controllable generation of nitric oxide by near-infrared-sensitized upconversion nanoparticles for tumor therapy. Adv Funct Mater 2015;25:3049–3056. Copyright 2015, John Wiley & Sons.*

upconversion nanoparticles (UCNPs) have been demonstrated to absorb NIR light and convert them into higher-energy UV/Vis photons, and this provides an effective means to induce UV-light-triggerable NO donors in the therapeutic window of NIR light [93, 94] (Fig. 16). Encouraged by these desired features of UCNPs, Zhang et al. developed an NIR-controlled NO-generating UCNP-based platform for tumor therapy [93]. By employing a photo-triggered NO donor (RBS) and anticancer drug (DOX), they demonstrated that a high concentration of NO produced by NIR light can directly kill the cancer cells, while a low concentration can significantly inhibit the Pgp expression in MDR cancer cells and reverse MDR in chemotherapy.

CONCLUSION

Several studies have established that the function of NO is highly dependent on its local concentration. Especially, anticancer effect mediated by NO requires high amount of NO concentration. However, administration of low-molecular-weight NO donors did not produce an optimal therapeutic concentration at the tumor tissue, owing to poor aqueous stability, off-target specificity, and uncontrolled NO release. Therefore, developing nanoplatforms that can deliver NO to the tumor sites in a spatiotemporal controlled fashion is highly essential to realize the full potential of NO. To date, a number of NO-releasing nanoplatforms have been developed by combining NO donors and functional nanomaterials as carriers. Both in vitro and in vivo studies have demonstrated high therapeutic efficacies for NO-releasing nanoplatforms compared with free NO donors. Furthermore, these nanoplatforms also permit to load anticancer therapeutic agents along with the NO donors. This combined delivery of anticancer drugs and NO using NO-releasing nanoplatforms demonstrated an enhanced in vivo antitumor efficacy by reversing Pgp-mediated MDR.

Acknowledgments

This research was supported by the Bio & Medical Technology Development Program of the National Research Foundation of Korea (NRF; NRF-2017M3A9F5030930), Basic Science Research Program through the NRF (2017R1E1A1A01074088), and Creative Materials Discovery Program through the National Research Foundation of Korea (NRF-2018M3D1A1058813) funded by the Ministry of Science, ICT & Future Planning.

References

[1] SoRelle R. Nobel prize awarded to scientists for nitric oxide discoveries. Circulation 1998;98:2365–6.

[2] Kim J, Saravanakumar G, Choi HW, Park D, Kim WJ. A platform for nitric oxide delivery. J Mater Chem B 2014;2:341–56.

[3] Carpenter AW, Schoenfisch MH. Nitric oxide release: Part II. Therapeutic applications. Chem Soc Rev 2012;41:3742–52.

[4] Chinje EC, Stanford IJ. Role of nitric oxide in growth of solid tumours: a balancing act. Essays Biochem 1997;32:61–72.

[5] Morbidelli L, Donnini S, Ziche M. Role of nitric oxide in tumor angiogenesis. Cancer Treatment Res 2004;117:155–67.

[6] Pacher P, Beckman JS, Liaudet L. Nitric oxide and peroxynitrite in health and disease. Phys Rev 2007;87:315–424.

[7] Hirst D, Robson T. Targeting nitric oxide for cancer therapy. J Pharm Pharmacol 2007;59:3–13 (prev p).

[8] Keefer LK. Fifty years of diazeniumdiolate research. From laboratory curiosity to broad-spectrum biomedical advances. ACS Chem Biol 2011;6:1147–55.

[9] Drago RS, Paulik FE. The reaction of nitrogen(II) oxide with diethylamine. J Am Chem Soc 1960;82:96–8.

[10] Maragos CM, Morley D, Wink DA, Dunams TM, Saavedra JE, Hoffman A, Bove AA, Isaac L, Hrabie JA, Keefer LK. Complexes of NO with nucleophiles as agents for the controlled biological release of nitric oxide. Vasorelaxant effects. J Med Chem 1991;34:3242–7.

[11] Sergio H. Nitric oxide for cancer therapy. Future Sci OA 2015;1:FSO44.

[12] Kim J, Lee Y, Singha K, Kim HW, Shin JH, Jo S, Han DK, Kim WJ. NONOates-polyethylenimine hydrogel for controlled nitric oxide release and cell proliferation modulation. Bioconjug Chem 2011;22:1031–8.

[13] Shami PJ, Saavedra JE, Wang LY, Bonifant CL, Diwan BA, Singh SV, Gu Y, Fox SD, Buzard GS, Citro ML, Waterhouse DJ, Davies KM, Ji X, Keefer LK. JS-K, a glutathione/glutathione S-transferase-activated nitric oxide donor of the diazeniumdiolate class with potent antineoplastic activity. Molecular. Cancer Ther 2003;2:409–17.

[14] Hogg N. The biochemistry and physiology of S-nitrosothiols. Annu Rev Pharmacol Toxicol 2002;42:585–600.

[15] Wang PG, Xian M, Tang X, Wu X, Wen Z, Cai T, Janczuk AJ. Nitric oxide donors: chemical activities and biological applications. Chem Rev 2002;102:1091–134.

[16] Miller MR, Megson IL. Recent developments in nitric oxide donor drugs. Br J Pharmacol 2007;151:305–21.

[17] Kleschyov AL, Oelze M, Daiber A, Huang Y, Mollnau H, Schulz E, Sydow K, Fichtlscherer B, Mulsch A, Munzel T. Does nitric oxide mediate the vasodilator activity of nitroglycerin? Circ Res 2003;93:e104–12.

[18] Bourassa J, DeGraff W, Kudo S, Wink DA, Mitchell JB, Ford PC. Photochemistry of Roussin's red salt, $Na_2[Fe_2S_2(NO)_4]$, and of Roussin's black salt, $NH_4[Fe_4S_3(NO)_7]$. In situ nitric oxide generation to sensitize γ-radiation induced cell death. J Am Chem Soc 1997;119:2853–60.

[19] Grossi L, D'Angelo S. Sodium nitroprusside: mechanism of NO release mediated by sulfhydryl-containing molecules. J Med Chem 2005;48:2522–626.

[20] Rose MJ, Mascharak PK. Photoactive ruthenium nitrosyls: effects of light and potential application as NO donors. Coordination Chem Rev 2008;252:2093–114.

[21] Salvatore S, Salvatore P, Giuseppe C, Sabrina C, Giuseppe C. Light-controlled nitric oxide generation from a novel self-assembled monolayer on a gold surface. Angew Chem Int Ed 2002;41:1914–7.

[22] Suzuki T, Nagae O, Kato Y, Nakagawa H, Fukuhara K, Miyata N. Photoinduced nitric oxide release from nitrobenzene derivatives. J Am Chem Soc 2005;127:11720–6.

[23] Sortino S, Marconi G, Condorelli G. New insight on the photoreactivity of the phototoxic anticancer flutamide: photochemical pathways selectively locked and unlocked by structural changes upon drug compartmentalization in phospholipid bilayer vesicles. Chem Commun 2001;13:1226–7.

[24] Namiki S, Arai T, Fujimori K. High-performance caged nitric oxide: a new molecular design, synthesis, and photochemical reaction. J Am Chem Soc 1997;119:3840–1.

[25] Ieda N, Hotta Y, Miyata N, Kimura K, Nakagawa H. Photomanipulation of vasodilation with a blue-light-controllable nitric oxide releaser. J Am Chem Soc 2014;136:7085–91.

[26] Albina JE, Reichner JS. Role of nitric oxide in mediation of macrophage cytotoxicity and apoptosis. Cancer Metastasis Rev 1998;17:39–53.

[27] Klimp AH, de Vries EG, Scherphof GL, Daemen T. A potential role of macrophage activation in the treatment of cancer. Crit Rev Oncol/Hematol 2002;44:143–61.

[28] Hibbs JB, Taintor RR, Vavrin Z. Macrophage cytotoxicity: role for L-arginine deiminase and amino nitrogen oxidation to nitrite. Science 1987;235:473–6.

[29] Stuehr DJ, Nathan CF. Nitric oxide: a macrophage product responsible for cytostasis and respiratory inhibition in tumor target cells. J Exp Med 1989;169:1543–55.

[30] Hibbs JB, Taintor RR, Vavrin Z, Rachil EM. Nitric oxide: a cytotoxic activated macrophage effector molecule. Biochem Biophys Res Commun 1988;157:87–94.

[31] Lin KT, Xue JY, Nomen M, Sper B, Wong PYK. Peroxynitrite-induced apoptosis in HL-60 cells. J Biol Chem 1990;265:7108–11.

[32] Rohwer N, Cramer T. Hypoxia-mediated drug resistance: novel insights on the functional interaction of HIFs and cell death pathways. Drug Resist Updates 2011;14:191–201.

[33] Wang GL, Semenza GL. Purification and characterization of hypoxia-inducible factor 1. J Biol Chem 1995;270:1230–7.

[34] Trédan O, Galmarini CM, Patel K, Tannock IF. Drug resistance and the solid tumor microenvironment. J Natl Cancer Inst 2007;99:1441–54.

[35] Comerford KM, Wallance TJ, Karhausen J, Louis NA, Montalto MC, Colgan SP. Hypoxia-inducible factor-1-dependent regulation of multidrug resistance (MDR1) gene. Cancer Res 2002;62:3387–94.

[36] Lv Y, Zhao S, Han J, Zheng L, Yang Z, Zhao L. Hypoxia-inducible factor-1α induces multidrug resistance protein in colon cancer. OncoTargets Ther 2015;8:1941–8.

[37] Li DW, Dong P, Wang F, Chen XW, Xu CZ, Zhou L. Hypoxia induced multidrug resistance of laryngeal cancer cells via hypoxia-inducible factor-1α. Asian Pac J Cancer Preven 2013;14:4853–8.

[38] Gottesman MM. Mechanisms of cancer drug resistance. Annu Rev Med 2002;53:615–27.

[39] Milane L, Duan Z, Amiji M. Role of hypoxia and glycolysis in the development of multi-drug resistance in human tumor cells and the establishment of an orthotopic multi-drug resistant tumor model in nude mice using hypoxic pre-conditioning. Cancer Cell Int 2011;11:3.

[40] Kondo A, Safaei R, Mishima M, Niedner H, Lin X, Howell SB. Hypoxia-induced enrichment and mutagenesis of cells that have lost DNA mismatch repair. Cancer Res 2001;61:7603–7.

[41] Yasuda H. Solid tumor physiology and hypoxia-induced chemo/radio-resistance: novel strategy for cancer therapy: Nitric oxide donor as a therapeutic enhancer. Nitric Oxide 2008;19:205–16.

[42] Yasuda H, Nakayama K, Watanabe M, Suzuki S, Fuji H, Okinaga S, Kanda A, Zayasu K, Sasaki T, Asada M, Suzuki T, Yoshida M, Yamanda S, Inoue D, Kaneta T, Kondo T, Takai Y, Sasaki H, Yanagihara K, Yamaya M. Nitroglycerin treatment may enhance chemosensitivity to docetaxel and carboplatin in patients with lung adenocarcinoma. Clin Cancer Res 2006;12:6748–57.

[43] Agani FH, Puchowicz M, Chavez JC, Pichiule P, LaManna J. Role of nitric oxide in the regulation of HIF-1alpha expression during hypoxia. Am J Physiol 2002;283:C178–86.

[44] Kim J, Yung BC, Kim WJ, Chen X. Combination of nitric oxide and drug delivery systems: tools for overcoming drug resistance in chemotherapy. J Control Release 2017;263:223–30.

[45] Thomas DD, Espey MG, Ridnour LA, Hofseth LJ, Mancardi D, Harris CC, Wink DA. hypoxic inducible factor 1 alpha, extracellular signal-regulated kinase, and p53 are regulated by distinct threshold concentrations of nitric oxide. Proc Natl Acad Sci U S A 2004;101:8894–9.

[46] Lowe SW, Ruley HE, Jacks T, Houseman DE. p53-dependent apoptosis modulates the cytotoxicity of anticancer agents. Cell 1993;74:957–67.

[47] Huerta S, Chilka S, Bonavida B. Nitric oxide donors: novel cancer therapeutics (Review). Int J Oncol 2008;33:909–27.

[48] Mattews NE, Adams MA, Maxwell LR, Gofton TE, Graham CH. Nitric oxide-mediated regulation of chemo-sensitivity in cancer cells. J Natl Cancer Inst 2001;93:1879–85.

[49] Azizadeh B, Yip HT, Blackwell KE, Horvath S, Calcaterra TC, Buga GM, Ignarro LJ, Wang MB. Nitric oxide improves cisplatin cytotoxicity in head and neck squamous cell carcinoma. Laryngoscope 2001;111(11 Pt 1):1896–900.

[50] Perrotta C, Bizzozero L, Falcone S, Rovere-Querini P, Prinetti A, Schuchman EH, Sonnino S, Manfredi AA, Clementi E. Nitric oxide boosts chemoimmunotherapy via inhibition of acid sphingomyelinase in a mouse model of melanoma. Cancer Res 2007;67:7559–64.

[51] Jia L, Schweizer J, Wang Y, Cerna C, Wong H, Revilla M. Effect of nitric oxide on cytotoxicity of Taxol: enhanced Taxol transcellular permeability. Biochem Pharmacol 2003;66:2193–9.

[52] Konovalova NP, Goncharova SA, Volkova LM, Rajewskaya TA, Eremenko LT, Korolev AM. Nitric oxide donor increases the efficiency of cytostatic therapy and retards the development of drug resistance. Nitric Oxide 2003;8:59–64.

[53] Moeller BJ, Richardson RA, Dewhirst MW. Hypoxia and radiotherapy, opportunities for improved outcomes in cancer treatment. Cancer Metastasis Rev 2007;26:241–8.

[54] Deschnes EE, Gray LH. Influence of oxygen tension on x-ray-induced chromosomal damage in Ehrlich ascites tumor cells irradiated in vitro and in vivo. Radiation Res 1959;11:115–46.

[55] Rockwell S, Dobrucki IT, Kim EY, Tucker Marrison S, Vu VT. Hypoxia and radiation therapy: past history, ongoing research, and future promise. Curr Mol Med 2009;9:442–58.

[56] Grimes DR, Partridge M. A mechanistic investigation of the oxygen fixation hypothesis and oxygen enhancement ratio. Biomed Physics Eng Express 2015;1:045209.

[57] Willers H, Azzoli CG, Santivasi WL, Xia F. Basic mechanisms of therapeutic resistance to radiation and chemotherapy in lung cancer. Cancer J 2013;19:200–7.

[58] Oronsky BT, Knox SJ, Scicinski J. Siz degrees separation: the oxygen effect in the development of radiosensitizer. Transl Oncol 2011;4:189–98.

[59] Oronsky BT, Knox SJ, Scicinski JJ. Is nitric oxide (NO) the last word in radiosensitization? A review. Transl Oncol 2012;5:66–71.

[60] Mitchell JB, Wink DA, DeGraff W, Gamson J, Keefer LK, Krishna MC. Hypoxic mammalian cell radiosensitization by nitric oxide. Cancer Res 1993;53:5845–8.

[61] Griffin RJ, Makepeace CM, Hur WJ, Song CW. Radiosensitization of hypoxic tumor cells in vitro by nitric oxide. Int J Radiation Oncol 1996;36:377–83.

[62] Jordan BF, Sonveaux P, Feron O, Grégoire V, Beghein N, Dessy C, Gallez B. Nitric oxide as a radiosensitizer: evidence for an intrinsic role in addition to its effect on oxygen delivery and consumption. Int J Cancer 2004;109:768–73.

[63] De Ridder M, Verellen D, Verovski V, Storme G. Hypoxic tumor cell radiosensitization through nitric oxide. Nitric Oxide 2008;19:164–9.

[64] Verovski VN, Van den Berge DL, Soete GA, Bols BL, Storme GA. Intrinsic radiosensitivity of human pancreatic tumour cells and the radiosensitising potency of the nitric oxide donor sodium nitroprusside. Br J Cancer 1996;74:377–83.

[65] Janssens MY, Verovsk VN, Van den Berge DL, Monsaert C, Storme GA. Radiosensitization of hypoxic tumour cells by S-nitroso-N-acetylpenicillamine implicates a bioreductive mechanism of nitric oxide generation. Br J Cancer 1999;79:1085–9.

[66] Swaroop GR, Malcon GP, Kelly PA, Ritchie I, Whittle IR. Effects of nitric oxide modulation on tumour blood flow and microvascular permeability in C6 glioma. Neuroreport 1998;9:2577–81.

[67] Jordan BF, Sonveaux P, Feron O, Grégoire V, Beghein N, Dessy C, Gallez B. Nitric oxide-mediated increase in tumor blood flow and oxygenation of tumors implanted in muscles stimulated by electric pulses. Int J Radiation Oncol 2003;55:1066–73.

[68] Singh S, Gupta AK. Nitric oxide: role in tumour biology and iNOS/NO-based anticancer therapies. Cancer Chemother Pharmacol 2011;67:1211–24.

[69] Howard-Flanders. Effect of nitric oxide on the radiosensitivity of bacteria. Nature 1957;180:1191–2.

[70] Jen MC, Serrano MC, van Lith R, Ameer GA. Polymer-based nitric oxide therapies: recent insights for biomedical applications. Adv Funct Mater 2012;22:239–60.

[71] de Souza GF, Yokoyama-Yasunaka JK, Seabra AB, Miguel DC, de Oliveira MG, Uliana SR. Leishmanicidal activity of primary S-nitrosothiols against Leishmania major and Leishmania amazonensis: implications for the treatment of cutaneous leishmaniasis. Nitric Oxide 2006;15:209–16.

[72] Riccio DA, Nugent JL, Schoenfisch MH. Stöber synthesis of nitric oxide-releasing S-nitrosothiol-modified silica particles. Chem Mater 2011;23:1727–35.

[73] Duong HTT, Kamarudin ZM, Erlich RB, Li Y, Jones MW, Kavallaris M, Boyer C, Davis TP. Intracellular nitric oxide delivery from stable NO-polymeric nanoparticle carriers. Chem Commun 2013;49:4190–2.

[74] Buck ME, Lynn DM. Azlactone-functionalized polymers as reactive platforms for the design of advanced materials: progress in the last ten years. Polymer Chem 2012;3:66–80.

[75] Fan W, Bu W, Zhang Z, Shen B, Zhang H, He Q, Ni D, Cui Z, Zhao K, Bu J, Du J, Liu J, Shi J. X-ray radiation-controlled NO-release for on-demand depth-independent hypoxic radiosensitization. Angew Chem Int Ed 2015;54:14026–30.

[76] Ma N, Xu H, An L, Li J, Sun Z, Zhang X. Radiation-sensitive diselenide block co-polymer micellar aggregates: toward the combination of radiotherapy and chemotherapy. Langmuir 2011;27:5874–8.

[77] Guo R, Tian Y, Wang Y, Yang W. Near-infrared laser-triggered nitric oxide nanogenerators for the reversal of multidrug resistance in cancer. Adv Funct Mater 2017;1606398.

[78] Chung MF, Liu HY, Lin KJ, Chia WT, Sung HW. A pH-responsive carrier system that generates NO bubbles to trigger drug release and reverse P-glycoprotein-mediated multidrug resistance. Angew Chem Int Ed 2015;54:9890–3.

[79] Chung MF, Chia WT, Liu HY, Hsiao CW, Hsiao HC, Yang CM, Sung HW. Inflammation-induced drug release by using a pH-responsive gas-generating hollow-microsphere system for the treatment of osteomyelitis. Adv Healthc Mater 2014;3:1854–61.

[80] Liu Q, Chen X, Jia J, Zhang W, Yang T, Wang L, Ma G. pH-responsive poly(D,L-lactic-co-glycolic acid) nanoparticles with rapid antigen release behavior promote immune response. ACS Nano 2015;9:4925–38.

[81] Tahara Y, Yoshikawa T, Sato H, Mori Y, Zahangir MH, Kishimura A, Mori T, Katayama Y. Encapsulation of a nitric oxide donor into a liposome to boost the enhanced permeation and retention (EPR) effect. Med Chem Commun 2017;8:415–21.

[82] Kumar V, Hong SY, Maciag AE, Saavedra JE, Adamson DH, Prud'homme RK, Keefer LK, Chakrapani H. Stabilization of the nitric oxide (NO) prodrugs and anticancer leads, PABA/NO and double JS-K, through incorporation into PEG-protected nanoparticles. Mol Pharmaceut 2010;7:291–8.

[83] Udupi V, Yu M, Malaviya S, Saavedra JE, Shami PJ. JS-K, a nitric oxide prodrug, induces cytochrome c release and caspase activation in HL-60 myeloid leukemia cells. Leukem Res 2006;30:1279–83.

[84] Nichols SP, Strom WL, Koh A, Schoenfisch MH. Local delivery of nitric oxide: targeted delivery of therapeutics to bone and connective tissues. Adv Drug Deliv Rev 2012;64:1177–88.

[85] Song Q, Tan S, Zhuang X, Guo Y, Zhao Y, Wu T, Ye Q, Si L, Zhang Z. Nitric oxide releasing D-α-tocopheryl polyethylene glycol succinate for enhancing antitumor activity of doxorubicin. Mol Pharmaceut 2014;11:4118–29.

[86] Deepagan VG, Ko H, Kwon S, Rao NV, Kim SK, Um W, Lee S, Min J, Lee J, Choi KY, Shin S, Suh M, Park JH. Intracellularly activatable nanovasodilators to enhance passive cancer targeting regime. Nano Lett 2018;18:2637–44.

[87] Ieda N, Hotta Y, Miyata N, Kimura K, Nakagawa H. Photomanipulation of vasodilation with a blue-light-controllable nitric oxide release. J Am Chem Soc 2014;136:7085–91.

[88] Yang L, Wu S, Lin B, Huang Tm Chen X, Yan X, Han S. A targetable nanogenerator of nitric oxide for light-triggered cytotoxicity. J Mater Chem B 2013;1:6115–22.

[89] Shishido-Silvia M, Ganzarolli de Oliveira M. Photosensitivity of aqueous sodium nitroprusside solutions: nitric oxide release versus cyanide toxicity. Prog React Kinetics Mech 2001;26:239–61.

[90] Fan J, He Q, Zhang F, Yang X, Wang Z, Lu N, Fan W, Lin L, Niu G, He N, Song J, Chen X. Light-responsive biodegradable nanomedicine overcomes multidrug resistance via NO-enhanced chemosensitization. ACS Appl Mater Interf 2016;8:13804–11.

[91] Tan L, Wan A, Zhu X, Li H. Nitric oxide release triggered by two-photon excited photoluminescence of engineered nanomaterials. Chem Commun 2014;50:5725–8.

[92] Garcia JV, Yang J, Shen D, Yao C, Li X, Wang R, Stucky GD, Zhao D, Ford PC, Zhang F. NIR-triggered release of caged nitric oxide using upconverting nanostructured materials. Small 2012;8:3800–5.

[93] Zhang X, Tian G, Yin W, Wang L, Zheng X, Yan L, Li J, Su H, Chen C, Gu Z, Zhao Y. Controllable generation of nitric oxide by near-infrared-sensitized upconversion nanoparticles for tumor therapy. Adv Funct Mater 2015;25:3049–56.

[94] Zhong Y, Tian G, Gu Z, Yang Y, Gu L, Zhao Y, Ma Y, Yao J. Elimination of photon quenching by a transition layer to fabricate a quenching-shield sandwich structure for 800 nm excited upconversion luminescence of Nd^{3+}-sensitized nanoparticles. Adv Mater 2014;26:2831–7.

Further Reading

Huang Z, Fu J, Zhang Y. Nitric oxide donor-based cancer therapy: advances and prospects. J Med Chem 2017;60:7617–35.

PART 2

Applications of Nitric Oxide in Inflammatory-Related Disorders

CHAPTER 11

The Role of Nitric Oxide on Vascular Dysfunction During Aging and Alzheimer's Disease

Massimo Venturelli*,[†]

*Department of Neurosciences, Biomedicine and Movement Sciences, University of Verona, Verona, Italy

[†]Department of Internal Medicine, University of Utah, Salt Lake City, United States

ABSTRACT

Nitric oxide (NO) is an endogenously synthesized free radical involved in a plethora of physiological phenomena affecting cardiovascular homeostasis and neural functions. Its unregulated bioavailability has been recognized as a key factor in neurodegenerative disorders, especially in relation to the mechanisms through which NO-mediated vascular impairment accentuates the effect of reactive oxygen species. Interestingly, the recent literature indicates the pivotal role of NO and oxidative stress to both early and advanced stages of neurodegenerative disorders, as well as promoting their progression. Alzheimer's disease (AD) is the most common form of dementia, characterized by extracellular amyloid (Aβ) plaques and intracellular neurofibrillary tangles coupled with reactive microgliosis and the loss of neurons and synapses in the cortex. However, the recent research supports the hypothesis of the pivotal role of NO depletion in the reduction of extracranial blood flow and impairment of cortical and peripheral circulation with a consequent detriment of cognitive function in humans with AD. It is worth of mention that AD, the leading form of dementia, continues to elude the scientific "armada" devoted on fighting this disruptive neurodegenerative disease, particularly with respect to its multifaceted origin. Although research on the biological mechanisms underlying the cause and the progression of AD is advancing, we are far to discover effective therapeutics or disease-modifying approaches. Currently, there is no absolute cure for AD: the few drugs available simply lessen the clinical symptoms. Therefore, new therapeutic approaches, including NO homeostasis, should be considered and might be useful as therapeutic targets.

Therapeutic Application of Nitric Oxide in Cancer and Inflammatory Disorders. https://doi.org/10.1016/B978-0-12-816545-4.00011-6

ABBREVIATIONS

Aβ	amyloid
AD	Alzheimer's disease
CNS	central nervous system
FA	femoral artery
ICA	internal carotid artery
MCI	mild cognitive impairment
ROS	reactive oxygen species

Conflict of Interest

No potential conflicts of interest were disclosed.

Nitric oxide (NO) is an endogenously synthesized free radical involved in a multitude of physiological phenomena affecting cardiovascular homeostasis and neural functions. In the cardiovascular system, this molecule is involved in the relaxation of smooth muscles of the vascular tissue. Interestingly, it also can intermediate synergistic and sometimes divergent physiological effects, which are likely associated with a number of local triggers including the associated courier through which it is mediating its physiological functions. Despite the well-established physiological effects of NO, its potential beneficial roles were tempered by the incongruence and variety of targeted clinical trials. On the other hand, in the past few decades, a great amount of data has been published about its multiple physiological roles in the normal functioning of the body. After the seminal study of Furchgott and coworkers [1–4] on the endothelium-derived relaxing factor, the role of NO on the central nervous system physiology has been deeply investigated. Its unregulated bioavailability was recognized as a key factor also in neurodegenerative disorders, especially in relation to the mechanisms through which NO-mediated vascular impairment accentuates the effect of reactive oxygen species (ROS). Interestingly, the recent literature indicates the pivotal role of NO and oxidative stress to both early and advanced stages of neurodegenerative disorders, as well as promoting their progression. During oxidative stress, NO and its reactive ancillary metabolites react with various molecular targets such as proteins, lipids, and nucleic acids, potentially triggering cellular dysfunctions.

Alzheimer's disease (AD) is the most common form of dementia, with a prevalence of ~24 million that is predicted to quadruplicate in the next decade [5]. The pathophysiological mechanisms of AD are well defined, including diffuse neuritic extracellular amyloid (Aβ) plaques and intracellular neurofibrillary tangles coupled with reactive microgliosis and the loss of neurons and synapses in the cortex. From the vascular point of view, the Aβ peptide accumulation in the tunica media and adventitia of cerebral blood vessels, a condition termed cerebral amyloid angiopathy, is associated with vessel smooth muscle cell degeneration, resulting in impaired cerebral circulation [6].

Apart from this direct effect of Aβ accumulation in the cortical neurons and vessels, there is increasing evidence that AD is associated with several dysregulated processes, which affect brain and systemic circulation, suggesting that vascular dysfunction may play a role in the pathogenesis of AD [7]. These pieces of evidence pose the question whether AD is an age-related neurodegenerative disorder with vascular consequences or a vascular disorder with neurodegenerative sequels [8]. Indeed, both aging and AD appear to be involved in the decline of the systemic and cerebral vascular functions [9]. However, the contribution of AD, *per se*, to vascular changes is still unclear. Additionally, the cardiovascular and the central nervous system (CNS) changes have been postulated to occur in parallel during the progression of AD [10].

In the brain of patients with AD, diffuse cortical changes have been demonstrated [11], coupled with a reduced perfusion of the temporoparietal association cortices, the mesial temporal structures, and the frontal association cortex [12]. However, the potential reduction of blood flow availability inward from the extracranial conduit arteries (i.e., internal carotid artery (ICA) and vertebral artery) may also contribute to the onset of AD [13, 14]. In this scenario, NO is considered the most important vasodilator factor responsible for the preservation of vasomotor function [15]. Indeed, reduced availability of NO in both cerebral and peripheral vessels results in major detrimental alterations of vascular function [15].

In a recent research, Venturelli and coworkers [16] have studied the mechanistic role of NO depletion in the reduction of extracranial blood flow and impairment of cortical and peripheral circulation in humans with AD. The main finding of this investigation was that the internal carotid blood flow, cortical perfusion, peripheral circulation, and systemic vascular function were reduced in healthy elderly in comparison with young controls and progressively further decreased in parallel to AD severity. Interestingly, these circulation impairments were associated with NO depletion (Fig. 1).

It should also be noted that, independent of neurodegeneration, reduced availability of NO, which is radically decreased in the aged population, results in major negative alterations of vascular function, including vasoconstriction and hypertension, leading to atherosclerosis. On the other hand, recent literature highlighted that AD is highly correlated to systemic vascular dysfunction [7]. Again, it is crucial to underline that the current literature suggests the key role of NO depletion in the pathogenesis of neurodegenerative disease [15]. Therefore, both age and AD appear to be implicated in the decline of the systemic and cerebral vascular functions likely because of endogenous NO reduced

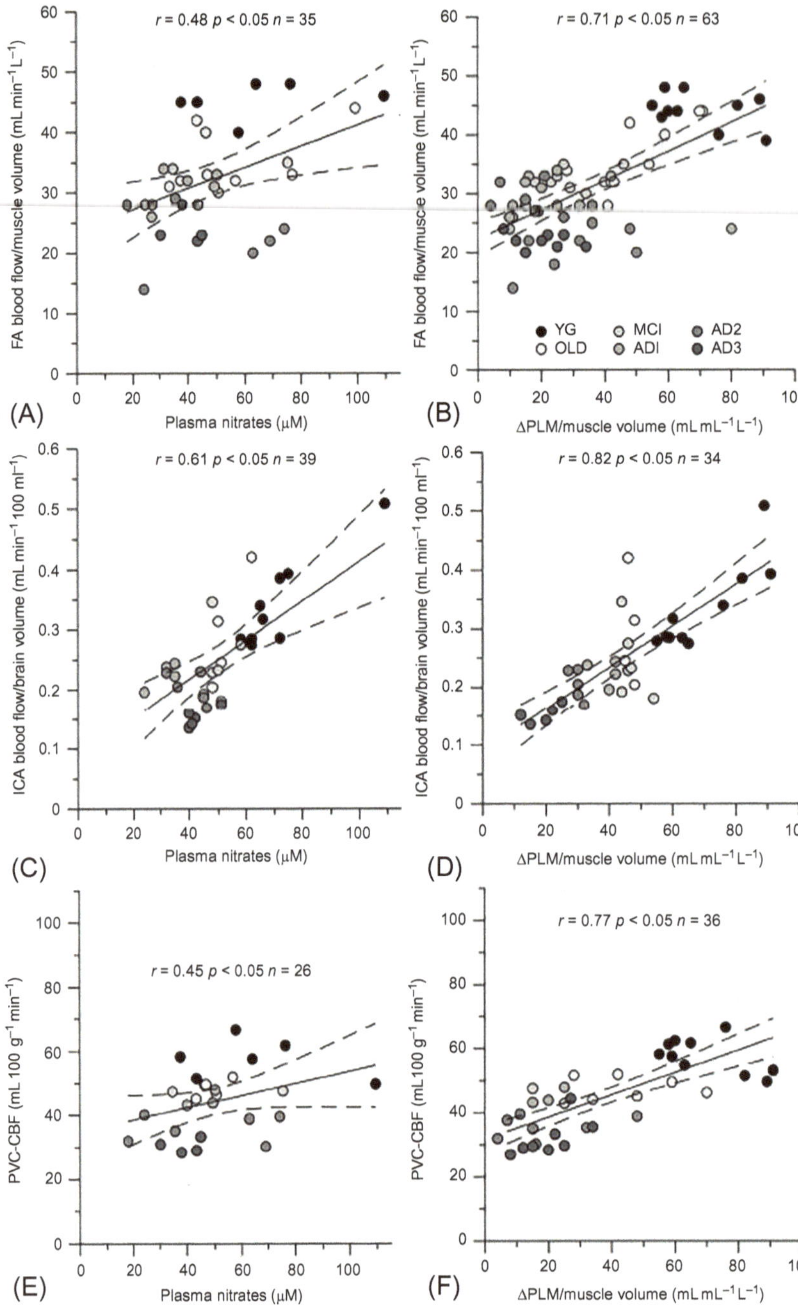

FIG. 1

Correlations between nitric oxide bioavailability; systemic vascular function; and cortical (PVC-CBF), internal carotid artery (ICA), and femoral artery (FA) blood flow normalized for brain and muscle volume, respectively. Nitric oxide (NO) bioavailability was determined by plasma levels of nitrates (panels A, C, and E) and passive limb movement-induced hyperemia normalized to muscle volume (Δ PLM; panels B, D, and F). Systemic vascular function determined via Δ PLM/muscle volume is represented in panels B, D, and F in healthy young (YG) and healthy old (OLD) controls and patients with mild cognitive impairment (MCI) and with different severity of AD, namely, mild AD (AD1), moderate AD (AD2), and severe AD (AD3). Each point represents a single subject; dashed lines represent interval of confidence. *Adapted from Venturelli M, Pedrinolla A, Boscolo Galazzo I, Fonte C, Smania N, Tamburin S, Muti E, Crispoltoni L, Stabile A, Pistilli A, Rende M, Pizzini FB, Schena F. Impact of nitric oxide bioavailability on the progressive cerebral and peripheral circulatory impairments during aging and Alzheimer's disease. Front Physiol 2018;9:169.*

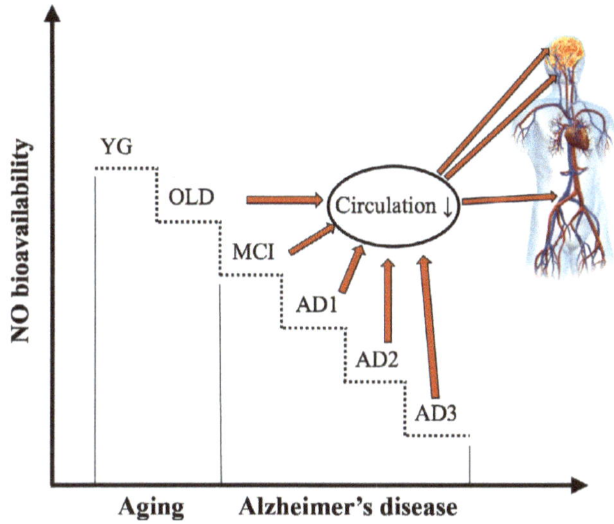

FIG. 2
Bioavailability of NO aging and AD: impact of nitric oxide bioavailability on the progressive cerebral and peripheral circulatory impairments during aging and Alzheimer's disease. *Adapted from Venturelli M, Pedrinolla A, Boscolo Galazzo I, Fonte C, Smania N, Tamburin S, Muti E, Crispoltoni L, Stabile A, Pistilli A, Rende M, Pizzini FB, Schena F. Impact of nitric oxide bioavailability on the progressive cerebral and peripheral circulatory impairments during aging and Alzheimer's disease. Front Physiol 2018;9:169.*

bioavailability. However, the recent knowledge in this field suggests a clear contribution of AD, *per se,* to this putative pathophysiological mechanism (Fig. 2).

In this scenario, there is new evidence that depletion in NO bioavailability is correlated with the reduction of cortical, extracranial, and peripheral blood flow in patients with AD. Indeed, depletion of NO and endothelial nitric oxide synthase enzymatic activity, as a major endogenous source of NO, are one of the mechanisms in the pathogenesis of endothelial dysfunction in both cerebral and peripheral blood vessels [15]. As previously reported, recent literature has underlined the key role of NO depletion in the early stage of neurodegenerative disorders and in their progression [15]. In support to this hypothesis, Merlini and coauthors [17] revealed that reduced NO bioavailability mediates cerebroarterial and peripheral dysfunction independently from cerebral amyloid angiopathy. Interestingly, the results of this study revealed that endothelium-dependent vasorelaxation was significantly impaired in both basilar and femoral arteries of 15-month-old Swedish arctic (SweArc) transgenic AD mice compared with that of age-matched wild-type and 6-month-old SweArc. This vascular impairment was accompanied by significantly reduced levels of cyclic GMP, demonstrating the central role of NO bioavailability in the pathogenesis and development of AD. Accordingly, in humans, the depletion in NO bioavailability is correlated with the reduction of cortical, extracranial, and peripheral blood flow during aging and in parallel to AD severity (Figs. 1 and 2).

Indeed, the recent scientific literature underlined that augmenting physical activity and fitness can protect NO bioavailability, attenuating the deleterious effects of advancing age on vascular function [18]. Therefore, particular attention on the determination of the physical activity level is needed in order to better describe the net effect of aging and AD to the systemic vascular function. Most importantly, physical activity needs to be mentioned as a potential nonpharmacological treatment for the abovementioned pathophysiological phenomena.

It is important to mention that NO is a free radical playing several positive regulatory functions at both the cellular and systemic levels. However, it is well established that elevated levels of free radicals have a plethora of deleterious effects on the vascular and nervous system during aging and AD, primarily associated with mitochondrial dysfunction. Indeed, Swerdlow and Khan [19] hypothesized the "mitochondrial cascade hypothesis" in AD, whereby mitochondrial dysfunction accumulates over the disease course, resulting in both symptoms and neuropathologic aspects of AD [20]. It is believed that mitochondrial dysfunction precedes Aβ formation, increasing ROS and oxidative stress, which, in turn, may facilitate overproduction of Aβ [21]. In AD, mitochondrial damage is characterized by decreased respiratory chain complex activities, where complexes III and IV are typically involved, causing ROS overproduction and reduced ATP synthesis [22–24]. In this regard, brain tissues are metabolically very active and are particularly susceptible to the damaging effects by ROS. In the case of AD, ROS have been reported within those brain regions, such as the cerebral cortex and hippocampus, which undergo selective neurodegeneration [25]. Interestingly, a large body of evidence shows that AD patients have oxidative metabolism dysfunction in both the central nervous system and peripheral tissues (i.e., vascular endothelial cells and platelets) suggesting that pathological changes coexist in the brain and nonneural tissues [21, 23]. Moreover, recent studies suggest that mitochondrial ROS overproduction contributes to accelerate the development of the senescent phenotype in endothelial cells, impairing regenerative and angiogenic capacity of the endothelium and promoting atherosclerosis by altering the secretion of cytokines, growth factors, and protease in the vascular wall [26, 27].

It is worth of mention that AD, the leading form of dementia, continues to elude the scientific "armada" devoted on fighting this disruptive neurodegenerative disease, particularly with respect to its multifaceted nature of its origin. Although research on the biological mechanisms underlying the cause and the progression of AD is advancing, we are far to discover effective therapeutics or disease-modifying approaches. Currently, there is no absolute cure for AD: the few drugs available simply alleviate some of the clinical symptoms. Therefore, new therapeutic approaches, including NO homeostasis, should be considered and might be useful as therapeutic targets.

References

[1] Matsunaga K, Furchgott RF. Interactions of light and sodium nitrite in producing relaxation of rabbit aorta. J Pharmacol Exp Ther 1989;248(2):687–95.

[2] Li J, Liu XJ, Furchgott RF. Blockade of nitric oxide-induced relaxation of rabbit aorta by cysteine and homocysteine. Zhongguo Yao Li Xue Bao 1997;18(1):11–20.

[3] Furchgott RF. Endothelium-derived relaxing factor: discovery, early studies, and identification as nitric oxide. Biosci Rep 1999;19(4):235–51.

[4] Furchgott RF. Endothelium-dependent relaxation, endothelium-derived relaxing factor and photorelaxation of blood vessels. Semin Perinatol 1991;15(1):11–5.

[5] Reitz C, Brayne C, Mayeux R. Epidemiology of Alzheimer disease. Nat Rev Neurol 2011; 7(3):137–52.

[6] Maier FC, Wehrl HF, Schmid AM, Mannheim JG, Wiehr S, Lerdkrai C, Calaminus C, Stahlschmidt A, Ye L, Burnet M, Stiller D, Sabri O, Reischl G, Staufenbiel M, Garaschuk O, Jucker M, Pichler BJ. Longitudinal PET-MRI reveals beta-amyloid deposition and rCBF dynamics and connects vascular amyloidosis to quantitative loss of perfusion. Nat Med 2014; 20(12):1485–92.

[7] Iturria-Medina Y, Sotero RC, Toussaint PJ, Mateos-Perez JM, Evans AC. Alzheimer's disease neuroimaging I. Early role of vascular dysregulation on late-onset Alzheimer's disease based on multifactorial data-driven analysis. Nat Commun 2016;7:11934.

[8] de la Torre JC. The vascular hypothesis of Alzheimer's disease: bench to bedside and beyond. Neurodegener Dis 2010;7(1–3):116–21.

[9] de la Torre JC. Cerebrovascular and cardiovascular pathology in Alzheimer's disease. Int Rev Neurobiol 2009;84:35–48.

[10] Picano E, Bruno RM, Ferrari GF, Bonuccelli U. Cognitive impairment and cardiovascular disease: so near, so far. Int J Cardiol 2014;175(1):21–9.

[11] Dallaire-Theroux C, Callahan BL, Potvin O, Saikali S, Duchesne S. Radiological-pathological correlation in Alzheimer's disease: systematic review of Antemortem magnetic resonance imaging findings. J Alzheimer's Dis 2017;57(2):575–601.

[12] Herholz K. Perfusion SPECT and FDG-PET. Int Psychogeriatr 2011;23(Suppl. 2):S25–31.

[13] Clark LR, Berman SE, Rivera-Rivera LA, Hoscheidt SM, Darst BF, Engelman CD, Rowley HA, Carlsson CM, Asthana S, Turski P, Wieben O, Johnson SC. Macrovascular and microvascular cerebral blood flow in adults at risk for Alzheimer's disease. Alzheimers Dement (Amst) 2017;7:48–55.

[14] Liu J, Zhu YS, Khan MA, Brunk E, Martin-Cook K, Weiner MF, Cullum CM, Lu H, Levine BD, Diaz-Arrastia R, Zhang R. Global brain hypoperfusion and oxygenation in amnestic mild cognitive impairment. Alzheimer's Dement 2014;10(2):162–70.

[15] Katusic ZS, Austin SA. Endothelial nitric oxide: protector of a healthy mind. Eur Heart J 2014;35(14):888–94.

[16] Venturelli M, Pedrinolla A, Boscolo Galazzo I, Fonte C, Smania N, Tamburin S, Muti E, Crispoltoni L, Stabile A, Pistilli A, Rende M, Pizzini FB, Schena F. Impact of nitric oxide bioavailability on the progressive cerebral and peripheral circulatory impairments during aging and Alzheimer's disease. Front Physiol 2018;9:169.

[17] Merlini M, Shi Y, Keller S, Savarese G, Akhmedov A, Derungs R, Spescha RD, Kulic L, Nitsch RM, Luscher TF, Camici GG. Reduced nitric oxide bioavailability mediates cerebroarterial dysfunction independent of cerebral amyloid angiopathy in a mouse model of Alzheimer's disease. Am J Physiol Heart Circ Physiol 2017;312(2):H232–8.

[18] Groot HJ, Rossman MJ, Garten RS, Wang E, Hoff J, Helgerud J, Richardson RS. The effect of physical activity on passive leg movement-induced vasodilation with age. Med Sci Sports Exerc 2016;48(8):1548–57.

[19] Swerdlow RH. Brain aging, Alzheimer's disease, and mitochondria. Biochim Biophys Acta 2011;1812(12):1630–9.

[20] Swerdlow RH, Burns JM, Khan SM. The Alzheimer's disease mitochondrial cascade hypothesis. J Alzheimer's Dis 2010;20(Suppl. 2):S265–79.

[21] Morris JK, Honea RA, Vidoni ED, Swerdlow RH, Burns JM. Is Alzheimer's disease a systemic disease? Biochim Biophys Acta 2014;1842(9):1340–9.

[22] Marques-Aleixo I, Oliveira PJ, Moreira PI, Magalhaes J, Ascensao A. Physical exercise as a possible strategy for brain protection: evidence from mitochondrial-mediated mechanisms. Prog Neurobiol 2012;99(2):149–62.

[23] Cadonic C, Sabbir MG, Albensi BC. Mechanisms of mitochondrial dysfunction in Alzheimer's disease. Mol Neurobiol 2016;53(9):6078–90.

[24] Pedrinolla A, Schena F, Venturelli M. Resilience to Alzheimer's disease: the role of physical activity. Curr Alzheimer Res 2017;14(5):546–53.

[25] Bhat AH, Dar KB, Anees S, Zargar MA, Masood A, Sofi MA, Ganie SA. Oxidative stress, mitochondrial dysfunction and neurodegenerative diseases: a mechanistic insight. Biomed Pharmacother 2015;74:101–10.

[26] Dai DF, Rabinovitch PS, Ungvari Z. Mitochondria and cardiovascular aging. Circ Res 2012;110(8):1109–24.

[27] El Assar M, Angulo J, Rodriguez-Manas L. Oxidative stress and vascular inflammation in aging. Free Radic Biol Med 2013;65:380–401.

CHAPTER 12

Carbonic Anhydrase Inhibitor—NO Donor Hybrids and Their Pharmacological Applications

Claudiu T. Supuran
University of Florence, Neurofarba Department, Section of Pharmaceutical and Nutraceutical Sciences, Firenze, Italy

ABSTRACT

Carbonic anhydrases (CAs, EC 4.2.1.1) are metalloproteins widespread in organisms all over the phylogenetic tree, with 15 different isoforms present in humans. Many of them are drug targets, and their inhibitors are used as diuretics, antiglaucoma, antiepileptic, antiobesity, and more recently antitumor agents. Many new classes of CA inhibitors (CAIs) were reported in the last period, among which sulfonamides, sulfamates, dithiocarbamates, coumarins, sulfocoumarins, and polyamines, together with hybrids, also incorporating NO-donating moieties. Sulfonamide CAI-NO donors belonging to several classes were shown to possess a significant antiglaucoma activity in animal models of the disease. The possibility to use such hybrid drugs in other pathologies such as cancer, neuropathic pain, arthritis, or cerebral ischemia was not yet explored, although several classes of CAIs were recently shown to be useful in these pathologies. CA IX, a membrane-bound, hypoxia-inducible enzyme, is highly expressed in solid tumors and the inflamed synovium of arthritis patients, showing restricted expression in normal tissues. The possibility to develop pharmacological inhibitors that selectively target this isoform, of the CAI/NO-donor type, may offer interesting pharmacological tools for exploring the role of NO and CA inhibition in various pathologies in need of effective therapeutic agents.

ABBREVIATIONS

CA	carbonic anhydrase
CAI	CA inhibitor
IOP	intraocular pressure
ISMN	isosorbide mononitrate
sGC	soluble guanylate cyclase

Therapeutic Application of Nitric Oxide in Cancer and Inflammatory Disorders. https://doi.org/10.1016/B978-0-12-816545-4.00012-8

Conflict of Interest

The author declares conflict of interest, being inventor on patents dealing with CA inhibitors with NO-releasing properties.

INTRODUCTION

Carbonic anhydrases (CAs, EC 4.2.1.1) are metalloenzymes acting as catalysts for the equilibration reaction between carbon dioxide and bicarbonate [1–5]. They usually have a Zn(II) ion at the active site that is crucial for catalysis. In the α-class mammalian CAs (also present in humans), the zinc ion is coordinated by three His residues and a water molecule or a hydroxide ion [1–5]. The catalysis occurs in two steps, by the mechanism shown below [6–9]:

H_2O

$$EZn^{2+} — OH^- + CO_2 \Leftrightarrow EZn^{2+} — HCO_{3^-} \Leftrightarrow EZn^{2+} - ¾OH_2 + HCO_{3^-} \quad (1)$$

$$EZn^{2+} — -OH_2 \Leftrightarrow EZn^{2+} — HO^- + H^+ \quad (2)$$

A nucleophilic attack of a zinc-bound hydroxide species of the enzyme on the substrate CO_2, bound in a hydrophobic pocket [8] and optimally orientated for the hydration reaction (Eq. 1), is the first step of the catalytic cycle. Bicarbonate formed in this way is thereafter replaced by an incoming water molecule, with the generation of the catalytically acid form of the enzyme, EZn^{2+}—OH_2 (Eq. 1). For the regeneration of the zinc hydroxide species, a proton transfer reaction occurs from the Zn(II)-bound water molecule to the external medium (Eq. 2), which is the rate-determining step, being generally assisted by active site residues acting as proton shuttles, such as a histidine residue placed in the middle of the cavity (His64, CA I numbering) [8–10]. Schematically, the CA catalytic inhibition and activation mechanisms are shown in Fig. 1.

The CA-catalyzed reaction transforms two neutral molecules, CO_2 and water, into a weak base (bicarbonate) and a strong acid (H^+ ions), and this is probably the reason why at least seven genetically distinct CA families are known to date (α-, β-, γ-, δ-, ζ-, η-, and θ-CAs), which are widespread in organisms all over the phylogenetic tree [1, 4, 11–13]. Both in prokaryotes and eukaryotes, these enzymes play crucial physiological functions connected with metabolic pathways in which CO_2 is involved (photosynthesis, carboxylation reactions, etc.) consisting of pH regulation; secretion of electrolytes; and metabolic processes such as glucogenesis, urea biosynthesis, and lipogenesis [1–5, 11–13]. CA inhibitors (CAIs) are in fact in clinical use for decades, mainly for the treatment of conditions such as edema [14, 15], glaucoma [16], epilepsy [17], obesity [18], or forms of cancers overexpressing some CA isoforms [19–21], though, more recently, the involvement of these enzymes and the possible use of their inhibitors for the management of neuropathic pain [22, 23], cerebral ischemia

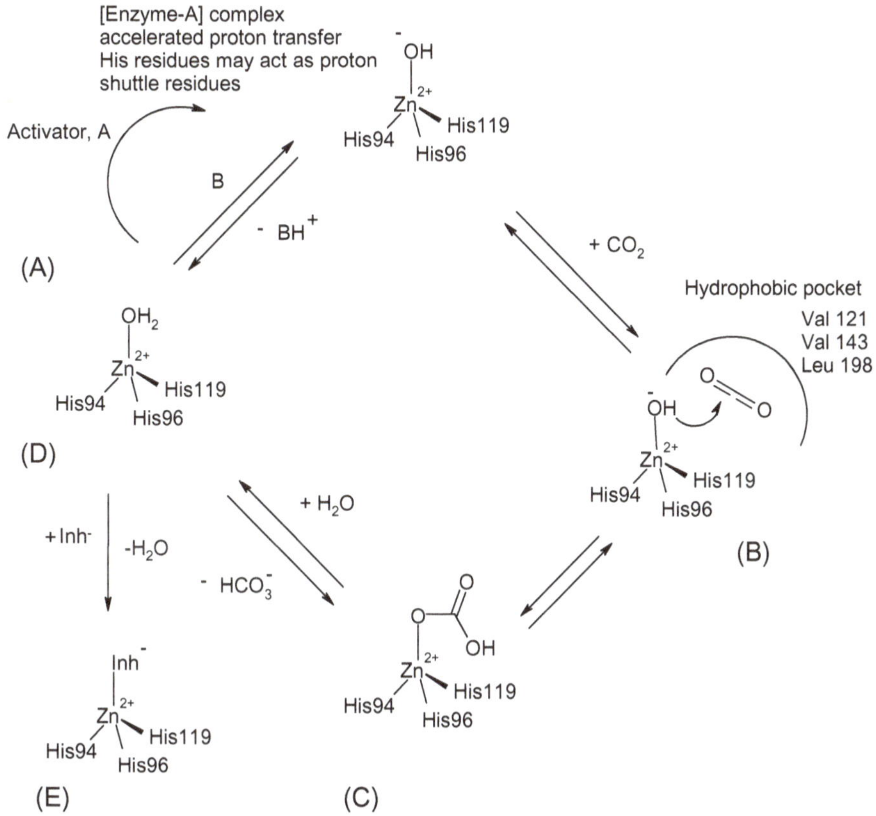

FIG. 1

Catalytic inhibition (with zinc binders) and activation mechanisms of α-CAs (hCA I amino acid numbering of the zinc ligands [20]). In the D to A step, proton shuttle residues (usually a His placed in the middle of the cavity, such as His64 in CA II) play a crucial role in the proton transfer processes. In E, an enzyme-inhibitor adduct is shown, whereas activators (A) may bind to the acidic or zinc hydroxide forms of the enzyme [7].

[24], and arthritis [25, 26] were also demonstrated. The role of CAs as metabolic enzymes, due to their involvement in several biosynthetic pathways mentioned above, started to be recognized only recently, mainly due to the role of some CA isoforms in the tumor metabolism [19, 21, 27, 28].

CA INHIBITORS

The only CA inhibition mechanism known for many years is the one illustrated in Fig. 1E, with the inhibitors binding to the metal ion in tetrahedral geometry or, more infrequently, by adding to the metal coordination sphere and generating trigonal bipyramidal geometries [1–5]. However, this situation changed dramatically in the last decade, with a large number of novel CA inhibition mechanisms reported (Fig. 2) and new classes of CAIs discovered, apart from the zinc binders.

Many classes of CAIs act as zinc binders (Fig. 1A). The classical ones are the sulfonamide and their isosteres (sulfamates and sulfamides) [1–5, 29], in which the zinc-binding group (ZBG) is a SO_2NH^- moiety. Such compounds also incorporate a scaffold that may interact with either the hydrophobic or the hydrophilic (or both) halves of the active site and with tails that bind toward the exit of the active site cavity, which is the most variable region of the various CA isoforms [1–6, 29]. Other classes of inhibitors that bind to the metal ion include carboxylates and hydroxamates [30, 31] and dithiocarbamates and isosteres [32, 33]. A second CA inhibition mechanism anchoring to the metal-ion-coordinated water molecule/hydroxide ion is shown in Fig. 2B. Such inhibitors incorporate an anchoring group (AG) that is hydrogen bonded to the zinc-coordinated nucleophile, eventually making additional hydrogen bonds with neighboring residues (Thr199 and Glu106) in the α-class enzymes. Phenols [34], polyamines [35], and sulfocoumarins [36] inhibit CAs by this mechanism. Similar to the zinc binders, the scaffold of this type of inhibitors may interact with one or both halves of the active site, and a tail may be also present for enhancing the isoform selectivity profile of the inhibitors [35, 36]. Situated even further away from the metal ion, at the entrance of the active site cavity, are the inhibitors that occlude the active site entrance, shown in Fig. 2C. Such compounds incorporate a sticky group (SG) that may be of the OH, amino, COOH, and other types, as well as the scaffold/tail, as the previous classes of inhibitors discussed above. This mechanism has been discovered for coumarins [37–39] being later shown that many other classes of structurally similar compounds bind in this manner to the enzyme [5, 39]. A fourth CA inhibition mechanism (Fig. 2D) was also documented by X-ray crystallography [40]. A benzoic acid derivative was observed bound outside the active site cavity, in a hydrophobic pocket adjacent to the entrance within the active site. The inhibition is achieved by blocking the proton shuttle residue (His64) in its out conformation, which leads to the collapse of the entire catalytic cycle [40]. Of all these classes of inhibitors mentioned above, only the sulfonamides and some sulfamates are clinically used [1–5]. The classical sulfonamide CAIs with antiglaucoma action are shown in Fig. 3, and they include the systemically acting drugs acetazolamide (**AAZ**), methazolamide (**MZA**), ethoxzolamide (**EZA**), and dichlorophenamide (**DCP**) [2, 3]. In the 1990s, two topically acting antiglaucoma sulfonamides were approved for clinical use, dorzolamide (**DZA**) and brinzolamide (**BRZ**) [2, 3]. Both of them, alone or in combination with β-blockers, are widely used for the management of glaucoma [16].

Indeed, glaucoma is the second leading cause of vision loss worldwide, after retinal degeneration, being estimated that as many as 70 million individuals may be affected [41, 42]. Glaucoma involves a progressive optic neuropathy exhibiting characteristic changes of the optic nerve head and visual fields, and it is

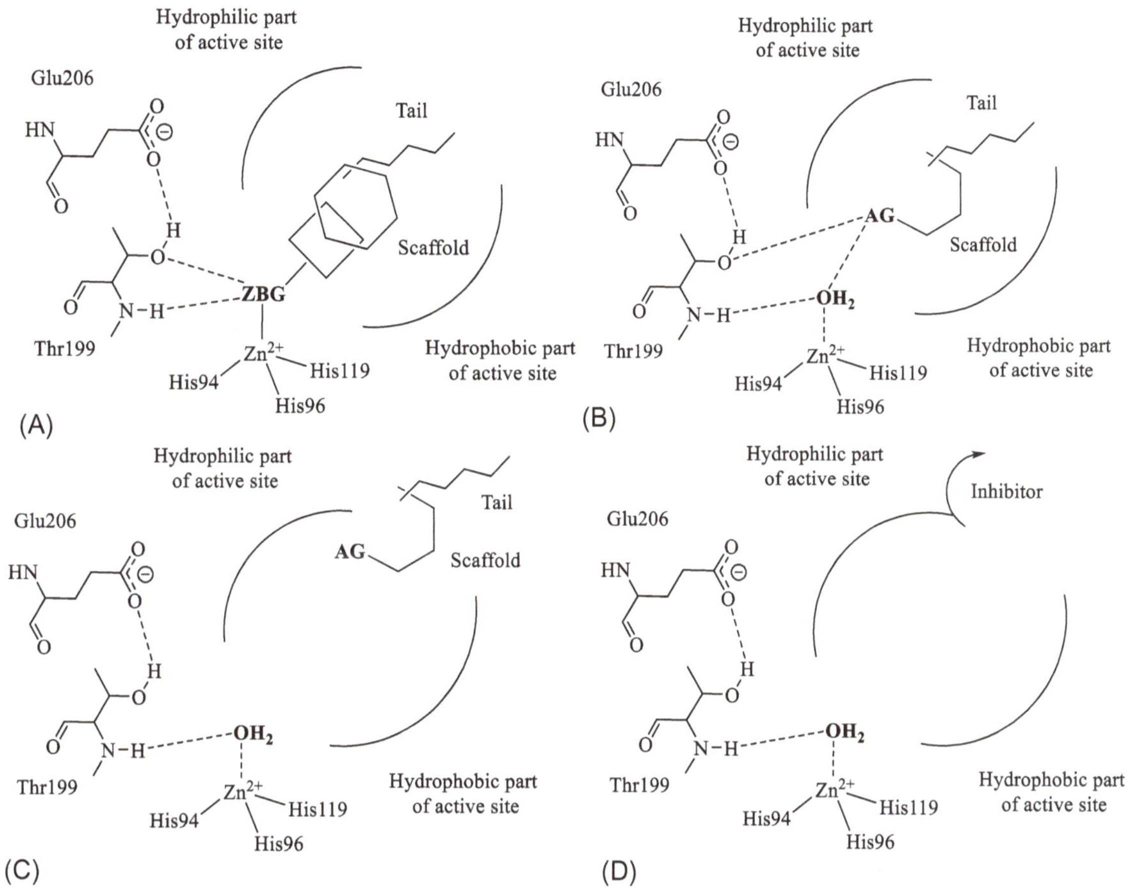

FIG. 2

CA inhibition mechanisms: (A) zinc binding, (B) anchoring to the metal-ion-coordinated water, (C) occlusion of the active site entrance, and (D) out of the active site binding [1–3].

commonly linked to elevated intraocular pressure (IOP). Treatment strategies to prevent glaucoma and the consequent irreversible loss of vision are based on the reduction of IOP by using topical hypotensive drugs, among which the CAIs are one of the most important pharmacological classes of agents [2, 42].

Nitric oxide (NO), a radical gas produced by the enzyme nitric oxide synthase (NOS), is involved in the aqueous humor outflow within the eye, local modulation of ocular blood flow, and retinal ganglion cell death by apoptosis [43–45]. It has been observed that hypertensive glaucoma patients have a decreased NO/cGMP content in the aqueous humor [43] and that several classes of NO donors decrease IOP in normal and pathological conditions [42–45].

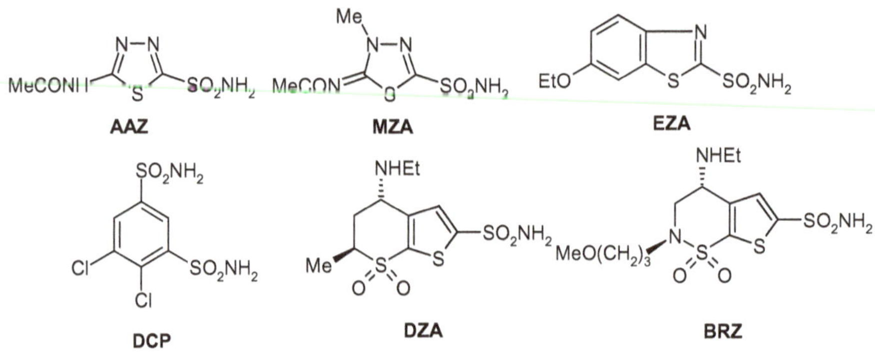

FIG. 3

Classical, clinically used CAIs: acetazolamide (AAZ), methazolamide (**MZA**), ethoxzolamide (**EZA**), dichlorophenamide (**DCP**), dorzolamide (**DZA**), and brinzolamide (**BRZ**).

Thus, the idea to design dual agents, incorporating both the CAI and the NO-donor moieties in the same molecule, appeared as an interesting working hypothesis for designing more effective antiglaucoma agents [46–49].

SULFONAMIDE CAI—NO DONOR HYBRIDS WITH ANTIGLAUCOMA ACTION

The investigation of NO-donating CAI, for the topical administration as potential antiglaucoma agents, started in 2009 by our group [42, 46]. Such agents, which would combine a dual mechanism of action in a single molecule, would be capable of enhancing the NO/soluble guanylate cyclase (sGC)/cGMP signaling while also inhibiting the CA enzymes involved in aqueous humor production within the ciliary bodies of the eye [42, 46]. Such NO-donating CAI represents an interesting new class of drugs, with improved activity over existing treatments, for reducing IOP more efficiently in hypertensive glaucoma patients compared with a single agent [36–49].

Several approaches have been used for obtaining such hybrid drugs. In the first one [46], the **DZA** scaffold has been used as a starting point, with the second amine moiety of the drug being decorated with nitrate esters (compounds **1–5**, Fig. 4).

The NO-donating moieties included in this first study were mainly alkyl nitrates, with the exception of the benzylic nitrate **2**. The dinitrate ester **5** was obtained to eventually measure the effect of increasing the theoretical deliverable amount of NO to eye tissues [46]. Tests carried out on compounds **1–5** to assess hydrolysis in a rabbit corneal homogenate documented that they

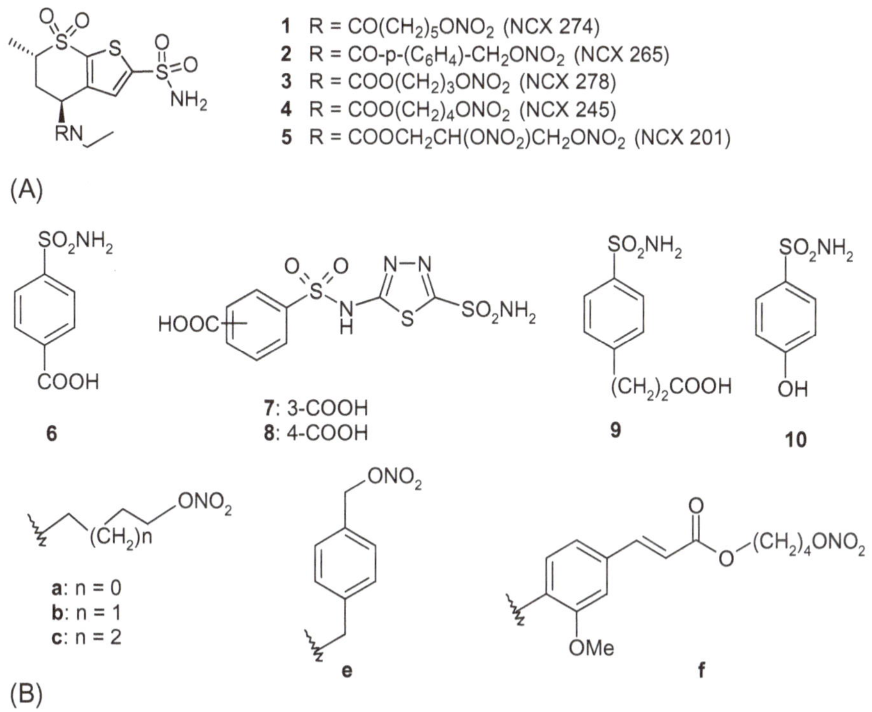

FIG. 4

Sulfonamide CAI/NO-donor hybrids. (A) Dorzolamide derivatives **1–5** [46]. (B) Hybrids incorporating aromatic/heterocyclic sulfonamides **6–10** and nitrate ester tails **a–e** [47].

were stable to eye esterases, thus suggesting that these compounds are novel, stable NO-donating CAI that exert their activity as such [46]. They were assayed for their inhibitory activity and selectivity on CA enzymes by exposing an increasing concentration of test drug to the human recombinant type I, II, and IV isoforms. Dorzolamide (**DZA**) was used as the reference standard, with CA II Ki = 9 nM and $Ki_{CAII}/Ki_{CAIV} = 4.7$ [46]. Compounds **1–5** inhibited CA II with the following order of potency: **3** ≥ **1** > **5** ≥ **2** ≥ **4** with Kis ranging from 13 to 76 nM. The alkyl nitrate linkers proved to be more selective for CA II than the benzyl nitrate **2**. Shortening the carbamate chain from four to three carbon atoms (compounds **4** and **3**) increased significantly both potency on CA II and selectivity over CA IV ($Ki_{CAII} = 13$ nM and $Ki_{CAII}/Ki_{CAIV} = 13.9$ for compound **3**). Notably, compounds **1** and **3** (Ki = 14 and 13 nM, respectively) were equipotent to dorzolamide (Ki = 9 nM) for CA II and comparably selective inhibitors of CA II over CA IV ($Ki_{CAII}/Ki_{CAIV} = 311$ and 13.9, respectively, for compounds **1** and **3**) to dorzolamide ($Ki_{CAII}/Ki_{CAIV} = 4.7$) [46]. The release of NO was clearly demonstrated for some of these compounds studied with 1, 3, and 5 being potent and effective stimulators of NO/sGC/cGMP signaling with an EC_{50} of 4.25 + 0.80 μM, 2.05 + 0.41 μM, and 0.37 + 0.05 μM, respectively, compared with the standard drug isosorbide mononitrate (ISMN), which gave an EC^{50} of 10.8 + 2.4 μM, whereas dorzolamide displayed no such effects [46]. The IOP lowering with some of these compounds was of 3.5–4.0 mm Hg, comparable with that of **DZA**, at 30–90 min post administration, in a hypertensive rabbit glaucoma model [46].

In a subsequent study [47], five different aromatic/heterocyclic sulfonamide scaffolds have been used to prepare new CAIs incorporating NO-donating moieties: 4-carboxybenzenesulfonamide **6**, 3- or 4-carboxybenzolamides **7** and **8**, 4-(2-carboxyethyl)-benzenesulfonamide **9**, and 4-hydroxybenzenesulfonamide **10**. The carboxylic or phenol moieties of **6–10** were derivatized to introduce NO-containing donor groups of types **a–e**, which incorporate either aliphatic C3–C5 moieties or aromatic nitrate esters of types **e** and **f** (ferulic acid derivatives) (Fig. 4B) [47]. The main target isoform for antiglaucoma sulfonamides hCA II was effectively inhibited by many of the hybrids, with K_Is in the range of 10–33 nM, in the same range as those of the systemically used antiglaucoma sulfonamides **AAZ** and **DCP** or the topically acting one **DZA** [47]. One of them (**7b**, incorporating the scaffold of sulfonamide **7** and the tail **b**, Fig. 5) was crystallized in complex with hCA II, allowing to evidence the binding mode to the enzyme (Fig. 5A) [47].

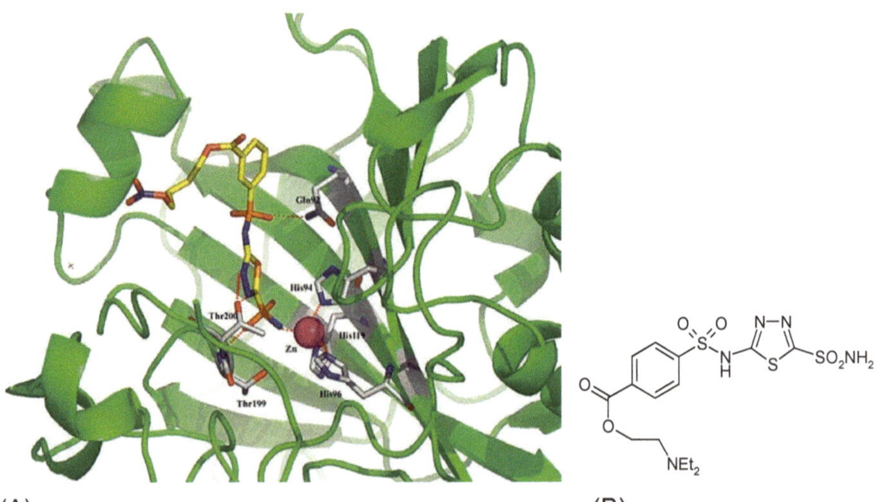

(A) (B)

FIG. 5

(A) View of compound **7b** (*yellow*) bound within the hCA II active site. The Zn(II) ion (*violet sphere*); its three His ligands (His94, 96, and 119); and residues Thr199, Thr200, and Gln92 with which the inhibitor interacts are shown. The protein scaffold is shown as green ribbon. (B) The chemical structure of the CAI **7b** [47].

As for other hCA II-sulfonamide adducts investigated earlier [1–3, 11], the deprotonated sulfonamide moiety (SO_2NH^-) of the inhibitor was coordinated to the Zn(II) ion from the enzyme active site, at a distance of 1.96Å. The same NH moiety made a hydrogen bond with the OH of Thr199 (of 2.86Å). Furthermore, the two endocyclic nitrogen atoms of the 1,3,4-thiadiazole ring participate in two hydrogen bonds (of 2.50–2.80Å) with the OH of Thr200 [47]. One oxygen of the secondary SO_2 moiety of inhibitor **7b** made a hydrogen bond (of 2.97Å) with the NH_2 of Gln92. However, due to the *meta*substituent of the phenyl moiety present in **7b**, the conformation of **7b** is different from that of other structurally related inhibitors [11] bound to the enzyme active site, as observed from Fig. 5. Such data demonstrated that the nitrate esters of the sulfonamides are stable in the presence of the enzyme and that they possess a similar CA inhibition mechanism to the nonderivatized sulfonamide CAIs [47–49].

However, the compound that subsequently was the most investigated one from the pharmacological viewpoint was **6a**, also called **NCX250** (Fig. 6), for which the antiglaucoma properties (IOP-lowering effects) were investigated in detail in rabbits with carbomer-induced glaucoma [48]. In this study, it has been demonstrated that the sulfonamide incorporating NO-donating moieties **NCX250** (which was a potent inhibitor of CA isoforms involved in aqueous humor secretion, i.e., CA II, IV, and XII [48], Fig. 6B) at concentrations of 0.5–2 % was highly effective in reducing IOP, with duration of action much longer than **DZA**, usually of 1 h [47, 48] of 2–4h post administration (Fig. 6A) [48]. Chronic administration of **NCX250** 2% on glaucomatous albino rabbits also showed an important reduction in IOP values already after the first week of treatment, with a regular decreasing trend during all treatment periods [48]. In the ophthalmic artery of the treated animals, both the systolic and diastolic velocities were significantly reduced in **NCX250**-treated eyes, in comparison with **DZA**-treated ones, suggesting a beneficial effect of **NCX250** on the blood supply to the optic nerve, which was not observed with **DZA** [48]. These results clearly indicated that in carbomer-induced glaucoma, the association of classical antiglaucomatous agents, such as the sulfonamide CA inhibitor dorzolamide with NO donors, significantly augment their effects on IOP lowering, increasing the ocular hemodynamics and decreasing the inflammatory processes and ocular apoptosis [48]. Such results suggested that not only does NO participate in the regulation of IOP in glaucoma but also this small molecule has antiapoptotic and anti-inflammatory effects, which were scarcely investigated in eye diseases until that date [48].

A series of sulfonamide—furazan/furoxan CAI hybrids—of types **11–26** (Fig. 7) were also prepared, using as NO donor the furazan ring [49] and not the nitrate esters as for the compounds **1–10** discussed above [45–48]. Many of these compounds were low-nanomolar inhibitors of isoforms hCA I, II, IX, and XII [49] (two of these isoforms, hCA II and XII, are involved in glaucoma, whereas hCA

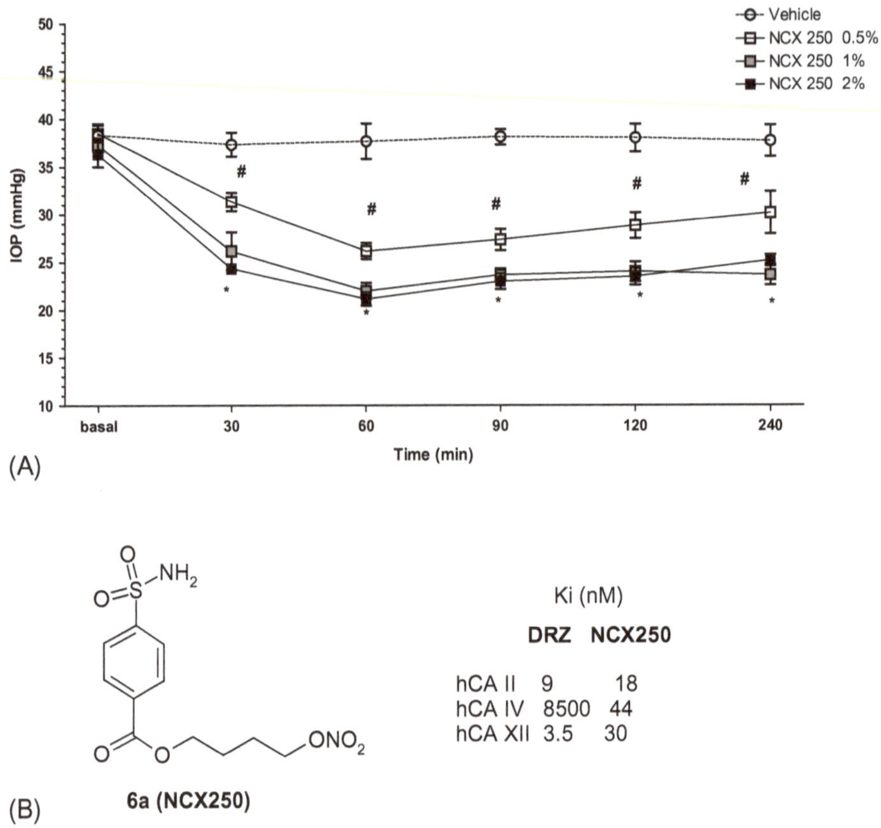

FIG. 6

(A) Effects of various concentrations of **NCX250** (compound **6a**) on IOP in carbomer-induced glaucoma in albino rabbits [48]. (B) Chemical structure of **NCX250** and its CA inhibitory properties in comparison with **DZA** as standard drug [48].

IX is a tumor-associated isoform [1–3]). Furthermore, as the derivatives containing nitrate ester moieties, the furazan/furoxan CAI hybrids also showed significant reduction of IOP in various animal models of glaucoma, making them attractive candidates for pharmacological studies.

CONCLUSIONS AND FURTHER DEVELOPMENTS

The CAI/NO-donor hybrids started to be investigated in the last decade as potential pharmacological agents for the treatment of diseases in which both the CAs and the NO signaling molecule are involved. Significant advances have been done in the synthesis of sulfonamide/NO-donor hybrids, which incorporate two types of NO-donating moiety: the nitrate esters and the furazan/

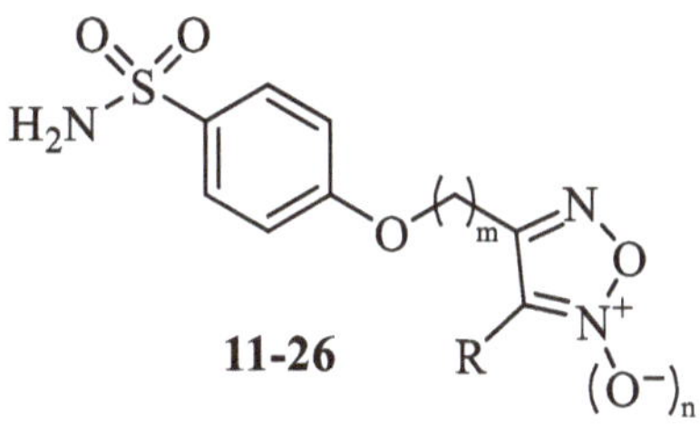

11 X = Br, R = Ph, $n = 0$, $m = 1$
12 X = Br, R = Ph, $n = 1$, $m = 1$
13 X = Br, R = $CONH_2$, $n = 0$, $m = 1$
14 X = Br, R = $CONH_2$, $n = 1$, $m = 1$
15 X = Br, R = CN, $n = 0$, $m = 1$
16 X = Br, R = CN, $n = 1$, $m = 1$
17 X = SO_2Ph, R = SO_2Ph, $n = 0$, $m = 0$
18 X = SO_2Ph, R = SO_2Ph, $n = 1$, $m = 0$
19 R = Ph, $n = 0$, $m = 1$
20 R = Ph, $n = 1$, $m = 1$
21 R = $CONH_2$, $n = 0$, $m = 1$
22 R = $CONH_2$, $n = 1$, $m = 1$
23 R = CN, $n = 0$, $m = 1$
24 R = CN, $n = 1$, $m = 1$
25 R = SO_2Ph, $n = 0$, $m = 0$
26 R = SO_2Ph, $n = 1$, $m = 0$

FIG. 7
Sulfonamide—furazan/furoxan CAI hybrids **11–26** [49].

furoxan ring systems [45–49]. Both types of donors are stable in solution, and they showed significant CA inhibitory properties against isoforms involved in various pathologies, such as CA I, II, IV, IX, and XII [45–49]. For at least several sulfonamide-nitrate ester hybrids, the X-ray crystal structures in adducts with CA II, the physiologically dominant isoform for glaucoma, were also reported, allowing the understanding at the molecular level of the inhibition mechanism and allowing for rational drug design campaigns [45, 46]. These hybrids were also shown to release NO in vivo, allowing thus for additive effects to the CA inhibition (produced by the sulfonamide fragment present in these molecules) and its IOP lowering, coupled with vasodilation and increased outflow of blood to the optical nerve (due to the NO), which make them of great interest as novel antiglaucoma agents [42–49]. However, of the wealth of CAI chemical classes, only the sulfonamides were investigated for the moment in order to obtain hybrids with NO-donating agents. Other classes of CAIs, such as the sulfamates, sulfamides, coumarins, sulfocoumarins, carboxylates, and hydroxamates, were not yet used for obtaining such hybrid pharmacological agents.

Apart from the use of the CAI/NO-donor hybrids for glaucoma, there could potentially be many other applications for such hybrid drugs, since in neuropathic pain [22, 23], cerebral ischemia [24], arthritis [25, 26], and many types of solid tumors [21, 27], sulfonamide/sulfamate CAIs were recently shown to possess efficient therapeutic values, by a novel mechanism of action compared with classically used pharmacological agents. Coupling the beneficial effects of

CA inhibition in these pathologies, with the effects of NO as a signaling molecule, may offer additional advantages for the development of novel therapeutic approaches in a multitude of diseases, some of which were mentioned in this chapter.

References

[1] Supuran CT. Structure and function of carbonic anhydrases. Biochem J 2016;473:2023–32.

[2] Supuran CT. Carbonic anhydrases: novel therapeutic applications for inhibitors and activators. Nat Rev Drug Discov 2008;7:168–81.

[3] Supuran CT. How many carbonic anhydrase inhibition mechanisms exist? J Enzyme Inhib Med Chem 2016;31:345–60.

[4] Xu Y, Feng L, Jeffrey PD, Shi Y, Morel FM. Structure and metal exchange in the cadmium carbonic anhydrase of marine diatoms. Nature 2008;452:56–61.

[5] Ferry JG. The gamma class of carbonic anhydrases. Biochim Biophys Acta 2010;1804:374–81.

[6] Macauley SR, Zimmerman SA, Apolinario EE, Evilia C, Hou YM, Ferry JG, Sowers KR. The archetype gamma-class carbonic anhydrase (Cam) contains iron when synthesized in vivo. Biochemistry 2009;48:817–9.

[7] Supuran CT. carbonic anhydrase activators. Future Med Chem 2018;10:561–73.

[8] Domsic JF, McKenna R. Sequestration of carbon dioxide by the hydrophobic pocket of the carbonic anhydrases. Biochim Biophys Acta 2010;1804:326–31.

[9] Tu CK, Silverman DN, Forsman C, Jonsson BH, Lindskog S. Role of histidine 64 in the catalytic mechanism of human carbonic anhydrase II studied with a site-specific mutant. Biochemistry 1989;28:7913–8.

[10] Briganti F, Mangani S, Orioli P, Scozzafava A, Vernaglione G, Supuran CT. Carbonic anhydrase activators: X-ray crystallographic and spectroscopic investigations for the interaction of isozymes I and II with histamine. Biochemistry 1997;36:10384–92.

[11] Alterio V, Di Fiore A, D'Ambrosio K, Supuran CT, De Simone G. Multiple binding modes of inhibitors to carbonic anhydrases: how to design specific drugs targeting 15 different isoforms? Chem Rev 2012;112:4421–68.

[12] Supuran CT. Carbonic anhydrases: from biomedical applications of the inhibitors and activators to biotechnological use for CO_2 capture. J Enzyme Inhib Med Chem 2013;28:229–30.

[13] Del Prete S, Vullo D, Fisher GM, Andrews KT, Poulsen SA, Capasso C, Supuran CT. Discovery of a new family of carbonic anhydrases in the malaria pathogen Plasmodium falciparum—the η-carbonic anhydrases. Bioorg Med Chem Lett 2014;24:4389–96.

[14] Supuran CT. Diuretics: from classical carbonic anhydrase inhibitors to novel applications of the sulfonamides. Curr Pharm Des 2008;14:641–8.

[15] Carta F, Supuran CT. Diuretics with carbonic anhydrase inhibitory action: a patent and literature review. Expert Opin Ther Pat 2005–2013;23:681–91.

[16] Masini E, Carta F, Scozzafava A, Supuran CT. Antiglaucoma carbonic anhydrase inhibitors: a patent review. Expert Opin Ther Pat 2013;23:705–16.

[17] Supuran CT. Structure-based drug discovery of carbonic anhydrase inhibitors. J Enzyme Inhib Med Chem 2012;27:759–72.

[18] Scozzafava A, Supuran CT, Carta F. Antiobesity carbonic anhydrase inhibitors: a literature and patent review. Expert Opin Ther Pat 2013;23:725–35.

[19] Neri D, Supuran CT. Interfering with pH regulation in tumours as a therapeutic strategy. Nat Rev Drug Discov 2011;10:767–77.

[20] Monti SM, Supuran CT, De Simone G. Anticancer carbonic anhydrase inhibitors: a patent review. Expert Opin Ther Pat 2008–2013;23:737–49.

[21] Supuran CT. Carbonic anhydrase inhibition and the management of hypoxic tumors. Metabolites 2017;7:E48.

[22] Carta F, Di Cesare Mannelli L, Pinard M, Ghelardini C, Scozzafava A, McKenna R, Supuran CT. A class of sulfonamide carbonic anhydrase inhibitors with neuropathic pain modulating effects. Bioorg Med Chem 2015;23:1828–40.

[23] Supuran CT. Carbonic anhydrase inhibition and the management of neuropathic pain. Expert Rev Neurother 2016;16:961–8.

[24] Di Cesare Mannelli L, Micheli L, Carta F, Cozzi A, Ghelardini C, Supuran CT. Carbonic anhydrase inhibition for the management of cerebral ischemia: in vivo evaluation of sulfonamide and coumarin inhibitors. J Enzyme Inhib Med Chem 2016;31:894–9.

[25] Margheri F, Ceruso M, Carta F, Laurenzana A, Maggi L, Lazzeri S, Simonini G, Annunziato F, Del Rosso M, Supuran CT, Cimaz R. Overexpression of the transmembrane carbonic anhydrase isoforms IX and XII in the inflamed synovium. J Enzyme Inhib Med Chem 2016;31 (sup4):60–3.

[26] Bua S, Di Cesare Mannelli L, Vullo D, Ghelardini C, Bartolucci G, Scozzafava A, Supuran CT, Carta F. Design and synthesis of novel nonsteroidal anti-inflammatory drugs and carbonic anhydrase inhibitors hybrids (NSAIDs-CAIs) for the treatment of rheumatoid arthritis. J Med Chem 2017;60:1159–70.

[27] Supuran CT. Carbonic anhydrases and metabolism. Metabolites 2018;8:E25.

[28] Supuran CT, Capasso C. An overview of the bacterial carbonic anhydrases. Metabolites 2017;7:E56.

[29] Carta F, Scozzafava A, Supuran CT. Sulfonamides (RSO2NH2): a patent review 2008-2012. Expert Opin Ther Pat 2012;22:747–58.

[30] Di Fiore A, Maresca A, Supuran CT, De Simone G. Hydroxamate represents a versatile zinc binding group for the development of new carbonic anhydrase inhibitors. Chem Commun 2012;48:8838–40.

[31] Supuran CT. Carbon-versus sulphur-based zinc binding groups for carbonic anhydrase inhibitors? J Enzyme Inhib Med Chem 2018;33:485–95.

[32] Maresca A, Carta F, Vullo D, Supuran CT. Dithiocarbamates strongly inhibit the β-class carbonic anhydrases from Mycobacterium tuberculosis. J Enzyme Inhib Med Chem 2013;28:407–11.

[33] Carta F, Aggarwal M, Maresca A, Scozzafava A, McKenna R, Masini E, Supuran CT. Dithiocarbamates strongly inhibit carbonic anhydrases and show antiglaucoma action in vivo. J Med Chem 2012;55:1721–30.

[34] Nair SK, Ludwig PA, Christianson DW. Two-site binding of phenol in the active site of human carbonic anhydrase II: structural implications for substrate association. J Am Chem Soc 1994;116:3659–60.

[35] Carta F, Temperini C, Innocenti A, Scozzafava A, Kaila K, Supuran CT. Polyamines inhibit carbonic anhydrases by anchoring to the zinc-coordinated water molecule. J Med Chem 2010;53:5511–22.

[36] Tars K, Vullo D, Kazaks A, Leitans J, Lends A, Grandane A, Zalubovskis R, Scozzafava A, Supuran CT. Sulfocoumarins (1,2-benzoxathiine-2,2-dioxides): a class of potent and isoform-selective inhibitors of tumor-associated carbonic anhydrases. J Med Chem 2013;56:293–300.

[37] Maresca A, Temperini C, Vu H, Pham NB, Poulsen SA, Scozzafava A, Quinn RJ, Supuran CT. Non-zinc mediated inhibition of carbonic anhydrases: coumarins are a new class of suicide inhibitors. J Am Chem Soc 2009;131:3057–62.

[38] Maresca A, Temperini C, Pochet L, Masereel B, Scozzafava A, Supuran CT. Deciphering the mechanism of carbonic anhydrase inhibition with coumarins and thiocoumarins. J Med Chem 2010;53:335–44.

[39] Bonneau A, Maresca A, Winum JY, Supuran CT. Metronidazole-coumarin conjugates and 3-cyano-7-hydroxy-coumarin act as isoform-selective carbonic anhydrase inhibitors. J Enzyme Inhib Med Chem 2013;28:397–401.

[40] D'Ambrosio K, Carradori S, Monti SM, Buonanno M, Secci D, Vullo D, Supuran CT, De Simone G. Out of the active site binding pocket for carbonic anhydrase inhibitors. ChemCommun 2015;51:302–5.

[41] Quigley HA, Nickells RW, Kerrigan LA, Pease ME, Thibault DJ, Zack DJ. Retinal ganglion cell death in experimental glaucoma and after axotomy occurs by apoptosis. Invest Ophthalmol Vis Sci 1995;36:774–86.

[42] Mincione F, Scozzafava A, Supuran CT. Antiglaucoma carbonic anhydrase inhibitors as ophthalomologic drugs. In: Supuran CT, Winum JY, editors. Drug design of zinc-enzyme inhibitors: functional, structural, and disease applications. Hoboken (NJ: Wiley; 2009. p. 139–54.

[43] Galassi F, Masini E, Giambene B, Fabrizi F, Uliva C, Bolla M, Ongini E. A topical nitric oxide-releasing dexamethasone derivative: effects on intraocular pressure and ocular haemodynamics in a rabbit glaucoma model. Br J Ophthalmol 2006;90:1414–9.

[44] Impagnatiello F, Borghi V, Gale DC, Batugo M, Gazzetta M, Brambilla S, Carreiro ST, Chong WK, Prasanna G, Chiroli V, Ongini E, Krauss AH. A dual acting compound with latanoprost amide and nitric oxide releasing properties, shows ocular hypotensive effects in rabbits and dogs. Exp Eye Res 2011;93:243–9.

[45] Krauss AH, Impagnatiello F, Toris CB, Gale DC, Prasanna G, Borghi V, Chiroli V, Chong WK, Carreiro ST, Ongini E. Ocular hypotensive activity of BOL-303259-X, a nitric oxide donating Prostaglandin F2α agonist, in preclinical models. Exp Eye Res 2011;93:250–5.

[46] Steele RM, Batugo MR, Benedini F, Biondi S, Borghi V, Carzaniga L, Impagnatiello F, Miglietta D, Chong WKM, Rajapakse R, Cecchi A, Temperini C, Supuran CT. Nitric oxide-donating carbonic anhydrase inhibitors for the treatment of open-angle glaucoma. Bioorg Med Chem Lett 2009;19:6565–70.

[47] Mincione F, Benedini F, Biondi S, Cecchi A, Temperini C, Formicola G, Pacileo I, Scozzafava A, Masini E, Supuran CT. Synthesis and crystallographic analysis of new sulfonamides incorporating NO-donating moieties with potent antiglaucoma action. Bioorg Med Chem Lett 2011;21:3216–21.

[48] Fabrizi F, Mincione F, Somma T, Scozzafava G, Galassi F, Masini E, Impagnatiello F, Supuran CT. A new approach to antiglaucoma drugs: carbonic anhydrase inhibitors with or without NO donating moieties. Mechanism of action and preliminary pharmacology. J Enzyme Inhib Med Chem 2012;27:138–47.

[49] Chegaev K, Lazzarato L, Tamboli Y, Boschi D, Blangetti M, Scozzafava A, Carta F, Masini E, Fruttero R, Supuran CT, Gasco A. Furazan and furoxan sulfonamides are strong α-carbonic anhydrase inhibitors and potential antiglaucoma agents. Bioorg Med Chem 2014;22:3913–21.

CHAPTER 13

Understanding the Roles of Nitric Oxide During Sepsis, an Inflammatory Disorder

Shikha Yadav*, Taru Verma†, Sanmoy Pathak*, Dipankar Nandi*,†
**Department of Biochemistry, Indian Institute of Science, Bangalore, India*
†Centre for Biosystems Science and Engineering, Indian Institute of Science, Bangalore, India

ABSTRACT

Inflammation is a cascade of host immune responses triggered by foreign agents and toxic stimuli including infections and tissue injury. It can be broadly classified into acute and chronic inflammation. Sepsis, an acute inflammatory disorder, is a leading cause of critical illness and mortality worldwide. It primarily results due to infections, involving Gram-negative and Gram-positive bacteria, which further results in organ dysfunction due to dysregulated host responses. A subset of sepsis patients develop circulatory and metabolic abnormalities, leading to hypotension and hyperlactatemia that is refractory to fluid infusion therapy, known as septic shock. There are multiple factors that are involved in the pathogenesis of sepsis: hyperinflammatory innate immunity and simultaneous adaptive immune suppression, leading to cytokine storm, lymphocyte anergy, and apoptosis. Oxidative stress is one of the hallmarks of sepsis, and nitric oxide (NO) plays a key role in the generation of reactive intermediates that contribute to sepsis-mediated organ dysfunction. NO has major roles to play in sepsis-mediated vascular dysfunction and immune modulation. Physiologically, NO is generated by three different isoforms of nitric oxide synthase (NOS): neuronal NOS (NOS1), inducible NOS (NOS2), and endothelial NOS (NOS3). Significantly, higher expressions of NOS2 and amounts of NO metabolites are observed in sepsis patients. Contextually, pharmacological studies involving in vivo animal models and clinical trials have been performed with inhibitors of NOS and NO donors. The present review highlights the diverse nature of past and recent studies on the roles of NO and the expression of different isoforms of NOS during sepsis. Further, studies will need to take into account various parameters along with careful monitoring of vascular hemodynamics before consensus is achieved regarding the diverse roles of NO during sepsis.

Therapeutic Application of Nitric Oxide in Cancer and Inflammatory Disorders. https://doi.org/10.1016/B978-0-12-816545-4.00013-X

ABBREVIATIONS

CLP	cecal ligation and puncture
DETA	diethylenetriamine
ICUs	intensive care units
L-NMMA	*N*-methylarginine
LPS	lipopolysaccharide
NADP	nicotinamide adenine diphosphate
NLRs	NOD-like receptors
NO	nitric oxide
NOD	nucleotide-binding oligomerization domain
NOS	nitric oxide synthase
O_2^-	superoxide
OH	hydroxyl radical
$ONOO^-$	peroxynitrite
PRRs	pathogen recognition receptors
qSOFA	Quick SOFA
ROS	reactive oxygen species
SIRS	systemic inflammatory response syndrome
SOFA	sequential organ failure assessment
TLRs	toll-like receptors
TRM	tissue-resident macrophage

Conflict of Interest

No potential conflicts of interest were disclosed.

INTRODUCTION

Inflammation is an essential complex immune response that is critical for survival during infection or injury. It is a cascade of protective immune responses triggered by foreign agents and toxic stimuli such as infections and tissue injury. It also maintains tissue homeostasis under a myriad of harmful conditions caused by cellular stress and tissue malfunction [1]. As explained in Fig. 1, based on various immune factors involved and the duration of the processes, inflammation can be broadly classified as acute and chronic inflammations [2]. Acute inflammation can last for few minutes to days and primarily occurs at the site of infection or tissue injury, involving a coordinated supply of plasma proteins and leukocytes. This type of response is best characterized in case of microbial infections, which are usually recognized by innate immune receptors such as

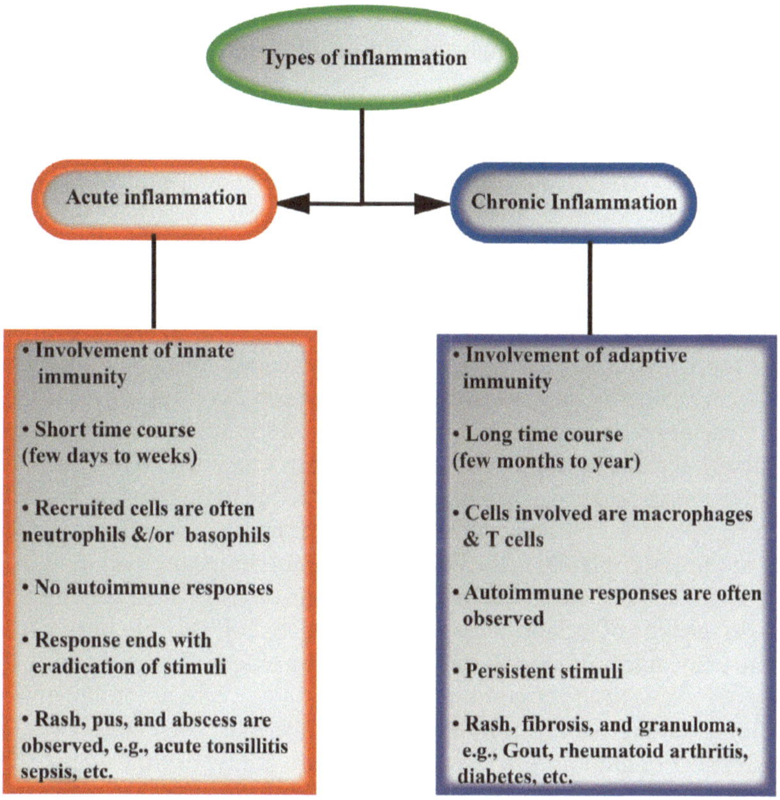

FIG. 1

Broad classification of inflammatory responses. The activation of acute inflammatory pathways is restricted to vascular changes with stereotype responses including vasodilation, increased vascular permeability, and extravasation of neutrophils and plasma proteins, ending with the elimination of the stimulus. Chronic inflammation is an ongoing inflammatory response because of persistent stimulus. The extent of response varies depending on the type of effector tissues and cells, leading to a variety of chronic inflammatory disorders.

toll-like receptors (TLRs) and nucleotide-binding oligomerization domain (NOD) protein-like receptors (NLRs) [3]. Infections are initially detected by tissue-resident macrophages (TRM), which further lead to the production of a variety of inflammatory mediators including chemokines, cytokines, nitric oxide (NO), reactive oxygen species (ROS), eicosanoids, vasoactive amines, and proteolytic enzymes. These inflammatory mediators act on local blood vessels to promote vasodilation, vascular permeability, and extravasation of leukocytes (neutrophils) and plasma proteins [4]. One of the most immediate effects of vasodilation is the local elicitation of the inflammatory responses through exudate formation by infiltration of neutrophils and plasma proteins while preventing the exit of erythrocytes. After reaching the site of infection, neutrophils can be activated in two ways. They are activated by either direct contact with the pathogens or indirectly through the actions of cytokines, produced by TRM. Neutrophils attempt to kill the foreign agents by producing noxious agents including NO, ROS, proteinases, cathepsins, and elastases [5]. Since these highly potent toxic effector molecules cannot discriminate between foreign and host targets, collateral host-tissue damage during inflammation is inescapable [6].The controlled acute inflammatory responses are beneficial in providing protection against infections, however, if uncontrolled, can become detrimental because of their tissue-damaging potential. A classic example of a dysregulated acute inflammatory response is sepsis.

In case the acute inflammatory response fails to completely eradicate the pathogen, a characteristic state of chronic inflammation develops. Chronic inflammation usually occurs when the actual infectious stimuli cannot be eliminated completely. It can also happen by the continuous exposure to chemicals, recurrent infections, and persistence of foreign agents due to incomplete breakdown of the compounds [7]. During chronic inflammation, the inflammatory cascades are not completely resolved, causing tissue damage, malfunction, and organ disorder [2]. The molecular and cellular bases of inflammation differ based on the type of inflammatory cells and organs [8]. During chronic inflammation, neutrophil infiltrate is often replaced by macrophages and even T cells in case of infections. If the combined actions of these cells are still insufficient, granulomas and tertiary lymphoid tissues are formed by these cells. The specific characteristics of granulomas can differ based on the type of the effector T cells [9]. The failure of macrophages to engulf or destroy these pathogenic agents or foreign bodies can also result in the formation of granulomas, such that these invaders are surrounded by macrophages, in a final attempt to protect the host [1]. Typical examples of localized chronic inflammation are gout and rheumatoid arthritis, where the progressive destruction of joints occurs due to persistent immune cell activation and the secretion of inflammatory mediators such as cytokines, NO, and ROS. [10–12]. In this chapter, our focus will be on sepsis, which, as explained earlier, is an example of a dysregulated acute inflammation.

SEPSIS

Sepsis is a leading cause of critical illness and mortality worldwide. It is a multifactorial disorder with severe impairment in the functions of the host immune responses. With increasing information, it is now clear that sepsis is a highly heterogeneous, complex syndrome caused by physiological, pathological, and biochemical abnormalities in response to infections. Sepsis occurs when an infection exceeds the barrier of local tissue containment and induces a series of dysregulated host immune responses. The persistent inflammation and immune suppression altogether lead to organ dysfunction. During sepsis, the immune system is simultaneously divided into two completely opposite directions of excessive inflammation and immune suppression [13].

In a recent comprehensive study from the year 1979 to 2015, sepsis and severe sepsis-induced hospital mortality were reported to be 17% and 26%, respectively. Based on the data from high-income countries, a tentative extrapolation suggests global estimates of 31.5 million sepsis and 19.4 million severe sepsis cases per year, with approximately 5.3 million deaths worldwide annually [14]. Additionally, patients suffering from chronic inflammatory diseases such as cancer and pulmonary disorders are at a higher risk of sepsis due to their predisposition to infection. Interestingly, the incidence of sepsis is higher in males than in females [15].The incidence of severe sepsis is higher in infants and the elderly population than in populations of other age groups. Neonatal sepsis is the major cause of morbidity and mortality of infants, with case fatality ranging from 5% to 60% [16]. Approximately 0.6 million of neonatal deaths are observed because of severe bacterial infections alone. Recently, a randomized double-blinded, placebo-controlled trial involving 4556 infants with an oral preparation of *Lactobacillus plantarum* along with fructooligosaccharide was conducted. The synbiotic trials showed a significant reduction in culture-positive and culture-negative sepsis and lower respiratory tract infections. All these findings suggest that a large proportion of neonatal sepsis in developing countries can be effectively prevented with the use of a synbiotic containing *L. plantarum* ATCC-202195 [17].

HISTORICAL PERSPECTIVES AND CURRENT DEFINITIONS

Since ancient times, sepsis has been recognized as a deadly menace. Hippocrates defined the term "sepsis" as a process by which flesh rots and wound putrefies. Later on, with the confirmation of the famous germ theory by Semmelweis, Pasteur, and others, sepsis was redefined as a systemic infection, often described as "blood poisoning" [18]. In fact, the term "sepsis" has persisted for over 2700 years with nearly the same meaning [19]. However, it was only in the early 1990s that sepsis was defined clinically by a group of opinion leaders and the first consensus definition of sepsis was released.

A 1991 consensus conference, based on the prevailing view, developed the initial definition of "sepsis." This was termed as Sepsis-1, and according to this, sepsis was defined as the combination of infection with two or more features of "systemic inflammatory response syndrome" (SIRS). The criteria for SIRS include

- altered body temperature ($>38°C$ or $<36°C$)
- increased heart rate (>90 beats/min)
- elevated respiratory rate (>22 breaths/min)
- abnormal white blood cells count ($>12 \times 10^9/L$ or $<4 \times 10^9/L$)

Based on these definitions, sepsis complicated by multiple organ dysfunction was termed as severe sepsis, which rapidly progressed to shock. Septic shock, in turn, was defined as persistent sepsis-induced hypotension, which could not be restored despite adequate fluid resuscitation [20]. These definitions were, thereafter, updated in 2001 (Sepsis-2) where the criteria for diagnosis of critical illness were broadened and it was highlighted that sepsis occurs due to the dysregulated host immune responses and not the infecting organism [21].

The ineffective clinical strategies in the past for the treatment of sepsis coupled with an increased understanding of the mechanisms behind sepsis pathogenesis led to the Third International Consensus Definitions for Sepsis and Septic Shock (Sepsis-3) in 2016. The current definition of sepsis and septic shock was developed to address the limitations of the previous definition, which overfocused on SIRS. Sepsis-3 defined sepsis as "life-threatening organ dysfunction caused by a dysregulated host immune response to an infection." Since the definition was centered on organ dysfunction, the committee suggested the use of an organ damage scoring system, namely, the sequential organ failure assessment (SOFA) as the main criterion for sepsis diagnosis. Accordingly, a patient with an acute change in the SOFA score of two points or greater met the criteria for sepsis. The task force also advised that the basal SOFA score be zero, unless the patient had a known history of acute or chronic organ dysfunctions before any signs of infection were detected [22].

SOFA Score

The SOFA score is based on six different independent scores, including respiratory, cardiovascular, hepatic, renal, coagulation, and neurological systems, and is used to check the patient's status or the extent of organ failure and dysfunction in intensive care units (ICUs). Generally, it is designed to predict the mortality but not the success or failure of medical interventions.

The concept of quick SOFA (qSOFA) was also introduced by the task force of Sepsis-3, keeping into consideration the limitations of using the SOFA score outside ICUs. Therefore, to identify the patients with the highest probability of poor outcome associated with sepsis, a bedside clinical score named as qSOFA was developed, which is based on at least two of the following clinical criteria:

- Respiratory rate of 22 breaths/min or greater
- Systolic blood pressure of 100 mmHg or less
- Altered mentation with Glasgow Coma Scale/Score of 14 or less. It is a neurological scale for recording a person's consciousness before and after the assessment

Septic shock is now redefined as a subset of sepsis in which profound circulatory and metabolic abnormalities occur, resulting in a higher risk of mortality than sepsis alone. Clinically, patients with septic shock are on vasopressors to maintain a mean arterial pressure of 65 mmHg or greater. Additionally, the serum lactate levels are greater than 2 mmol/L (>18 mg/dL) in the absence of hypovolemia with in-hospital mortality rates of more than 40%. The 2016 report also discredits the misleading model that sepsis follows a linear continuum from SIRS through severe sepsis to septic shock, and declared the term severe sepsis redundant and unnecessary [22]. The definitions of sepsis defined by Sepsis-2 and Sepsis-3 are schematically represented in Fig. 2.

The medical community has become divided over the clinical worth of the new criteria for diagnosing sepsis. The Sepsis-3 criteria do not recommend any change in the treatment approach, especially regarding the antibiotic and fluid

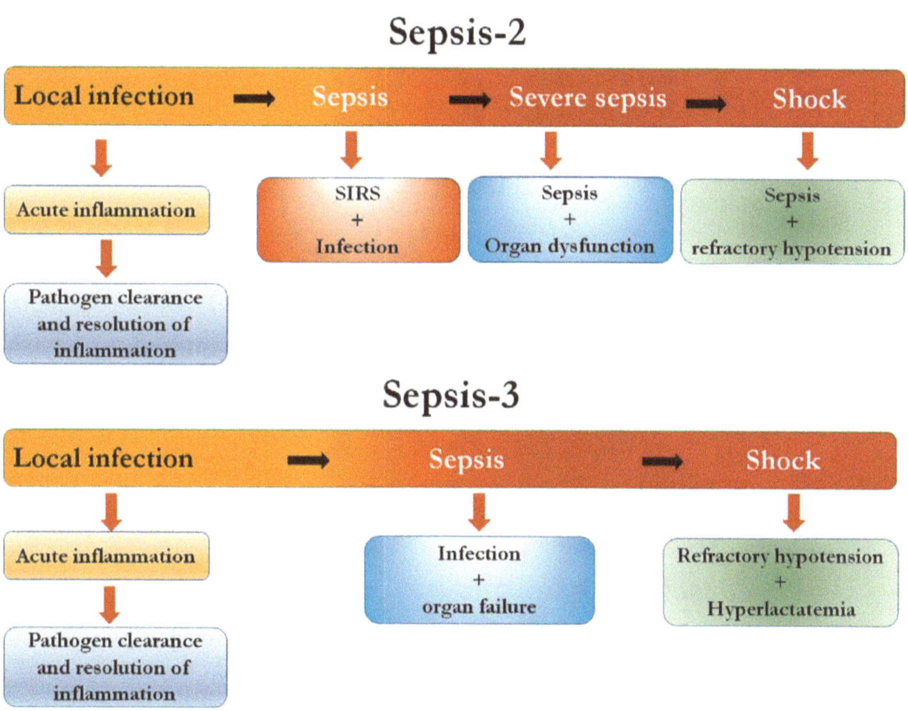

FIG. 2

A schematic representation of the definitions of sepsis by international consensus. The previous model (Sepsis-2) summarizes that sepsis follows a linear trajectory from infection-induced SIRS to severe sepsis (sepsis plus multiorgan failure (MOFs)), followed by the septic shock (hypotension). The revised definition of sepsis and septic shock was developed in 2016. Sepsis-3, unlike the previous definitions, which focused on SIRS, defined sepsis as an organ dysfunction caused by a dysregulated immune response. The term "severe sepsis" was declared redundant and unnecessary. In addition, septic shock is now defined as a subset of sepsis with profound metabolic and circulatory disorder associated with higher risk of mortality than with sepsis alone.

administration and vasopressor support, but the definitions have neglected the importance of early identification of sepsis before the organ failure sets in [23]. Following the new definitions, the earlier categorizations of disease severity and the subsequent mortality due to infection that progressed from sepsis, to severe sepsis, to shock have now been reduced to simple infection, sepsis, and septic shock.

The drop in blood pressure and lactatemia has allowed doctors to identify patients with different disease severities and prognoses since a long time. This has been summarized below:

- *Cryptic shock*: Defined as concentration of lactate ≥4 mmol/L without any arterial hypotension or the use of vasopressors
- *Septic shock*: Sepsis-induced hypotension that persists despite adequate fluid resuscitation and can manifest clinically in two ways
- *Vasoplegic shock*: Normal serum lactate levels but hypotension refractory to fluid therapy
- *Shock with tissue dysoxia*: Hyperlactatemia along with hypotension refractory to fluid therapy, associated with higher mortality compared with the other forms

According to Sepsis-3, only the last group (shock with tissue dysoxia) is classified as septic shock. The other phenotypic expressions of septic shock severity have been ignored in the new definitions as cryptic shock and vasoplegic shock are not considered. These categories are now classified as sepsis.

It is important to highlight that there is not a pathophysiological condition that is unique to sepsis and that the diagnosis of infection results from the crossing of three components: systemic manifestations, appearance of organ dysfunction, and microbiological evidence. Notably, no specific biomarkers for sepsis exist till date [23]. An overview of the definitions of sepsis and the diagnosing parameters are listed in Table 1.

Causative Organisms

Sepsis, often, occurs as a result of community-acquired and health-care-associated infections. For example, pneumonia is the most common cause of sepsis and accounts for almost half of the cases, followed by intra-abdominal and urinary tract infections. Blood cultures are positive in only one-third of the cases [24]. Based on past records, the most frequently encountered pathogens are the Gram-positive *Staphylococcus aureus* and *Streptococcus pneumoniae*, whereas *Escherichia coli*, *Klebsiella* species, and *Pseudomonas aeruginosa* are the common Gram-negative isolates [24, 25]. In a study involving 14,000 ICU patients from 75 countries, microbiological blood cultures were positive for 70% of patients, out of which the majority of the sepsis patients had Gram-

Table 1 Overview of the different sepsis definitions in order of their occurrence

Definition	Condition	Diagnosing Criteria
Sepsis-1 (1991)	Sepsis	SIRS + confirmed/suspected infection
	Severe sepsis	Sepsis + organ dysfunction
	Shock	Sepsis + persistent hypotension despite fluid resuscitation
Sepsis-2 (2001) Definitions remained the same as Sepsis-1; more parameters were added to the diagnosis	• Edema • Hyperglycemia in the absence of diabetes	>20 mL/kg in 24 h > 120 mg/dL glucose
	Inflammation markers	Increased serum C-reactive protein (CRP) and plasma procalcitonin
	Hemodynamic variables	Cardiac index >3.5 L/min/m2
	Organ dysfunction markers	• Arterial hypoxemia • Altered consciousness • Acute oliguria (<0.5 mL/kg/h) • Increased creatinine (>0.5 mg/dL), • Coagulation disorders • Thrombocytopenia (<100,000/μL) • Hyperbilirubinemia (>4 mg/dL or 70 mmol/L)
	Tissue perfusion indicators	• Hyperlactatemia (>1 mmol/L) • Reduced capillary refill, mottled skin
Sepsis-3 (2016)	Sepsis	• qSOFA score—respiratory rate > 22 breaths/min • Systolic arterial pressure < 100 mmHg • Altered mentation
	Shock	• Arterial hypotension requiring vasopressors for maintaining mean arterial pressure > 65 mmHg • Hyperlactatemia (>18 mg/dL or 2 mmol/L) despite fluid resuscitation

negative bacterial infections followed by gram-positive bacterial infections and few incidences of fungal infections [26].

Due to advancements in critical care and improved surgical interventions, the fatality rates of sepsis have decreased. However, the absolute mortality caused by sepsis tends to increase because of increasing numbers of patients with sepsis [27]. In addition, patients who survive the early phase of sepsis and are discharged from the hospitals are at a higher risk of sepsis-induced long-term mortality (within a year of discharge from the hospital). Sepsis survivors often have a poor quality of life owing to the cognitive disabilities, cardiovascular diseases, and secondary infections, altogether indicating the significant impact of sepsis on the health care and the society [28]. Although the clinical manifestations of sepsis are highly variable, depending on the site of infection, causative agent, and the underlying health status of the patients, broadly, sepsis-induced pathophysiology can be explained as follows.

Hyperinflammation

Due to the loss of local-regional containment of infection, the body is systemically exposed to infectious agents and damaged tissue products. These lead to the induction of SIRS, through the recognition of pathogen-associated molecular patterns (PAMPs) by the ubiquitously present pathogen recognition receptors (PRRs) on innate immune cells and on various nonprofessional immune cells. Currently, four different classes of PRRs have been identified:

- TLRs
- C-type lectin receptors
- Cytosolic proteins such as retinoic acid-inducible gene (RIG)-I-like receptors (RLRs)
- NODs

One of the most potent microbial inducer implicated in the pathogenesis of sepsis and septic shock is lipopolysaccharide (LPS). LPS is the notable bacterial endotoxin, involved in the induction of dysregulated hyperinflammatory responses upon recognition by innate immune cells through PRRs [29]. PRRs can also recognize endogenous molecules released from damaged tissues and cells, called damage-associated molecular patterns (DAMPs) such as high-mobility group box 1, adenosine triphosphate, and nucleic acids. PRRs, upon recognition of their ligands, upregulate the transcription of various inflammatory proteins such as cytokines, chemokines, antimicrobial peptides, and other proteins involved in the regulation of inflammation pathway [30]. During sepsis, the systemic activation of innate immune cells by PAMPs and DAMPs results in sustained and drastic activation of the inflammatory responses. Sepsis is often characterized by an excessive release of proinflammatory cytokines such as IL-1, TNFα, and IL-17 over a relatively smaller duration of 2–3 days collectively known as "cytokine storm." The progression to a state of persistent inflammation can further result in dysregulated immune responses with ongoing catabolic metabolism and cellular stress. These altogether hamper tissue homeostasis and induce organ dysfunction leading to multiorgan failure (MOF) due to deterioration in the ability of immune system to clear infections and heal the injured tissues [13].

Immune Suppression

Until recently, it was thought that the initial hyperinflammatory phase (SIRS) of sepsis was followed by a counteractive, compensatory antiinflammatory response syndrome or immunosuppressive phase, which began several days later. However, it is now evident that both the proinflammatory and antiinflammatory responses occur early and coexist. The net effect of these two competing processes is manifested predominantly by the early hyperinflammatory responses [13]. During sepsis, the functions of both the innate and the adaptive immune effector cells are impaired. The expression of the genes involved in the regulation of the adaptive immune response is highly downregulated, which results in sustained complications [31]. It is demonstrated that in ongoing inflammation-induced organ damage and dysfunction, in conjunction with recurrent secondary nosocomial infections, the recovery of sepsis patients is impaired leading to mortality [13].

INNATE IMMUNITY AND IMMUNE SUPPRESSION

Sepsis-induced immune suppression is mediated directly or indirectly by the impaired functions of virtually all kinds of innate and adaptive immune cells. Some of the key innate immune cells with impaired functions responsible for the immunosuppressive effects are discussed below.

Neutrophils

These are the most prevalent innate immune cells comprising the majority of the bone marrow cellularity and are essential for the control of invading pathogens at an early stage. Neutrophils are also the first responders to migrate to the site of infection. Under a normal scenario, circulating neutrophils die within 24h after their release from the bone marrow. However, in sepsis patients, neutrophil apoptosis is delayed because of those patients that show a remarkable increase in the number of circulating neutrophils with varying degrees of maturation [32]. It is also evident from the experimental studies on various mouse models and human patient samples that during sepsis, there are profound defects in the chemotactic ability and recruitment of neutrophils to the site of infection. Neutrophils demonstrate clear deficits in oxidative burst with lesser production of ROS, RNI, and antimicrobial proteins and cytokines leading to infectious complications [33]. More importantly, multiple studies with human patients have implicated the roles of persistent neutrophil dysfunction in the development of hospital-acquired infections [34–36]. A vast majority of patients undergoing long-term, sepsis-induced mortality have ongoing secondary, nosocomial infections, indicating that the neutrophil-mediated bacterial clearance could serve as a potential therapeutic target.

In addition to the conventional ways of phagocytosis, both the degranulation and oxidative burst to eliminate pathogens, neutrophils can also eradicate a wide range of infectious organisms by forming neutrophil extracellular traps by a process called as NETosis. This process involves the release of antimicrobial proteins such as cathepsin G, proteinase 3, myeloperoxidase, elastase, lactoferrin, gelatinase, and calprotectin as well as histone proteins H1, H2A, H2B, H3, and H4, all embedded in the neutrophil chromatin material. The chromatin fibers promote physical containment, whereas the granular proteins confer antimicrobial activity [37].

MONOCYTES AND MACROPHAGES

Under normal conditions, after the clearance of the infectious agents, impaired tissue cells and leukocytes undergo apoptosis or are engulfed by macrophages and removed from the inflamed site to resolve inflammation. During this process, macrophages further trigger the production of IL-10 and TGF-β resulting in the healing of the injured tissue [38]. The hallmark antiinflammatory cytokine, IL-10, suppresses the production of IL-6 and IFNγ and inhibits the proinflammatory signaling, by stimulating the production of soluble TNF receptor and IL-1 receptor antagonist. However, the roles of monocyte subpopulations during sepsis have been the subject of intense investigations. For decades, it has been apparent that myeloid cells undergo "endotoxin tolerance," a state in which they respond with a reduced production of proinflammatory cytokines and enhanced potency to produce the antiinflammatory cytokine IL-10 upon LPS stimulation. Furthermore, reduced human leukocyte antigen-DR isotype (HLA-DR) expression on monocytes, a hallmark of sepsis-related immune suppression, directly correlates with severity and mortality in sepsis [31]. Currently, the reduced monocyte HLA-DR expression is considered as a surrogate marker of "monocyte anergy." The mechanisms responsible for the reduced ability of monocytes to secrete pro-inflammatory cytokines, TNFα, IL-1, IL-6, and IL-12, upon LPS stimulation are still not clearly defined [39]. Analysis of mRNA expression levels demonstrated increased levels of inhibitory signaling molecules and antiinflammatory cytokines with reduced levels of pro-inflammatory cytokines (IL-6 and TNFα) and chemokine receptor protein mRNAs [40]. These aspects may contribute to the lower lymphocyte proliferation probably due to poor antigen presentation [41]. Recent studies highlight the roles of epigenetic regulations during endotoxin tolerance, through histone protein modification and chromatin remodeling [42]. Interestingly, IFNγ-mediated monocyte stimulation can abrogate endotoxin tolerance by TLR-induced chromatin remodeling and restores the cytokine release by monocytes [43].

ADAPTIVE IMMUNITY AND IMMUNE SUPPRESSION

Unlike neutrophils and myeloid cells, which show impaired functions during sepsis with delayed apoptosis, circulating lymphocytes undergo significant apoptosis. The elementary observation of a profound depletion in the number of $CD4^+$ and $CD8^+$ T cells and B cells due to enhanced apoptosis in dying sepsis patients was the basis for sepsis-induced immune suppression [44]. Sepsis-induced apoptosis occurs in all age groups, including neonatal, pediatric, and adult populations, in response to various classes of microorganisms [44–46]. Lymphocyte apoptosis not only leads to the reduction in the number of functioning T cells and B cells but also impacts the function of surviving cells that uptake the apoptotic cells. Macrophages and DCs that engulf the apoptotic lymphocytes undergo a state of anergy further contributing to immune tolerance [47].

SEPSIS-INDUCED METABOLIC DEFECTS

Sepsis-induced metabolic defects, i.e., a shift from oxidative phosphorylation to glycolysis, lead to the accumulation of metabolites, which further act as modulators of epigenetic enzymes. For example, in some patients, decreased levels of intracellular acetyl CoA can lead to reduced histone acetylation. Decreased histone acetylation further inhibits the gene transcription resulting into immunoparalysis. In contrast, highly activated innate immune cells show higher succinate accumulation, which stabilizes the transcription factor hypoxia-inducible factor-1α (HIF-1α). Increased HIF-1α along with the inhibition of demethylases leads to the upregulation of cytokines and increased inflammation [48].

Sepsis and Neuroimmunology

On the basis of advances in the field of neuroimmunology, it is also evident that during inflammation, the immune and the nervous systems communicate with each other [49, 50]. Accordingly, sepsis is also defined as a neuroendocrine disorder [51]. The vagus nerve signaling, through cholinergic receptors expressed by phagocytes, has crucial roles in the regulation of inflammation. This is also known as the cholinergic antiinflammatory pathway, which dampens the cytokine production by macrophages [52]. Inflammation is under neuronal control by the autonomic nervous system, which signals through acetylcholine and its receptors and inhibits the synthesis of proinflammatory cytokines, termed as "inflammatory reflex" [53]. It has been shown that the branches of the vagus nerve in the spleen are crucial for the suppression of cytokine synthesis during sepsis [54]. In addition to cholinergic receptors, the autonomic nervous system also signals through catecholamines and its adrenergic receptors expressed on leukocytes and all other tissues and cell types. The early inflammatory phase of sepsis is characterized by high circulating amounts of catecholamines such as epinephrine and norepinephrine. The activation of adrenergic receptors on the immune cells triggers cytokine response in an NFκB-dependent manner [55].

Sepsis-Induced Mortality

Historically, sepsis-induced death occurred in a biphasic manner, where the - sepsis-induced mortality of patients often occurs within few days (early mortality) and several weeks later (late-phase mortality), after the onset of infection due to cytokine-mediated hyperinflammatory and antiinflammatory responses, respectively, as shown in Fig. 3. Subsequent advances in intensive care and specific interventions have led to a drop in sepsis-induced biphasic mortality during the initial periods. However, based on the recent trends in sepsis outcomes, as depicted in Fig. 3, the majority of the sepsis survivors succumb to long-term mortality (60–90 days post the onset of sepsis), mounting a trimodal distribution for sepsis-induced deaths. Hence, according to the trimodal distribution curve, sepsis-induced death can occur at three phases: early-phase (few days), late-phase (few weeks), and long-term mortality (several months to years) after the onset of sepsis [13]. Sepsis-associated long-term mortality occurs because of ongoing immunologic, biochemical, and physiological aberrations post sepsis that impair the recovery of sepsis patients. Historically, most of the clinical research to treat sepsis focused on ameliorating the

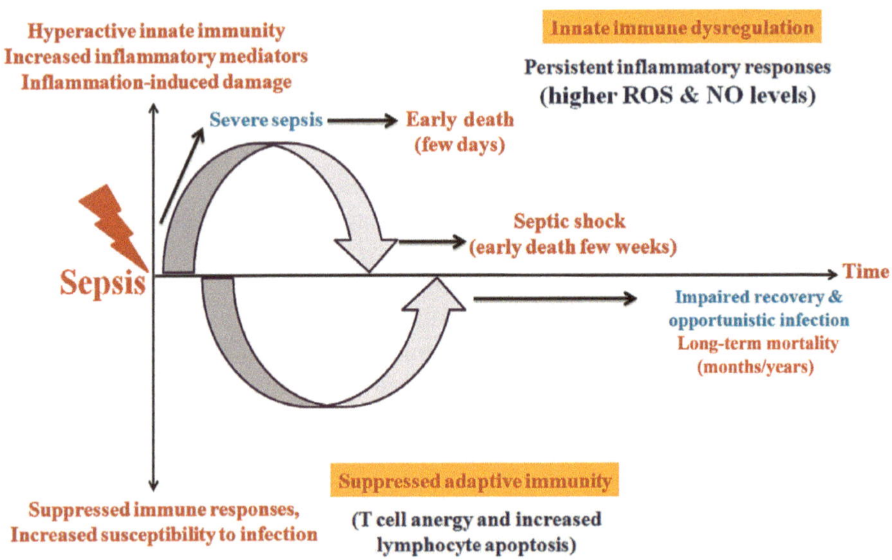

FIG. 3

A schematic overview of sepsis-associated immune dysregulation and death. Studies from the tissue samples of sepsis and critically ill patients have demonstrated that the overwhelming inflammatory response driven by the hyperactive innate and the suppressed adaptive immunity lead to persistent organ damage and dysfunction that may culminate in the death of the patients. Unregulated inflammatory responses can lead to organ dysfunction and sepsis-induced early mortality; the process is largely amended due to improvement in patient management protocols. However, the vast majority of elderly group of patients with comorbid conditions often succumb to sepsis-associated long-term mortality because of persistent derangement in cellular functions of innate and adaptive immune cells.

hyperinflammatory responses. In contrast, with the recent advances in the understanding of the sepsis pathogenesis, investigations have now shifted toward immune suppression and underlying innate and adaptive immune dysregulations. Also, new insights obtained from the tissue samples of sepsis and severely injured trauma patients have identified coexisting inflammatory and antiinflammatory responses driven by the dysregulated innate and suppressed adaptive immunities. Altogether, sepsis alters the innate and adaptive immune responses for a sustained period of time, culminating in persistent inflammation, chronic catabolism, and immune suppression. All these factors correlate with long-term sepsis-induced mortality due to ongoing tissue injury with poor tissue regeneration and recurrent and persistent secondary, nosocomial infections [13].

Models to Study Sepsis

The discovery of novel therapeutics to treat sepsis, apart from antibiotics, fluids, and vasopressors, poses a significant challenge to clinicians and internists, despite extensive research in this area. As it is not ethically feasible to test each new potential therapeutic agent directly in human patients, researchers have resorted to the use of different animal models as platforms for drug testing and to study the pathophysiology of sepsis. These models can provide levels of sophistication that are more advanced than any cell- or tissue-culture-based systems. The frequently used animal models based on the agents used for initiating sepsis and their shortcomings are described below:

- Exogenous administration of toxins such as LPS resulting in the development of LPS-induced endotoxic shock model
- Exogenously administered live bacterial infection (monobacterial) model
- Alterations of the host endogenous protective barrier to induce colonic permeability and bacterial translocation by cecal ligation and puncture (CLP) (polymicrobial) model

The use of animal models of sepsis, however, has also faced extensive criticism, particularly in the way they relate to human disease progression. Critics cite numerous differences between humans and mouse models, particularly at the genomic level. Some of the reasons behind the failure of mouse model of sepsis to accurately predict the outcome of clinical trials in humans are the following:

- Most laboratory mice, compared with humans, are relatively more resistant to the widely used endotoxin LPS. The dose of LPS that leads to death of approximately half of mice called LD_{50} is about 1000- to 10,000-fold higher in mice (1–25 mg/kg) than the dose of LPS required to induce severe illness in humans (2–4 ng/kg) [56, 57]. An improvement in

LPS-induced endotoxic shock model in animals would be selecting the animals that are sensitive to LPS like humans; for example, large animals like sheep mimics remarkable changes in pulmonary arterial pressure, cardiac output, and lung microvasculature upon infusion of LPS at the rate of 9 ng/kg of body weight per hour [58]. Unlike infection-induced sepsis, eradication of the infectious organism is not crucial for survival in the acute endotoxemia model. However, an overwhelming inflammatory response generated by massive activation of the innate immune system induces death in this model of sepsis. During endotoxemia, systemic cytokine levels show a transient and very rapid induction. However, in a clinical setting of sepsis, prolonged increase in systemic cytokine level is seen with cytokine concentration several orders of magnitude lower than in the endotoxemia model. Therefore, antiinflammatory therapies are like anti-TNFα therapy that is highly beneficial in murine models of endotoxemia but not in the case of human sepsis [56, 59].

- There is a difference in the timings of development of the disease. The onset and progression of sepsis to MOF in all the animal models of sepsis occur rapidly within hours to days, whereas in humans, it occurs gradually and can take days to several weeks:
- Experimental animals are rarely provided with supportive interventions, in contrast to standard therapies for the treatment of sepsis in humans such as antibiotics and vasopressor therapy, mechanical ventilation, and fluid resuscitation.
- Most of the mouse studies are done using young, healthy mice without any comorbid conditions. While in humans, sepsis usually occurs in neonates and at an older age with one or more comorbidities [60].

In contrast, other researchers argue that there exist several similarities between humans and animal models and stress on the importance of choosing the appropriate model to draw comparisons with human sepsis. However, it cannot be ignored that sufficient differences exist between the mouse and human immunology, and therefore, one must always keep in mind that any given response in a mouse may not translate exactly into humans. While this necessitates exercising caution while interpreting data obtained in mice, these animals have and will continue to be the primary models in contributing to our understanding of the human immunology in health and disease. The relatively low cost, easy handling, and availability of genetically modified strains are added benefits of working with mice models. Other small mammals like rats, guinea pigs, and rabbits have also been used to study different aspects of sepsis. Rabbits are large enough to endure and survive protocols that demand the use of several samples like blood samples. Owing to their larger size, dogs, cats, pigs, and sheep have been used for protocols that demand surgery and blood sampling and for procedures that require implantation of devices for real-time monitoring of physiological parameters. Assessing the cardiac output, ventilatory parameters, and pulmonary arterial pressure is also relatively easy in these larger-sized animals in comparison with smaller animals like mice and guinea pigs. The anatomy and physiology of pigs is remarkably similar to that of humans. Nonhuman primates like monkeys are the most similar to humans, but working with such animals is quite challenging. Due to extensive ethical issues, studies on nonhuman primates are reserved for preclinical studies only [61].

Treatment of sepsis has improved steadily. The progressive improvement in treatment of sepsis is based on the understanding of the underlying mechanisms behind the failure of clinical trials over the past decades. Based on this, current potential therapeutics to modulate sepsis-induced derangement in immune responses are granulocyte-monocyte colony-stimulating factor; recombinants IL-17, IL-7, and IFNγ; etc. Sepsis-induced pathophysiology primarily occurs through molecular alterations that further disrupt cellular functions. Many different molecular mediators have been linked to the pathophysiology of sepsis. The major molecular alterations that can be considered as central hubs during sepsis, which impact cellular phenotype with pleiotropic effects and connection with various pathways, are complement activation, coagulopathy, redox imbalance, etc. Abundant literature from clinical and experimental studies has demonstrated the existence of overwhelming oxidative stress in sepsis patients [62, 63]. Oxidative stress results from uncontrolled production of ROS and reactive nitrogen species. One of the key protein responsible for excessive oxidative stress during sepsis is inducible nitric oxide synthase 2 (NOS2), which is overexpressed in activated immune cells such as macrophages and neutrophils [64]. Some of the oxidizing molecules like NO, hydrogen peroxide (H_2O_2) and hydroxyl radical (OH) act together to promote cellular dysfunction and organ failure.

NITRIC OXIDE

Nitric oxide is a highly reactive signaling molecule with a half-life of ~ 1 s. It is a small, neutral, hydrophobic molecule, capable of permeating cell membranes and cellular compartments. It is involved in the regulation of several activities from neurotransmission and vascular hemodynamics by stimulating the NO-sensitive guanylyl cyclase to gene transcription and translation by binding to iron-responsive elements and posttranslational modification of proteins by ADP-ribosylation [65]. Several of the target molecules of NO are regulatory molecule themselves, for example, transcription factors and enzymes due to which NO exhibits heterogeneous and diverse phenotypic effects [66]. Depending on its concentration and cellular compartment, NO can either have beneficial or detrimental effects. At physiological concentrations, it can regulate cell-cell interactions and immune responses, whereas at pathologically higher concentrations, it can have detrimental effects due to increased oxidation of biomolecules like nucleic acids, proteins, lipids, and prosthetic groups, which may further contribute to inflammation-induced damages [67].

NO can be synthesized by three different isoforms of NO synthases (NOS): calcium-dependent, constitutively expressed, neuronal NOS (NOS1); endothelial NOS (NOS3); and calcium-independent, inducible NOS (NOS2), which is expressed in a variety of immune and nonimmune cells upon activation. NOS converts L-arginine to L-citrulline by utilizing molecular oxygen and reduced nicotinamide adenine dinucleotide phosphate (NADPH) as cosubstrate through a complex set of oxidation reactions, as explained in Fig. 4. Flavin adenine dinucleotide (FAD), flavin mononucleotide (FMN), and 5,6,7, 8-tetrahydro L-biopterin (BH_4) are used as cofactors by all isoenzymes [68]. All three isoforms of NOS are active as homodimers and require calmodulin, heme, and Zn^+ for their dimerization [68]. A functional homodimer of NOS transfers electrons from NADPH, via FAD and FMN, in the carboxy-terminal reductase domain to the heme-containing amino-terminal oxygenase domain. The oxygenase domain binds with the essential cofactor BH_4, molecular oxygen and L-arginine [65]. Although three isoforms of NOS catalyze the same reaction, they differ with respect to their regulations, amplitude and durations of the production of NO, and cellular distributions. The expression of NOS2 by cytokines is independent of the intracellular calcium concentration and is primarily regulated by de novo synthesis and stability of the mRNA [69]. In contrast, NOS1 and NOS3 exist in cells as preformed proteins, and their activity is dependent on the increase in the intracellular calcium concentration. A high concentration of intracellular calcium (200–400 nM) is required for calmodulin binding and, thus, catalytic activities of NOS1 and NOS3. Due to differences in amino acid sequence and structure of the calmodulin-binding site of NOS2, calmodulin binds to NOS2 at an extremely low intracellular concentration of calcium (below 40 nM) [65]. The significance of NOS and L-arginine-mediated pathway in inflammation is described in Fig. 4.

NO and Oxidative Stress

NO is a labile molecule and is quickly scavenged in the biological system. Under normal physiological conditions, intracellular NO can react and further oxidize to form nitrogen dioxide (NO_2), partially stable nitrite (NO_2^-), and nitrate (NO_3^-) anions. Hence, nitrite is now recognized as a reservoir of an NO-like bioactive molecule in blood that can be recycled back to form NO and other bioactive nitrogen oxides [67]. All of these oxidized states of NO can either act as antimicrobial agents or induce signaling events in the host [69]. During steady state of equilibrium, nitrite with *S*-nitrosothiols has been

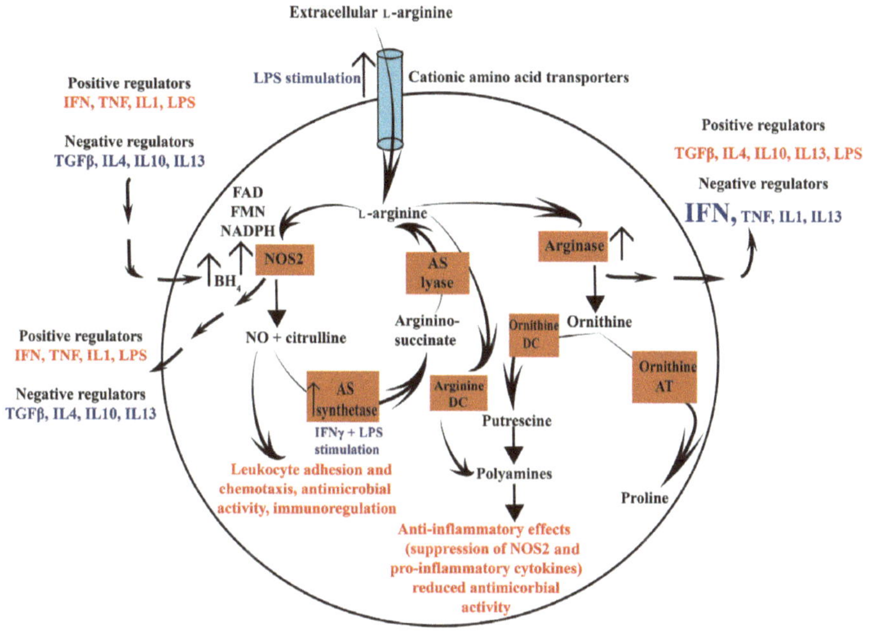

FIG. 4

Importance of L-arginine-mediated pathways and its roles during inflammation. The activity of NOS2 in macrophages is regulated by cytokines or bacterial components, e.g., LPS, which modulate the uptake of extracellular L-arginine by cationic amino acid transporters, the synthesis of cofactors like tetrahydrobiopterin (BH_4), the expression and stability of NOS2 mRNA and protein, the enzymatic recycling of citrulline to arginine, and the depletion of arginine by arginase and arginine decarboxylase (DC). The availability of extracellular L-arginine is strongly modulated by the activity of arginase, which can be secreted into the extracellular space and breaks down arginine into urea and ornithine. There are two isoforms of arginase, cytosolic arginase I and mitochondrial arginase II. Antiinflammatory cytokines like IL-4, IL-10, and TGFβ increase the activity of arginases. On the other hand, pro-inflammatory cytokines like IFNγ and TNFα increase the activity of NOS2. Pro- and antiinflammatory cytokines can also induce or suppress the synthesis of BH_4, respectively, a positive regulator of NOS2 enzyme. The induction of arginase by anti-inflammatory cytokines prior to the activation of NOS2 by pro-inflammatory cytokines can inhibit the production of NO due to the depletion of substrate (arginine). This is independent of IL-10, IL-13, or TGFβ-mediated inhibition of transcription and translation of *NOS2* gene. However, in case arginase and NOS2 are coinduced, for example, by LPS, the production of NO is not impaired because of ~3000-fold higher affinity of NOS2 enzyme for arginine compared with the arginase enzyme. Further, macrophages and vascular smooth muscle cells expressing NOS enzymes can regenerate arginine from citrulline by argininosuccinate (AS) synthetase, and IFNγ and LPS can induce the activity of AS synthetase. Polyamines such as putrescine, produced by ornithine DC and arginine DC, are antiinflammatory and act as immunosuppressants. This further downregulates the proinflammatory effects of NO. Besides, increased arginase activity in the absence of NOS2 through ornithine aminotransferase (AT) leads to higher amounts of proline required for collagen synthesis and contribute to tissue fibrosis. Since arginine is an essential amino acid, required for the microbial growth, depletion of arginine by NOS2 may indirectly contribute to its antimicrobial activity. Importantly, the arginase inhibitor L-hydroxyarginine, which inhibits the activity of arginase enzyme in macrophages (increased NOS2 activity) and parasites (prevent their replication), is used for killing intracellular parasites like *Leishmania*.

shown to increase cGMP levels in tissues [67]. Nitrite can be reduced back into NO by a variety of host enzymes such as hemoglobin and xanthine oxidoreductase. In addition, the nitrite reductase activity of mammalian tissues is also contributed by the mitochondrial electron transport chain, protonation, and deoxyhemoglobin. Since the mitochondria are the major source of cellular oxygen consumption via oxidative phosphorylation, mitochondrial dysfunction during sepsis can also contribute to excessive ROS production [70]. Pertinent to immune cell activation and mitochondrial dysfunction, sepsis is characterized by an excess of ROS and RNS. Importantly, NO can also react with superoxide (O_2^-) to form a highly potent oxidizing molecule, known as peroxynitrite ($ONOO^-$). $ONOO^-$ can react with cysteine and thiol groups of proteins to promote nitration of protein tyrosine residues that affects the normal functioning of proteins [71]. $ONOO^-$ exists in equilibrium with its protonated form peroxynitrous acid (ONOOH). Under physiologically relevant pH conditions, both the species coexist and react differently. Under alkaline conditions, ONOOH adopts a cis configuration and degrades into inactive NO_3^- anion; however, during acidic conditions, it isomerizes to transconfiguration and yields hydroxyl radical and nitrogen dioxide. In general, $ONOO^-$ is associated with the pathophysiology of a variety of inflammatory disorders including atherosclerosis, hypertension, neurodegeneration, cancer, and sepsis [71]. The pathological effects of $ONOO^-$ are mediated by inhibition of the beneficial actions of NO, protein nitration, or both. In fact, $ONOO^-$ and protein nitration are considered to be good biomarkers of inflammatory disease progression. Due to enhanced oxidative stress observed during vascular dysfunction and inflammation, an increased degradation or consumption of NO by its reaction with O_2^- occurs and leads to the formation of $ONOO^-$. $ONOO^-$ oxidizes BH_4 due to which the dimerization of NOS is inhibited and its affinity for its substrate L-arginine is reduced. Consequently, reduction of oxygen and the production of NO by NOS are uncoupled, known as NOS uncoupling. Under these conditions, the NO generating functional homodimer of NOS is converted into dysfunctional O_2^--producing monomers of NOS, which further contributes to enhanced O_2^- production [65]. Thus, based on the local conditions, NO either can behave as a toxic-free radical and further potentiate the toxicity of circulating oxygen radicals or can ameliorate and scavenge the O_2^- [12, 71]. The roles of multiple pathways involved in the oxidative stress and their interactions with NO are explained in Fig. 5.

Roles of NO in Inflammation

Profound physiological changes are observed during inflammation including vasodilation, fever, and activation of the immune system. In addition to the regulation of these changes by diverse kinds of inflammatory mediators, such as cytokines, proteolytic enzymes, lipids, vasoactive amines, and peptides, NO governs a broad spectrum of these processes and plays an important role during

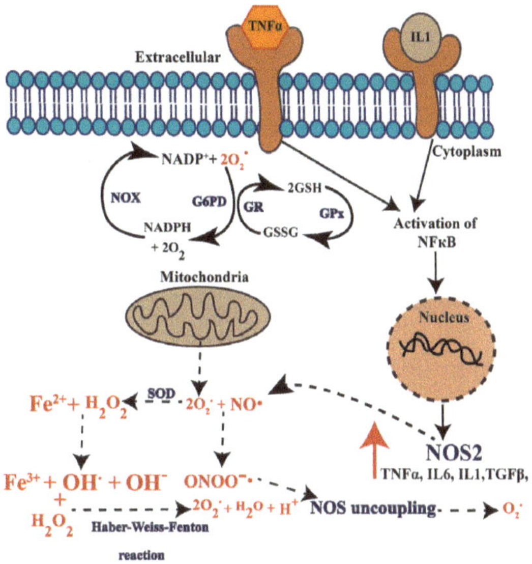

FIG. 5

Multiple pathways are involved in the generation of oxidative stress. The principal sources of intracellular ROS during inflammatory conditions are NADPH oxidase (NOX), mitochondria, and cytokine-mediated activation of NFκB pathway. NOX-mediated induction of O_2^- in immune cells contributes to oxidative burst, involved in killing of intracellular pathogens. Significant amounts of cellular ROS are formed by mitochondria during oxidative phosphorylation, which is further increased during inflammatory conditions. In addition, proinflammatory cytokine-mediated induction of NOS2 leads to increased amounts of NO. O_2^- and NO can further react to form the highly potent oxidizing agent $ONOO^-$. The excess of $ONOO^-$ during oxidative stress results in NOS uncoupling that further contributes to O_2^- formation and oxidative damage. Intracellular iron amounts can further lead to the formation of superoxide through the Haber-Weiss-Fenton reaction. Since free radicals are formed as a by-product of beneficial cellular processes, cells have evolved with intracellular antioxidant to maintain the redox equilibrium, such as glutathione (GSH) and NADPH.

inflammation. NO is involved in the neurotransmission and regulation of vascular hemodynamics through the activation of cGMP. Physiologically, NO plays a key role in the production of cytokines, differentiation, proliferation, and apoptosis of immune cells as well as inhibition of adhesion and recruitment of leukocytes to the site of inflammation [66, 72]. Hyperactive NOS1, upon stimulation with a huge calcium influx in neuronal cells, is associated with neuronal cell death in cerebrovascular stroke [73]. Higher levels of NO can also lead to neuronal death due to energy depletion, obstruction of mitochondrial respiration, and glycolysis [74]. Increased NOS2 activity is observed in a broad spectrum of inflammatory diseases in humans [75, 76]. The majority of inflammatory lesions are characterized by the accumulation of activated macrophages and neutrophils with significant amounts of NOS2-derived

NO [75]. Also, nonspecific allograft rejection is likely to be associated with higher amounts of NOS2-derived NO [77]. In contrast, the protective roles of NO in a variety of experimental autoimmune models are also known [66]. For example, NOS2 negatively regulates the function of autoimmune T_H1 cells and protects the host from sequelae of immunopathologic responses [66]. Furthermore, NOS2-mediated host protective responses during infection-induced inflammation involves inhibition of tissue fibrosis and the termination of the immune response by apoptosis of activated $CD4^+$ cells [78].

NOS3-derived NO released toward the vascular lumen is a potent inhibitor of platelet aggregation and adhesion to the walls of blood vessels [65]. Endothelial NO protects against the onset of atherosclerosis by preventing the cytokine-induced apoptosis of endothelial cells and inducing the remodeling of angiogenesis [79]. Out of the three distinct isoforms of NOS, it is the inducible NOS2, which is a predominant form expressed in a wide variety of cell types during inflammation, producing much larger amounts of NO compared with the other two isoforms. The production of NO in models of chronic localized inflammation is usually associated with the induction of maladaptive processes. Evidently, elevated levels of NO_2^- are observed in the synovial fluid of arthritis patients, and chronic administration of an NOS inhibitor reduces it significantly. Also increased production of NO_2^- by pancreatic islet cells upon stimulation with IL-1β correlated with the development of experimental diabetes and was abrogated upon application of an NOS inhibitor [12].

Roles of NO in Sepsis and Its Therapeutic Implications

Disordered leukocyte trafficking, activation of coagulation, and endothelial damage during sepsis altogether affect the integrity and function of the microcirculatory system. Collectively, the microcirculatory dysfunction is a critical event in the development of sepsis, leading to sepsis-induced circulatory defects. NO plays central and multifaceted roles in maintaining the microcirculatory integrity and function [80].

NO and Vascular Hemodynamics in Sepsis

In 1991, elevated levels of plasma NO_2^- and NO_3^-, the by-products of NO, were observed in sepsis patients for the first time and have been implicated in sepsis-induced vascular manifestations [81]. Later, several reports described the roles of NO in the manifestation of dysregulated vascular hemodynamic during sepsis [82–85]. Also, studies using animal models have demonstrated the role of NOS2 in mediating vasodilation and hypotension, leading to septic shock [86, 87]. In fact, a profound increase in the amounts of NOS2-derived NO is counterbalanced by the suppression of endothelial NOS3 expression in the heart, lungs, and aorta of rats during endotoxemia [88]. Contextually, sepsis-related changes in the metabolism of NO have promoted pharmacological studies with the aim to reduce and to increase NO in order to improve sepsis-induced hemodynamic manifestations. It is relevant to point out the failure of clinical trials with the nonspecific NOS inhibitor, 546C88, for the treatment of sepsis [89, 90]. Initial studies with the administration of NOS inhibitors in animal models of sepsis reversed sepsis-induced hypotension [91]. These findings subsequently led to clinical trials that targeted NOS using L-NMA. In comparison with untreated/mocktreated groups, increased mortality in patients treated with L-NMA was observed. Thus, these clinical trials had to be terminated prematurely [85]. Subsequent reports suggested that inhibiting NOS functions dysregulates the microcirculation leading to lower organ perfusion, cardiac output, and accelerated mortality [85]. The inhibition of NOS during endotoxemia results in an increased hepatic injury [92]. Subsequent studies demonstrated that increased hepatic injury upon NOS inhibition is partially due to O_2^- and OH^- radicals, indicating the organ-protective role of NO by scavenging free radicals [93]. NO also plays cytoprotective roles and prevents renal and intestinal injuries during endotoxemia [94, 95]. Supportingly, a recent study of *Salmonella typhimurium*-induced peritonitis model of sepsis has shown NOS2 as a critical factor required for the initiation of the host innate immune responses at the site of infection. Consequently, the lack of NOS2 leads to increased organ damage and reduced survival [96].

During pathological inflammatory conditions such as sepsis, NO induces hypotension due to systemic vasodilation. Simultaneously, it also exhibits organ beneficial effects by maintaining end-organ perfusion [80]. Efforts have been made to improve the deranged lower microcirculation during sepsis by increasing the availability of NO through exogenous NO donors. The application of DETA-NO attenuates the adhesion of platelets to the endothelium during LPS-induced endotoxemia [97]. In agreement, reduced survival of $Nos2^{-/-}$ mice compared with wild-type mice, during *S. typhimurium* peritonitis-induced sepsis, is significantly increased upon DETA-NO supplementation [96]. Also, LPS-induced manifestation in the rat hepatic microcirculation is improved by the early application of the NO donor, sodium nitroprusside. Linsidomine or molsidomine (SIN-1) improves the cardiac index, mesenteric blood flow, and myocardial dysfunction in the early phase of endotoxic shock in fluid-resuscitated dogs. Altogether, the combination of fluid infusion therapy along with an NO donor and an NOS2 inhibitor might improve locally deranged microcirculation under the condition of heterogeneous NOS expression [98]. Across clinical and preclinical studies with various models of sepsis, administration of NO donors is associated with improved pulmonary pressure and reduced organ damage and mortality [99]. List of NO-modulating drugs and their preclinical and clinical trials is listed in Table 2.

Table 2 List of nonclinical and clinical trials performed to investigate the effects of NOS inhibitors and NO donors during sepsis and septic shock

Experimental Designs (Preclinical Studies)	Modulation of NO	Effects
Intravenous administration of group B *Streptococcus* in piglets	Sodium nitroprusside (NO donor)	Reduced mean arterial pressure. No effect on cardiac index, pulmonary arterial pressure, and pulmonary vascular resistance index [100]
Intravenous administration of group B *Streptococcus* in piglets	Nitroglycerin (NO donor)	Reduced cardiac index. No effect on mean arterial pressure, pulmonary arterial pressure, and pulmonary vascular resistance index [100]
Intraperitoneal administration of cecal slurry in rats	Aminoguanidine (iNOS inhibitor)	Reduced stroke volume in nonseptic rats. Increased heart rate in septic rats. Reduced blood flow in nonseptic rats [101]
Administration of *P. aeruginosa* in sheep via inhalation	BBS-2 (iNOS inhibitor)	Proper pulmonary gas exchange was observed. Reduced obstruction of air passages [102]
Administration of *P. aeruginosa* in pigs intravenously	L-NIL (iNOS inhibitor) + tempol (superoxide scavenger)	Reduced renal, cardiovascular, and coagulation abnormalities [103]
Intraperitoneal administration of *S. typhimurium* in mice	DETA NONOate (NO donor)	Improved recruitment of neutrophils at site of infection. Reduction of bacterial load in the peritoneum [96]
Patients with septic shock	L-NMMA (NOS inhibitor)	Reduction in cardiac output. Increase in blood pressure. Improved vascular tone [104]
Patients with septic shock	L-NAME (NOS inhibitor)	Improvement in mean arterial pressure. Better vascular resistance [105]
Septic shock patients	Nitroglycerin (NO donor)	Decrease in mean arterial pressure temporarily. Increased microvascular flow after a short time [106]
Septic shock patients	546C88 (NOS inhibitor)	Increase in mortality [90]
Severe sepsis patients	Nitroglycerin (NO donor)	Higher mortality. No change in microcirculation [107]

As explained before the neutrophils, a major cell type of the innate immune system, they are often the first cells to be recruited to the site of infection and play significant roles in pathogen clearance. They are equipped to generate large amounts of oxidants, proteases, and antimicrobial proteins [108]. Although neutrophils are known to be important during sepsis, their efficacy with respect to the pathogenesis of sepsis is controversial. The uncontrolled neutrophil activation during sepsis may lead to inflammation-induced host-tissue damage and multiorgan failure [109]. In the contrary, defects in the ability of neutrophils to migrate to the sites of infection are often associated with increased sepsis severity and mortality [35]. Also, the lack of adequate neutrophil response at the site of infection leads to higher lung tissue damage in peritonitis-induced septic shock in rabbits [110]. Neutropenia and low neutrophil storage pool in the bone marrow of neonates are associated with neonatal sepsis [34]. These studies demonstrate the relevance of neutrophil migration and function during sepsis.

NO and Neutrophil Migration During Sepsis

Neutrophils of septic patients express high amounts of *Nos2* mRNA [111]. Further increased amounts of NO upon sepsis lower the neutrophil migration by inhibiting the process of rolling and adherence to the blood vessel endothelium [112, 113]. Conversely, NO is also reported to increase the amounts of IL-8, a potent chemoattractant and activator of neutrophils [114]. In a CLP model of sepsis, increased amount of TNFα is correlated to the reduced neutrophil migration to the affected site. This reduction in neutrophil migration during CLP is not observed in mice lacking TLR4 or NOS2 [35]. However, in case of peritonitis-induced sepsis upon intraperitoneal *S. typhimurium* infection, the lack of NOS2 inhibits the neutrophil infiltration in the peritoneal cavity [96]. These studies illustrate that the roles of NOS2 may vary depending on the cellular location, temporal expression, and type of the inflammatory response.

EMERGING CONCEPTS OF NO-BASED THERAPIES DURING SEPSIS

Single-nucleotide polymorphisms of NOS3 (E298D) and NOS2A (exon 22), either individually or in combination, are associated with increased plasma concentration of NO metabolites in sepsis patients, which in turn is associated with increased hemodynamic failure and mortality [115]. Reduced bioavailability of NO in sepsis patient and increased expression of an endogenous NOS inhibitor, like dimethylarginine dimethylaminohydrolase 2 in immune cells of septic patients, are associated with altered antimicrobial activity and increased sepsis severity [116]. In the contrary, reduced expression of NOS2 in mouse macrophages upon LPS stimulation through the inhibition of heme oxygenase-1 and increased lysosomal degradation of NOS2 results in reduced myocardial damage and sepsis-induced cardiac dysfunction [117]. Noteworthily, the regulation and induction of NOS2 in human and mouse macrophages differ significantly, and there is a substantial controversy as to whether human macrophages express NOS2 [118]. It is reported that, unlike mouse macrophages that show significant induction of NOS2 upon IFNγ and LPS stimulation, human macrophages lack the induction of NOS2 under similar conditions. However, it is shown that other cytokines such as IFNα/β, IL-4 along with IL-23, and several other chemokines are much more efficient inducers of NOS2 in human macrophages [119].

The levels of NO during sepsis are critical and of particular interest, basically because of two reasons. First, NO regulates systemic blood circulation by smooth muscle relaxation and vasodilation. Therefore, excessive production of NO results in severe hypotension and septic shock. On the contrary, NO is also essential for tissue oxygenation and the supply of nutrients to tissues and organs by regulating the homeostasis in the context of microcirculation.

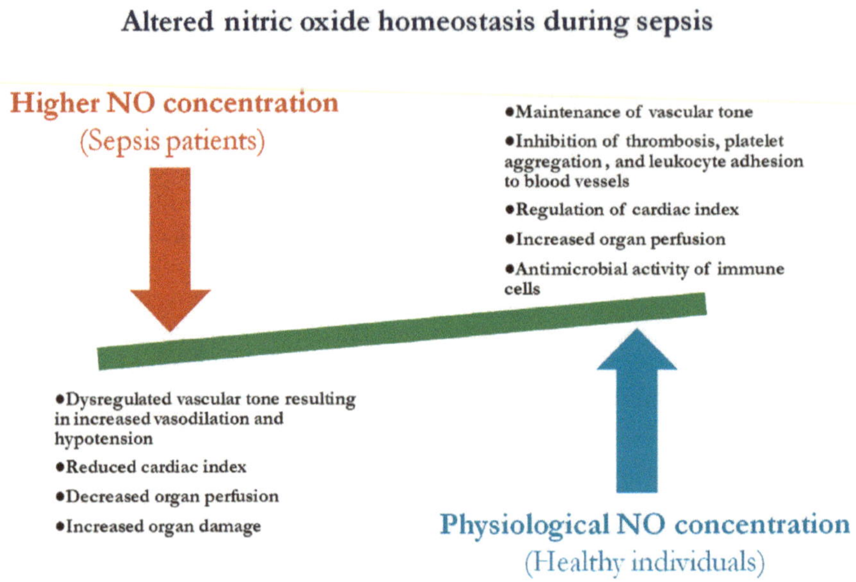

FIG. 6

Roles of altered NO homeostasis in the pathogenesis of sepsis. NO is a pleiotropic molecule with varied and pivotal roles in sepsis. In a healthy individual, at physiological concentrations, NO regulates the macrocirculation by modulating the cardiac output and vasodilation as well as the microcirculation by controlling end-organ perfusion, platelet aggregation, vascular pressure, etc. NO also exhibits antimicrobial activity and synchronizes immune cell function.

NO maintains the microcirculation by dilation and regulation of the vascular tone and by inhibiting the platelet aggregation and leukocyte attachment to the endothelium of blood vessel. Second, NO has antimicrobial effects and lowers microbial growth during infections [120]. Hence, during sepsis, NO acts like a double-edged sword where excessive production of NO on one hand can lead to hypotension, whereas the lack of NO on the other hand can lead to reduced organ perfusion and poor resistance to infection [82–84, 121]. The roles of NO and its dysregulated homeostasis in pathophysiology of sepsis are described in Fig. 6.

Several issues that need clarification and consideration for the establishment of improved, effective, and successful clinical trials with NO-based therapeutics are as follows: Dose of NO-modulating drugs should be tittered based on cardiopulmonary monitoring, age of patient, and phase of sepsis. Also, the administration of NOS2-specific inhibitor should be taken into consideration than nonspecific NOS inhibitors. It was observed in the ovine model of sepsis that endothelial NOS3 is induced transiently at early time point (8–12 h) post injury, whereas inducible NOS2 is induced profoundly at later time point (12h) post injury, without any significant induction of NOS1 [122]. In

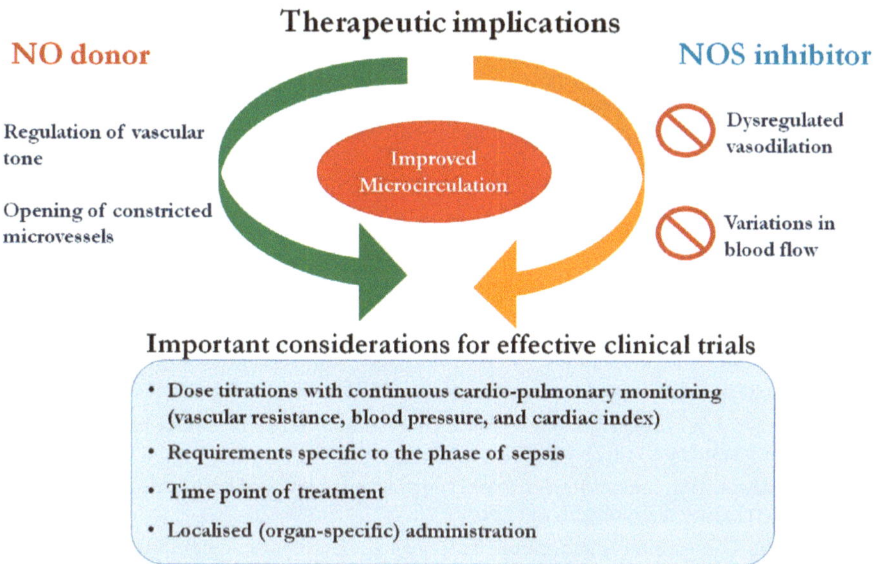

FIG. 7

Hypothetical mechanism to improve NO-based therapeutics during sepsis. An approach to stabilize the microcirculatory blood flow and the vascular hemodynamic parameters during sepsis may consist of a combination of fluid infusion therapy along with nitric oxide donor and a selective NOS2 inhibitor. On the other hand, NO donors might supplement the physiological concentration of endothelial NO and improve the microcirculation, whereas the selective inhibition of NOS2 might prevent the excessive and heterogeneous production of NO.

contrast to the treatment with nonselective NOS inhibitors or the inhibition of a selective NOS isoform at the wrong time point, sepsis patients should be treated with selective NOS inhibitors at the time of their maximum activity. The effects of NO-modulatory drugs in the pathogenesis of sepsis are described in Fig. 7.

CONCLUSION

In conclusion, sepsis is a complex, dysregulated inflammatory disorder. Delineating the composite roles of NO in the regulation of inflammatory responses, macro- and microcirculation, and immune pathology will be critical in exploring its therapeutic efficacies. Since NO exhibits heterogeneous effects during sepsis, future trials with administration of NO-modulating compounds will need to be carefully planned, keeping various parameters in mind. Comprehensive studies with different in vivo models are essential before the concept of NO-modulating compounds can be extended to trials with human sepsis patients.

Acknowledgments

We thank all the previous and current members of the laboratory for their support. Grants from CSIR and DBT have allowed us to set up and investigate mice models of sepsis. Our studies were possible due to the infrastructural support by the DBT-IISc program and central facilities in IISc, which are greatly appreciated.

References

[1] Medzhitov R. Origin and physiological roles of inflammation. Nature 2008;454 (7203):428–35.

[2] Arulselvan P, Fard MT, Tan WS, Gothai S, Fakurazi S, Norhaizan ME, Kumar SS. Role of antioxidants and natural products in inflammation. Oxid Med Cell Longev 2016;2016:5276130.

[3] Barton GM. A calculated response: control of inflammation by the innate immune system. J Clin Invest 2008;118(2):413–20.

[4] Markiewski MM, Lambris JD. The role of complement in inflammatory diseases from behind the scenes into the spotlight. Am J Pathol 2007;171(3):715–27.

[5] Nathan C. Neutrophils and immunity: challenges and opportunities. Nat Rev Immunol 2006;6(3):173–82.

[6] Nathan C. Points of control in inflammation. Nature 2002;420(6917):846–52.

[7] Ferguson LR. Chronic inflammation and mutagenesis. Mutat Res 2010;690(1–2):3–11.

[8] Weber A, Boege Y, Reisinger F, Heikenwalder M. Chronic liver inflammation and hepatocellular carcinoma: persistence matters. Swiss Med Wkly 2011;141:w13197.

[9] Drayton DL, Liao S, Mounzer RH, Ruddle NH. Lymphoid organ development: from ontogeny to neogenesis. Nat Immunol 2006;7(4):344–53.

[10] Martinon F, Petrilli V, Mayor A, Tardivel A, Tschopp J. Gout-associated uric acid crystals activate the NALP3 inflammasome. Nature 2006;440(7081):237–41.

[11] Choy E. Understanding the dynamics: pathways involved in the pathogenesis of rheumatoid arthritis. Rheumatology 2012;51:v3–v11.

[12] Wong JM, Billiar TR. Regulation and function of inducible nitric oxide synthase during sepsis and acute inflammation. Adv Pharmacol 1995;34:155–70.

[13] Delano MJ, Ward PA. The immune system's role in sepsis progression, resolution, and long-term outcome. Immunol Rev 2016;274(1):330–53.

[14] Fleischmann C, Scherag A, Adhikari NK, Hartog CS, Tsaganos T, Schlattmann P, Angus DC, Reinhart K, T. International Forum of Acute Care. Assessment of global incidence and mortality of hospital-treated sepsis. Current estimates and limitations. Am J Respir Crit Care Med 2016;193(3):259–72.

[15] Angus DC, Linde-Zwirble WT, Lidicker J, Clermont G, Carcillo J, Pinsky MR. Epidemiology of severe sepsis in the United States: analysis of incidence, outcome, and associated costs of care. Crit Care Med 2001;29(7):1303–10.

[16] Thaver D, Zaidi AK. Burden of neonatal infections in developing countries: a review of evidence from community-based studies. Pediatr Infect Dis J 2009;28:S3–9.

[17] Panigrahi P, Parida S, Nanda NC, Satpathy R, Pradhan L, Chandel DS, Baccaglini L, Mohapatra A, Mohapatra SS, Misra PR, Chaudhry R, Chen HH, Johnson JA, Morris JG, Paneth N, Gewolb IH. A randomized synbiotic trial to prevent sepsis among infants in rural India. Nature 2017;548(7668):407–12.

[18] Funk DJ, Parrillo JE, Kumar A. Sepsis and septic shock: a history. Crit Care Clin 2009;25 (1):83–101.

[19] Geroulanos S, Douka ET. Historical perspective of the word "sepsis" Intensive Care Med 2006;32(12):2077.

[20] Bone RC, Sibbald WJ, Sprung CL. The ACCP-SCCM consensus conference on sepsis and organ failure. Chest 1992;101(6):1481–3.

[21] Levy MM, Fink MP, Marshall JC, Abraham E, Angus D, Cook D, Cohen J, Opal SM, Vincent JL, Ramsay G. SCCM/ESICM/ACCP/ATS/SIS, 2001 SCCM/ESICM/ACCP/ATS/SIS international Sepsis definitions conference. Crit Care Med 2003;31(4):1250–6.

[22] Singer M, Deutschman CS, Seymour CW, Shankar-Hari M, Annane D, Bauer M, Bellomo R, Bernard GR, Chiche JD, Coopersmith CM, Hotchkiss RS, Levy MM, Marshall JC, Martin GS, Opal SM, Rubenfeld GD, van der Poll T, Vincent JL, Angus DC. The third international consensus definitions for sepsis and septic shock (Sepsis-3). JAMA 2016;315(8):801–10.

[23] Carneiro AH, Povoa P, Gomes JA. Dear Sepsis-3, we are sorry to say that we don't like you. Rev Bras Ter Intensiva 2017;29(1):4–8.

[24] Ranieri VM, Thompson BT, Barie PS, Dhainaut JF, Douglas IS, Finfer S, Gardlund B, Marshall JC, Rhodes A, Artigas A, Payen D, Tenhunen J, Al-Khalidi HR, Thompson V, Janes J, Macias WL, Vangerow B, Williams MD, P.-S.S. Group. Drotrecogin alfa (activated) in adults with septic shock. N Engl J Med 2012;366(22):2055–64.

[25] Opal SM, Garber GE, LaRosa SP, Maki DG, Freebairn RC, Kinasewitz GT, Dhainaut JF, Yan SB, Williams MD, Graham DE, Nelson DR, Levy H, Bernard GR. Systemic host responses in severe sepsis analyzed by causative microorganism and treatment effects of drotrecogin alfa (activated). Clin Infect Dis 2003;37(1):50–8.

[26] Vincent JL, Rello J, Marshall J, Silva E, Anzueto A, Martin CD, Moreno R, Lipman J, Gomersall C, Sakr Y, Reinhart K, E.I.G.o. investigators. International study of the prevalence and outcomes of infection in intensive care units. JAMA 2009;302(21):2323–9.

[27] Gaieski DF, Edwards JM, Kallan MJ, Carr BG. Benchmarking the incidence and mortality of severe sepsis in the United States. Crit Care Med 2013;41(5):1167–74.

[28] Yende S, Austin S, Rhodes A, Finfer S, Opal S, Thompson T, Bozza FA, LaRosa SP, Ranieri VM, Angus DC. Long-term quality of life among survivors of severe sepsis: analyses of two international trials. Crit Care Med 2016;44(8):1461–7.

[29] Opal SM. Endotoxins and other sepsis triggers. Contrib Nephrol 2010;167:14–24.

[30] Takeuchi O, Akira S. Pattern recognition receptors and inflammation. Cell 2010;140 (6):805–20.

[31] Hotchkiss RS, Monneret G, Payen D. Sepsis-induced immunosuppression: from cellular dysfunctions to immunotherapy. Nat Rev Immunol 2013;13(12):862–74.

[32] Drifte G, Dunn-Siegrist I, Tissieres P, Pugin J. Innate immune functions of immature neutrophils in patients with sepsis and severe systemic inflammatory response syndrome. Crit Care Med 2013;41(3):820–32.

[33] Alves-Filho JC, Spiller F, Cunha FQ. Neutrophil paralysis in sepsis. Shock 2010;34:15–21.

[34] Christensen RD, Rothstein G. Exhaustion of mature marrow neutrophils in neonates with sepsis. J Pediatr 1980;96(2):316–8.

[35] Alves-Filho JC, de Freitas A, Spiller F, Souto FO, Cunha FQ. The role of neutrophils in severe sepsis. Shock 2008;30:3–9.

[36] Morris AC, Brittan M, Wilkinson TS, McAuley DF, Antonelli J, McCulloch C, Barr LC, McDonald NA, Dhaliwal K, Jones RO, Mackellar A, Haslett C, Hay AW, Swann DG, Anderson N, Laurenson IF, Davidson DJ, Rossi AG, Walsh TS, Simpson AJ. C5a-mediated neutrophil dysfunction is RhoA-dependent and predicts infection in critically ill patients. Blood 2011;117(19):5178–88.

[37] Brinkmann V, Reichard U, Goosmann C, Fauler B, Uhlemann Y, Weiss DS, Weinrauch Y, Zychlinsky A. Neutrophil extracellular traps kill bacteria. Science 2004;303(5663):1532–5.

[38] Kolb JP, Oguin 3rd TH, Oberst A, Martinez J. Programmed cell death and inflammation: winter is coming. Trends Immunol 2017;38:705–18.

[39] Cavaillon JM, Adib-Conquy M. Bench-to-bedside review: endotoxin tolerance as a model of leukocyte reprogramming in sepsis. Crit Care 2006;10(5):233.

[40] Ishii M, Wen H, Corsa CA, Liu T, Coelho AL, Allen RM, Carson WFt, Cavassani KA, Li X, Lukacs NW, Hogaboam CM, Dou Y, Kunkel SL. Epigenetic regulation of the alternatively activated macrophage phenotype. Blood 2009;114(15):3244–54.

[41] Cazalis MA, Friggeri A, Cave L, Demaret J, Barbalat V, Cerrato E, Lepape A, Pachot A, Monneret G, Venet F. Decreased HLA-DR antigen-associated invariant chain (CD74) mRNA expression predicts mortality after septic shock. Crit Care 2013;17(6):R287.

[42] Carson WF, Cavassani KA, Dou Y, Kunkel SL. Epigenetic regulation of immune cell functions during post-septic immunosuppression. Epigenetics 2011;6(3):273–83.

[43] Chen J, Ivashkiv LB. IFN-gamma abrogates endotoxin tolerance by facilitating toll-like receptor-induced chromatin remodeling. Proc Natl Acad Sci of U S A 2010;107(45): 19438–43.

[44] Hotchkiss RS, Swanson PE, Freeman BD, Tinsley KW, Cobb JP, Matuschak GM, Buchman TG, Karl IE. Apoptotic cell death in patients with sepsis, shock, and multiple organ dysfunction. Crit Care Med 1999;27(7):1230–51.

[45] Toti P, De Felice C, Occhini R, Schuerfeld K, Stumpo M, Epistolato MC, Vatti R, Buonocore G. Spleen depletion in neonatal sepsis and chorioamnionitis. Am J Clin Pathol 2004;122(5): 765–71.

[46] Felmet KA, Hall MW, Clark RS, Jaffe R, Carcillo JA. Prolonged lymphopenia, lymphoid depletion, and hypoprolactinemia in children with nosocomial sepsis and multiple organ failure. J Immunol 2005;174(6):3765–72.

[47] Voll RE, Herrmann M, Roth EA, Stach C, Kalden JR, Girkontaite I. Immunosuppressive effects of apoptotic cells. Nature 1997;390(6658):350–1.

[48] van der Poll T, van de Veerdonk FL, Scicluna BP, Netea MG. The immunopathology of sepsis and potential therapeutic targets. Nat Rev Immunol 2017;17(7):407–20.

[49] Elenkov IJ, Wilder RL, Chrousos GP, Vizi ES. The sympathetic nerve—an integrative interface between two supersystems: the brain and the immune system. Pharmacol Rev 2000;52(4): 595–638.

[50] Sternberg EM. Neural regulation of innate immunity: a coordinated nonspecific host response to pathogens. Nat Rev Immunol 2006;6(4):318–28.

[51] Munford RS, Tracey KJ. Is severe sepsis a neuroendocrine disease? Mol Med 2002;8(8): 437–42.

[52] Wang H, Yu M, Ochani M, Amella CA, Tanovic M, Susarla S, Li JH, Wang H, Yang H, Ulloa L, Al-Abed Y, Czura CJ, Tracey KJ. Nicotinic acetylcholine receptor alpha7 subunit is an essential regulator of inflammation. Nature 2003;421(6921):384–8.

[53] Tracey KJ. The inflammatory reflex. Nature 2002;420(6917):853–9.

[54] Huston JM, Ochani M, Rosas-Ballina M, Liao H, Ochani K, Pavlov VA, Gallowitsch-Puerta M, Ashok M, Czura CJ, Foxwell B, Tracey KJ, Ulloa L. Splenectomy inactivates the cholinergic antiinflammatory pathway during lethal endotoxemia and polymicrobial sepsis. J Ex Med 2006;203(7):1623–8.

[55] Flierl MA, Rittirsch D, Nadeau BA, Chen AJ, Sarma JV, Zetoune FS, McGuire SR, List RP, Day DE, Hoesel LM, Gao H, Van Rooijen N, Huber-Lang MS, Neubig RR, Ward PA. Phagocyte-derived catecholamines enhance acute inflammatory injury. Nature 2007;449 (7163):721–5.

[56] Taveira da Silva AM, Kaulbach HC, Chuidian FS, Lambert DR, Suffredini AF, Danner RL. Brief report: shock and multiple-organ dysfunction after self-administration of Salmonella endotoxin. N Engl J Med 1993;328(20):1457–60.

[57] Barber AE, Coyle SM, Fischer E, Smith C, van der Poll T, Shires GT, Lowry SF. Influence of hypercortisolemia on soluble tumor necrosis factor receptor II and interleukin-1 receptor antagonist responses to endotoxin in human beings. Surgery 1995;118(2):406–10.

[58] Traber DL, Redl H, Schlag G, Herndon DN, Kimura R, Prien T, Traber LD. Cardiopulmonary responses to continuous administration of endotoxin. Am J Physiol 1988;254(5):H833–9. Pt 2.

[59] Warren HS, Fitting C, Hoff E, Adib-Conquy M, Beasley-Topliffe L, Tesini B, Liang X, Valentine C, Hellman J, Hayden D, Cavaillon JM. Resilience to bacterial infection: difference between species could be due to proteins in serum. J Infect Dis 2010;201(2):223–32.

[60] Buras JA, Holzmann B, Sitkovsky M. Animal models of sepsis: setting the stage. Nat Rev Drug Discov 2005;4(10):854–65.

[61] Fink MP. Animal models of sepsis. Virulence 2014;5(1):143–53.

[62] Ogilvie AC, Groeneveld AB, Straub JP, Thijs LG. Plasma lipid peroxides and antioxidants in human septic shock. Intensive Care Med 1991;17(1):40–4.

[63] Santos SS, Brunialti MK, Rigato O, Machado FR, Silva E, Salomao R. Generation of nitric oxide and reactive oxygen species by neutrophils and monocytes from septic patients and association with outcomes. Shock 2012;38(1):18–23.

[64] Andrades ME, Morina A, Spasic S, Spasojevic I. Bench-to-bedside review: sepsis—from the redox point of view. Crit Care 2011;15(5):230.

[65] U. Forstermann, W.C. Sessa, Nitric oxide synthases: regulation and function, Eur Heart J 33 (7) (2012) 829–37, 837a-837d.

[66] Bogdan C. Nitric oxide and the immune response. Nat Immunol 2001;2(10):907–16.

[67] Bryan NS. Nitric oxide enhancement strategies. Future Sci 2015;1(1):FSO48.

[68] Stuehr DJ, Kwon NS, Nathan CF. FAD and GSH participate in macrophage synthesis of nitric oxide. Biochem Biophys Res Commun 1990;168(2):558–65.

[69] MacMicking J, Xie QW, Nathan C. Nitric oxide and macrophage function. Annu Rev Immunol 1997;15:323–50.

[70] Galley HF. Bench-to-bedside review: targeting antioxidants to mitochondria in sepsis. Crit Care 2010;14(4):230.

[71] Radi R. Oxygen radicals, nitric oxide, and peroxynitrite: redox pathways in molecular medicine. Proc Natl Acad Sci U S A 2018;115(23):5839–48.

[72] Hickey MJ. Role of inducible nitric oxide synthase in the regulation of leucocyte recruitment. Clin Sci 2001;100(1):1–12.

[73] Lipton SA, Choi YB, Pan ZH, Lei SZ, Chen HS, Sucher NJ, Loscalzo J, Singel DJ, Stamler JS. A redox-based mechanism for the neuroprotective and neurodestructive effects of nitric oxide and related nitroso-compounds. Nature 1993;364(6438):626–32.

[74] Brown GC. Nitric oxide and neuronal death. Nitric Oxide 2010;23(3):153–65.

[75] Kroncke KD, Kolb-Bachofen V, Berschick B, Burkart V, Kolb H. Activated macrophages kill pancreatic syngeneic islet cells via arginine-dependent nitric oxide generation. Biochem Biophys Res Commun 1991;175(3):752–8.

[76] Weinberg JB. Nitric oxide production and nitric oxide synthase type 2 expression by human mononuclear phagocytes: a review. Mol Med 1998;4(9):557–91.

[77] Langrehr JM, Hoffman RA, Billiar TR, Lee KK, Schraut WH, Simmons RL. Nitric oxide synthesis in the in vivo allograft response: a possible regulatory mechanism. Surgery 1991;110(2):335–42.

[78] Dalton DK, Haynes L, Chu CQ, Swain SL, Wittmer S. Interferon gamma eliminates responding CD4 T cells during mycobacterial infection by inducing apoptosis of activated CD4 T cells. J Exp Med 2000;192(1):117–22.

[79] Forstermann U. Oxidative stress in vascular disease: causes, defense mechanisms and potential therapies. Nat Clin Practice Cardiovasc Med 2008;5(6):338–49.

[80] Trzeciak S, Cinel I, Phillip Dellinger R, Shapiro NI, Arnold RC, Parrillo JE, Hollenberg SM. Resuscitating the microcirculation in sepsis: the central role of nitric oxide, emerging concepts for novel therapies, and challenges for clinical trials. Acad Emerg Med 2008;15 (5):399–413.

[81] Ochoa JB, Udekwu AO, Billiar TR, Curran RD, Cerra FB, Simmons RL, Peitzman AB. Nitrogen oxide levels in patients after trauma and during sepsis. Ann Surg 1991;214(5):621–6.

[82] Wang P, Ba ZF, Chaudry IH. Nitric oxide. To block or enhance its production during sepsis? Arch Surg 1994;129(11):1137–42.

[83] Wolkow PP. Involvement and dual effects of nitric oxide in septic shock. Inflamm Res 1998;47(4):152–66.

[84] Parratt JR. Nitric oxide in sepsis and endotoxaemia. J Antimicrob Chemother 1998;41:31–9.

[85] Cobb JP, Hotchkiss RS, Swanson PE, Chang K, Qiu Y, Laubach VE, Karl IE, Buchman TG. Inducible nitric oxide synthase (iNOS) gene deficiency increases the mortality of sepsis in mice. Surgery 1999;126(2):438–42.

[86] Thiemermann C, Vane J. Inhibition of nitric oxide synthesis reduces the hypotension induced by bacterial lipopolysaccharides in the rat in vivo. Eur J Pharmacol 1990;182(3):591–5.

[87] Kilbourn RG, Gross SS, Jubran A, Adams J, Griffith OW, Levi R, Lodato RF. NG-methyl-L-arginine inhibits tumor necrosis factor-induced hypotension: implications for the involvement of nitric oxide. Proc Natl Acad Sci U S A 1990;87(9):3629–32.

[88] Liu SF, Adcock IM, Old RW, Barnes PJ, Evans TW. Differential regulation of the constitutive and inducible nitric oxide synthase mRNA by lipopolysaccharide treatment in vivo in the rat. Crit Care Med 1996;24(7):1219–25.

[89] Grover R, Zaccardelli D, Colice G, Guntupalli K, Watson D, Vincent JL. An open-label dose escalation study of the nitric oxide synthase inhibitor, N(G)-methyl-L-arginine hydrochloride (546C88), in patients with septic shock. Glaxo Wellcome international septic shock study group. Crit Care Med 1999;27(5):913–22.

[90] Lopez A, Lorente JA, Steingrub J, Bakker J, McLuckie A, Willatts S, Brockway M, Anzueto A, Holzapfel L, Breen D, Silverman MS, Takala J, Donaldson J, Arneson C, Grove G, Grossman S, Grover R. Multiple-center, randomized, placebo-controlled, double-blind study of the nitric oxide synthase inhibitor 546C88: effect on survival in patients with septic shock. Crit Care Med 2004;32(1):21–30.

[91] Petros A, Bennett D, Vallance P. Effect of nitric oxide synthase inhibitors on hypotension in patients with septic shock. Lancet 1991;338(8782–8783):1557–8.

[92] Billiar TR, Curran RD, Harbrecht BG, Stuehr DJ, Demetris AJ, Simmons RL. Modulation of nitrogen oxide synthesis in vivo: NG-monomethyl-L-arginine inhibits endotoxin-induced nitrate/nitrate biosynthesis while promoting hepatic damage. J Leukoc Biol 1990;48(6): 565–9.

[93] Harbrecht BG, Billiar TR, Stadler J, Demetris AJ, Ochoa J, Curran RD, Simmons RL. Inhibition of nitric oxide synthesis during endotoxemia promotes intrahepatic thrombosis and an oxygen radical-mediated hepatic injury. J Leukoc Biol 1992;52(4):390–4.

[94] Boughton-Smith NK, Hutcheson IR, Deakin AM, Whittle BJ, Moncada S. Protective effect of S-nitroso-N-acetyl-penicillamine in endotoxin-induced acute intestinal damage in the rat. Eur J Pharmacol 1990;191(3):485–8.

[95] Shultz PJ, Raij L. Endogenously synthesized nitric oxide prevents endotoxin-induced glomerular thrombosis. J Clin Invest 1992;90(5):1718–25.

[96] Yadav S, Pathak S, Sarikhani M, Majumdar S, Ray S, Chandrasekar BS, Adiga V, Sundaresan NR, Nandi D. Nitric oxide synthase 2 enhances the survival of mice during

Salmonella typhimurium infection-induced sepsis by increasing reactive oxygen species, inflammatory cytokines and recruitment of neutrophils to the peritoneal cavity. Free Radic Biol Med 2018;116:73–87.

[97] Cerwinka WH, Cooper D, Krieglstein CF, Feelisch M, Granger DN. Nitric oxide modulates endotoxin-induced platelet-endothelial cell adhesion in intestinal venules. Am J Physiol Heart Circ Physiol 2002;282(3):H1111–7.

[98] Lupp C, Baasner S, Ince C, Nocken F, Stover JF, Westphal M. Differentiated control of deranged nitric oxide metabolism: a therapeutic option in sepsis? Crit Care 2013; 17(3):311.

[99] Lamontagne F, Meade M, Ondiveeran HK, Lesur O, Fox-Robichaud AE. Nitric oxide donors in sepsis: a systematic review of clinical and in vivo preclinical data. Shock 2008;30 (6):653–9.

[100] Rudinsky BF, Komar KJ, Strates E, Meadow WL. Neither nitroglycerin nor nitroprusside selectively reduces sepsis-induced pulmonary hypertension in piglets. Crit Care Med 1987;15 (12):1127–30.

[101] Alden KJ, Motew SJ, Sharma AC, Ferguson JL. Effect of aminoguanidine on plasma nitric oxide by-products and blood flow during chronic peritoneal sepsis. Shock 1998;9 (4):289–95.

[102] Enkhbaatar P, Murakami K, Traber LD, Cox R, Parkinson JF, Westphal M, Esechie A, Morita N, Maybauer MO, Maybauer DM, Burke AS, Schmalstieg FC, Hawkins HK, Herndon DN, Traber DL. The inhibition of inducible nitric oxide synthase in ovine sepsis model. Shock 2006;25(5):522–7.

[103] Matejovic M, Krouzecky A, Rokyta R, Radej J, Kralova H, Treska V, Radermacher P, Novak I. Effects of combining inducible nitric oxide synthase inhibitor and radical scavenger during porcine bacteremia. Shock 2007;27(1):61–8.

[104] Petros A, Lamb G, Leone A, Moncada S, Bennett D, Vallance P. Effects of a nitric oxide synthase inhibitor in humans with septic shock. Cardiovasc Res 1994;28(1):34–9.

[105] Avontuur JA, Tutein Nolthenius RP, Buijk SL, Kanhai KJ, Bruining HA. Effect of L-NAME, an inhibitor of nitric oxide synthesis, on cardiopulmonary function in human septic shock. Chest 1998;113(6):1640–6.

[106] Spronk PE, Ince C, Gardien MJ, Mathura KR, Oudemans-van Straaten HM, Zandstra DF. Nitroglycerin in septic shock after intravascular volume resuscitation. Lancet 2002;360 (9343):1395–6.

[107] Boerma EC, Koopmans M, Konijn A, Kaiferova K, Bakker AJ, van Roon EN, Buter H, Bruins N, Egbers PH, Gerritsen RT, Koetsier PM, Kingma WP, Kuiper MA, Ince C. Effects of nitroglycerin on sublingual microcirculatory blood flow in patients with severe sepsis/septic shock after a strict resuscitation protocol: a double-blind randomized placebo controlled trial. Crit Care Med 2010;38(1):93–100.

[108] Kolaczkowska E, Kubes P. Neutrophil recruitment and function in health and inflammation. Nat Rev Immunol 2013;13(3):159–75.

[109] Weiss SJ. Tissue destruction by neutrophils. N Engl J Med 1989;320(6):365–76.

[110] Matute-Bello G, Frevert CW, Kajikawa O, Skerrett SJ, Goodman RB, Park DR, Martin TR. Septic shock and acute lung injury in rabbits with peritonitis: failure of the neutrophil response to localized infection. Am J Respir Crit Care Med 2001;163(1):234–43.

[111] Tsukahara Y, Morisaki T, Kojima M, Uchiyama A, Tanaka M. iNOS expression by activated neutrophils from patients with sepsis. ANZ J Surg 2001;71(1):15–20.

[112] Benjamim CF, Silva JS, Fortes ZB, Oliveira MA, Ferreira SH, Cunha FQ. Inhibition of leukocyte rolling by nitric oxide during sepsis leads to reduced migration of active microbicidal neutrophils. Infect Immun 2002;70(7):3602–10.

[113] Crosara-Alberto DP, Darini AL, Inoue RY, Silva JS, Ferreira SH, Cunha FQ. Involvement of NO in the failure of neutrophil migration in sepsis induced by Staphylococcus aureus. Br J Pharmacol 2002;136(5):645–58.

[114] Sparkman L, Boggaram V. Nitric oxide increases IL-8 gene transcription and mRNA stability to enhance IL-8 gene expression in lung epithelial cells. Am J Physiol Lung Cell Mol Physiol 2004;287(4):L764–73.

[115] Martin G, Asensi V, Montes AH, Collazos J, Alvarez V, Perez-Is L, Carton JA, Taboada F, Valle-Garay E. Endothelial (NOS3 E298D) and inducible (NOS2 exon 22) nitric oxide synthase polymorphisms, as well as plasma NOx, influence sepsis development. Nitric Oxide 2014;42:79–86.

[116] Winkler MS, Kluge S, Holzmann M, Moritz E, Robbe L, Bauer A, Zahrte C, Priefler M, Schwedhelm E, Boger RH, Goetz AE, Nierhaus A, Zoellner C. Markers of nitric oxide are associated with sepsis severity: an observational study. Crit Care 2017;21(1):189.

[117] Jia L, Wang Y, Wang Y, Ma Y, Shen J, Fu Z, Wu Y, Su S, Zhang Y, Cai Z, Wang J, Xiang M. Heme oxygenase-1 in macrophages drives septic cardiac dysfunction via suppressing lysosomal degradation of inducible nitric oxide synthase. Circ Res 2018;122(11):1532–44.

[118] Schneemann M, Schoedon G. Species differences in macrophage NO production are important. Nat Immunol 2002;3(2):102.

[119] Mestas J, Hughes CC. Of mice and not men: differences between mouse and human immunology. J Immunol 2004;172(5):2731–8.

[120] Nathan C, Shiloh MU. Reactive oxygen and nitrogen intermediates in the relationship between mammalian hosts and microbial pathogens. Proc Natl Acad Sci U S A 2000;97(16):8841–8.

[121] Cobb JP. Use of nitric oxide synthase inhibitors to treat septic shock: the light has changed from yellow to red. Crit Care Med 1999;27(5):855–6.

[122] Lange M, Connelly R, Traber DL, Hamahata A, Nakano Y, Esechie A, Jonkam C, von Borzyskowski S, Traber LD, Schmalstieg FC, Herndon DN, Enkhbaatar P. Time course of nitric oxide synthases, nitrosative stress, and poly(ADP ribosylation) in an ovine sepsis model. Crit Care 2010;14(4):R129.

CHAPTER 14

Nanoporous Materials: New Generation of Nitric Oxide Donors

Rosana Vieira Pinto*,†, Moisés Luzia Pinto*

*CERENA, Departamento de Engenharia Química, Instituto Superior Técnico, Universidade de Lisboa, Lisboa, Portugal

†CQB and CQE, Departamento de Química e Bioquímica, Faculdade de Ciências, Universidade de Lisboa, Ed. C8, Campo Grande, Lisboa, Portugal

ABSTRACT

The ability of porous solids to store significant amounts of small gas molecules at the interior of their nanopores has been explored in recent years for the storage/release of nitric oxide (NO) for potential therapeutic applications, since this molecule plays a wide range of physiological and pathophysiological functions associated with cardiovascular homeostasis, immune response to infections, wound repair, tumor biology, and pathology. Indeed, the delivery of NO in a controllable manner only at specific sites of the human body can lead to therapies for many diseases. However, due to its short half-life, its rapid diffusion, and its high reactivity, significant challenges remain before the full potential of NO therapy is achieved. Conventional molecular donors that release NO at a known rate are well known, but their application in actual therapies has been limited since they are unstable and soluble in physiological media impairing their direct application to localized NO delivery. Thus, solid materials that deliver NO locally to the surrounding milieu may be a good solution to develop new products and therapies. Several materials have been proposed as NO-releasing systems, including polymer dressings, nanoparticle platforms, liposomes, dendrimers, and more recently porous materials. In this context, nanoporous materials are a very interesting alternative because they can store and release NO at target sites in the human body. Porous materials such as zeolites, clays, titanosilicates, and metal-organic frameworks (MOFs) with a variety of architectures and chemical compositions have been recently considered as NO donors, and this chapter contributes with an overview of the solids explored so far. The key parameters for designing an adequate structure, its mechanism for NO binding, and the first indicators of its potential therapeutic action are discussed, and possible further developments are suggested.

Therapeutic Application of Nitric Oxide in Cancer and Inflammatory Disorders. https://doi.org/10.1016/B978-0-12-816545-4.00014-1

ABBREVIATIONS

CO	carbon monoxide
H_2S	hydrogen sulfide
MOFs	metal-organic frameworks
MSN	mesoporous silica nanoparticles
NO	nitric oxide
NO_2^-	nitrite
NONOates	*N*-diazeniumdiolates
NOS	nitric oxide synthases
PAMAM	alkyl-chain-modified poly(amidoamine)
PTFE	polytetrafluoroethylene
RSNOs	*S*-nitrosothiols
SPER/NO	spermine/nitric oxide

Conflict of Interest

No potential conflicts of interest were disclosed.

INTRODUCTION

Nitric oxide (NO) is a signaling molecule involved in key physiological and regulatory processes, generated endogenously by isoform nitric oxide synthases (NOS) through the sequential catalysis of L-arginine [1]. Due to its free diffusion through aqueous solutions and cell membranes, NO rapidly interacts with numerous intracellular molecules being advantageous over most of the classical cellular messengers [2]. The NO generated from NOS has an essential role in the maintenance of vascular homeostasis [3], in neurotransmission [4], and in the wound healing process [5] and is involved in the immune response to infections [6], angiogenesis, and apoptosis [7, 8]. Deficiencies in NO biosynthesis, i.e., inappropriate local NO levels, have been linked to a number of disease states [2], and thus, several strategies for delivering NO are of growing interest for potential therapeutic applications. The NO-releasing treatment generally relies on the direct or indirect NO release (i.e., by metabolic activity, biotransformation, or redox activation) and on the use of mediators that increase NO bioactivity [9].

A great number of molecular NO donors have been proposed to chemically stabilize and release controllable amounts of NO during a precise period of time to trigger specific biological responses [10, 11]. Although many classes of NO donors exist, e.g., *S*-nitrosothiols (RSNOs), *N*-diazeniumdiolates (NONOates), metal-NO complexes, organic nitrates/nitrites, and nitrobenzene derivatives, only few have been transposed to the clinical practice [10–12]. For example, due to the high solubility of some donors, NO is released before reaching the target tissue [13]. Consequently, higher amounts of donor need to be provided to achieve the desired therapeutic effect, but additional chemical reactions may occur, generating reactive nitrogen species capable to induce cell toxicity [11]. Additionally, delivering NO in a temporally and spatially controlled manner is challenging because the release of a burst amount in the first few minutes can occur. Other concerns about these current homogenous donors are their decomposition products that can be potentially toxic/carcinogenic (e.g., nitrosamines) [11] and their storage instability since molecular donors decompose readily in the presence of moisture, light, and temperature, requiring very specific storage environments [14].

Inhaled NO gas or its topical application is another approach for exogenous NO therapy that has been used to cure respiratory diseases and skin wounds [15, 16]; however, it requires the use of pressurized gas tanks or plasma-chemical generators, which are costly and difficult to handle due to security reasons.

In recent years, much attention has been paid toward the synthesis of NO-releasing platforms, which are an innovative strategy that is focused on the combination of NO with drug delivery systems, giving promising features comparing with the previous native NO donors. Two main approaches have been proposed for the NO loading of the systems: (1) linking NO covalently to substrates (i.e., by substrate functionalization with RSNOs and NONOates) and (2) NO adsorption (chemisorption and/or physisorption) inside of porous solids bearing metal active sites by NO coordination to those sites [14]. Typical biomaterials used as carriers for NO are based on polymers [17], nanoparticles [18], and silica particles, and within the porous materials, zeolites [19] and metal-organic frameworks (MOFs) [20] are the most studied. The NO chemisorption is a recent and interesting method that allows releasing pure NO in the target site, being of special interest for topical therapeutic applications since its direct release avoids potential toxic by-products [14]. Based on this approach, we have been exploring and designing new porous structures for NO storage/release and evaluating their potentialities in the therapeutic field. This chapter shall contribute with the description of the state of the art, based on a literature survey, regarding the explored porous structures so far, including a discussion on the ideal features for designing these materials, the mechanisms for NO loading, and the preliminary therapeutic studies using these new NO donors performed until now. Moreover, a summarized description of the established NO-delivering platforms and their comparison with NO-releasing porous materials is reviewed, and possible further developments are suggested. With this work, we hope to encourage other researchers working in the development of new therapies based on NO to consider porous materials as possible vehicles for NO delivery.

NITRIC OXIDE-RELEASING PLATFORMS

The lack of suitable NO donors is a significant barrier to the use of NO as a therapeutic agent. Thus, it is highly necessary to design new NO-releasing strategies and to discover new technologies that store and release adequate NO amounts directly to the specific target, for improved therapeutic output and reduced side effects. However, filling all those requirements has been challenging.

Small molecules (generally described as molecules with formula weights <600) that release NO, which are well defined, and some organic nitrates have been used for decades for therapeutic purposes. Nevertheless, these NO donors generate serious undesired effects that can compromise their efficiency. Since for many therapeutic applications a more localized release of NO would be preferred, several researchers have been focusing on the design of NO-releasing platforms, which conjugate NO donors with carriers that stand up as new methods to address the above limitations. The different platforms that have been engineered to store and release NO are based of polymers, nanoparticles, liposomes, dendrimers, and more recently porous materials. Fig. 1 represents an overview of the main NO-releasing materials explored so far, which will be briefly described in the following paragraphs.

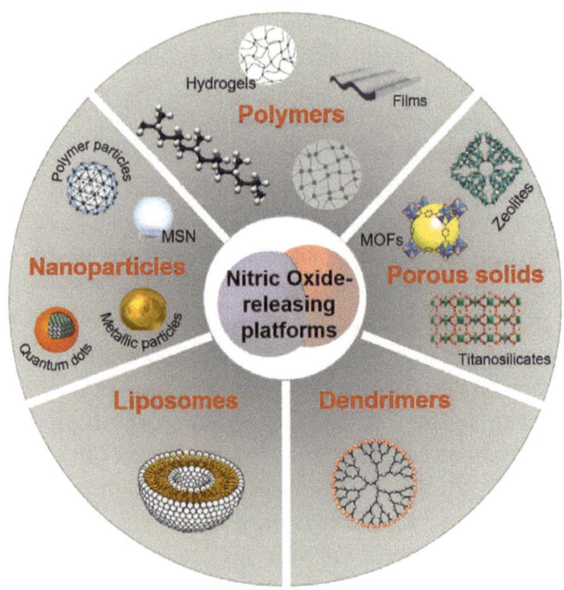

FIG. 1
A schematic summary of the different platforms that have been engineered to store and release NO.

Polymers

The authors first used matrix structures for functionalization with NO-releasing molecular systems [21]. These materials present great biocompatibility and exhibit interesting similarities with living tissues due to their softness and their high capacity to retain high amounts of water, being of special interest for topical applications [22]. The use of polymer-based materials for NO release adds another level of versatility, since they could be used as coatings, films, or ointments for a variety of therapeutic applications and its properties may be easily manipulated according to the therapeutic need. Typical NO-releasing polymers contain specific moieties to act as NO donors (i.e., RSNOs, nitrosamines, or NONOates), overcoming the limitations of the therapeutic use of conventional NO donors and allowing a large control over the kinetics of NO release. The most common preparation strategy is based on dispersing thiol- or amine-containing compounds in polymers, followed by exposure to NO gas to convert the parent polymer materials into NO donors. These NO donors allow a prolonged NO release, which can last for days [23] or even weeks [24]. Polymers functionalized with *N*-based diazeniumdiolate present great storage stability and a localized controllable NO release under physiological conditions. Still, *N*-diazeniumdiolate compounds may form potential toxic nitrosamines (a well-known class of carcinogens [25]) upon decomposition, which limit their use as therapeutic agents [26]. Different donors, such as *S*-nitroso and *C*-nitroso compounds, have been used for polymer functionalization [27, 28]; however, the therapeutic potential of *S*-nitroso compounds is limited due to their rapid and unpredictable decomposition (which release NO). Another feature to consider in the development of polymer-based NO donors is the hydrophobic/hydrophilic index of the polymer substrates. In most of the cases, NO release follows a water-substitution mechanism, and the hydrophobic/hydrophilic characteristics of the polymer will strongly affect the water penetration into the matrix and its contact with the NO donor segments to trigger NO releasing.

Nanomaterials

Nanomaterials applied to NO therapeutic actions led to the creation of an alternative strategy that may be used in multiple therapeutic applications. Nanoscale particles provide a higher interaction with the tissues and an enhanced penetration comparing with the previously described NO-releasing therapies. Moreover, their small size offers the advantage of easy application and storage.

Mesoporous silica nanoparticles (MSN) are well-known biocompatible systems that have been used as NO carriers [29–31]. This approach, which uses *N*-diazeniumdiolate as NO donor, allows a high physicochemical flexibility, achieving a final wide range of NO storage (0.05–1.78 $mmol\,g^{-1}$), being possible to be prepared with different particle sizes (d = 20–500 nm), different NO released amounts (10–5500 ppb mg^{-1}), and/or different NO release periods (up to 30 h) by varying the synthetic precursors and synthesis procedures [32, 33].

Another inorganic nanoparticles, such as gold nanoparticles [33–36], silver nanoparticles [37], quantum dots [38, 39], and iron oxide nanoparticles [40, 41], present unique optical, magnetic, and electronic properties that may afford an interesting dual functionality for both diagnostic and therapy. Their common preparation is based on their surface functionalization with diazeniumdiolate or S-nitrosothiol groups [26, 35, 40]. However, some of these systems are limited due to their low water solubility, low NO storage capacity, and short duration of release [12, 36]. For quantum dots, in vivo applications require long periods of wavelength excitation for efficient tissue penetration that may cause damage of the cells [42, 43]. Another concern for this type of materials is their long-term toxicity and their biodistribution and location in vivo that still need further investigation.

Liposomes

Liposomes are spherical vesicles formed by an aqueous core surrounded by synthetic or natural phospholipid bilayers, which can encapsulate both hydrophilic and hydrophobic NO donors or NO gas to be used in the therapeutic field [44–47]. These vesicles are of special interest for anticancer treatments since the lower pH inherent to the tumor tissues increases the permeability of the lipid layers triggering a local higher NO release. The main techniques used to form these carriers are thin-film hydration, solvent injection, reverse-phase evaporation, membrane extrusion, and microfluidic technology [48, 49]. These platforms have as main advantage the protection of the spontaneous NO release of the hydrophilic NO donors, ensuring a slow NO donor decomposition and consequently a prolonged NO release. For example, encapsulating SPER/NO (a homogeneous NO donor belonging to NONOate family) in liposomes resulted in lengthy durations of NO release with a half-life of 162 min, while free SPER/NO has a $t_{1/2}$ of only 35 min [50]. Moreover, by varying the liposome system, it is possible to store up to 9 μmol mL^{-1} of NO [46].

Unfortunately, it is hard to control the NO release rate by the liposomes, which may compromise the positive therapeutic effect.

Dendrimers

Dendrimers are monodispersing macromolecules with a tridimensional structure that consists of a central core surrounded by a highly branched corona with reactive surface groups. Their accessible functional groups at the periphery allow a multiple functionalization of the structure, which have attracted much attention for their use as NO carriers. Dendrimers can be synthesized by two different methods: (1) the divergent method, starting by the core site where repeat monomers are branched outward from there, or (2) the convergent method, where the synthesis inward from the surface group toward the core [51, 52]. Schoenfisch's group developed NO-releasing dendrimers by generating NONOates on primary and secondary amine-derived dendrimers, reaching improved NO storage levels with the structure containing secondary amines (up to 3.5 mmol mg^{-1}) [53]. Also, it is possible to control the NO release kinetics by using different dendrimer sizes (generations). For example, comparing (NO)-releasing alkyl-chain-modified poly(amidoamine) (PAMAM) dendrimers with different generations (from 1 to 4), the NO storage capacity is equivalent ($\approx$1 μmol mg^{-1}); however, the $t_{½}$ can vary from 20 to 30 min and the total time of NO release from 7 to 10 h [54].

Porous Materials

Porous materials, such as zeolites, titanosilicates, and MOFs with a variety of structural architectures and chemical compositions, have been recently considered as NO donors, being one of the most promising vehicles for practical applications of nitric oxide in the biomedical fields [51]. Due to their interesting porous framework, which can accommodate a wide range of metal ions within the pores, they allow NO to be stored by chemisorption (coordination of NO) or physisorption, by exposing the solid to NO gas [55–57]. Their NO release is triggered in the presence of water that replaces NO on the metal centers and diffuses out of the porous structure (Fig. 2).

This attractive NO-releasing approach is advantageous over the small molecular donors since it provides a highly efficient packing of NO within the solid and exhibits good stability for safe storage and a controlled delivery of pure NO at the target tissues, avoiding thus the leaching of undesirable species [14]. Due to

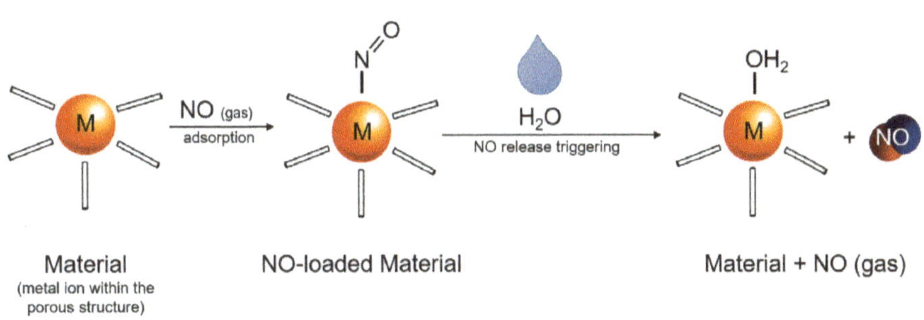

FIG. 2
Representation of the process for NO storage inside the structure of porous materials. When the solid is exposed to the gas, NO is coordinated to the metal ion. The release of NO is triggered by contact with water.

its novelty and to the fact that there is no detailed literature review regarding these materials as NO carriers, the following sections will describe the porous materials explored for this purpose and the biological studies conducted so far with these new NO donors.

As a summary of this section, Table 1 provides a comparison among the mentioned NO-releasing platforms by showing the key NO storage properties from several examples of the developed NO donors.

POROUS MATERIALS

Porous materials are a vast class of materials that cover a wide variety of structures [68]. These structures that can be organic, inorganic, and inorganic-organic hybrids may vary essentially in the pore size, in the percentage of porosity, and in the presence (or absence) of interconnectivity between them. In addition, they may have well-defined pore sizes or be structurally disordered with a wide range of different pore sizes. The possibility of engineering and manipulating the pore network is of scientific and technological importance because the pore walls are able to interact with ions and molecules and the pore space can store liquids, solid particles, or gas molecules [69].

These porous solids have been used in numerous applications, being gas adsorption as one of the most important ranging energy, environmental, and medical applications [68]. For this application, the preferred porous materials are the nanoporous (also called microporous solids, with pore size <2 nm) having a slightly larger diameter than the common gaseous molecules. Although not yet widely explored, gas storage in materials for biomedical applications has emerged in the last years, being increasingly closer to commercialization. The dominant and the most exploited gas for therapeutic applications is NO since its mechanisms of biological action are already well known, but there is a scope for developing carriers for carbon monoxide (CO) and hydrogen sulfide (H_2S), which also play an important role in the human body [70–72]. Unlike other applications where the most important feature to consider is high storage capacity, in the case of nanoporous materials for NO storage/release, it is necessary to combine various characteristics in a unique material to achieve the desired therapeutic effect (Fig. 3).

One of the key requirements for developing a delivery system is the control over the NO release kinetics under aqueous and physiological media. Since NO is a highly toxic gas, it is vital to obtain a slow release over extended periods and achieve a control in the concentration. Depending on the desired application, concentrations in nanomolar or micromolar levels may be preferred, which means that very small concentrations may not be sufficient to achieve the positive biological effect. As mentioned before, the release of NO from the porous

Table 1 Comparison of the NO storage properties from relevant examples of developed platforms, under physiological conditions.

NO-Releasing Materials	NO Donor	NO Storage Capacity ($mmol\,g^{-1}$)	t½ [NO] (min)[a]	t (Total) (h)[b]	Ref.
Polymers					
S-Nitrosothiol-based polyphosphazene coating	*S*-Nitrosothiol	0.89 ± 0.09	NA	336	[58]
Poly(organophosphazene) (POP-EtCys-SH)	*S*-Nitrosothiol	0.55 ± 0.04	NA	24	[59]
Polymethacrylate microbead	*N*-Diazeniumdiolate	1.99	56	NA	[60]
Poly(vinyl chloride) (12% wt) microfibers	PROLI/NO (1.2% wt)	0.38 ± 0.10	5055 ± 1078	NA	[61]
Nanoparticles					
Gold nanoparticles (C6-HD)	*N*-Diazeniumdiolate	0.09	68	600	[35]
PLGA particle-SNAP	SNAP	0.3–0.6	NA	240–336	[62]
Fumed silica particles	*N*-Diazeniumdiolate	0.56	43	10	[29]
Silica particles 150 nm diameter	*N*-Diazeniumdiolate	1.30 ± 0.11	40.7 ± 11.0	16.7 ± 1.4	[63]
Liposomes					
Liposome-DPTA/NO-SPER/NO	DPTA/NO + SPER/NO (50:50)	0.29 ± 0.03	996 ± 126	51.2 ± 2.1	[50]
Liposome-DPTA/NO	DPTA/NO	0.26 ± 0.05	1224 ± 162	65.9 ± 1.8	[50]
Liposome-SPER/NO	SPER/NO	7.73 ± 0.71	2718 ± 276	168.2 ± 17.0	[46]
Dendrimers					
Alkyl-chain-modified G1 PAMAM	*N*-Diazeniumdiolate	1.06 ± 0.10	21 ± 5	7 ± 2	[64]
DAB-C7–16/NO	*N*-Diazeniumdiolate	3.4	77	14	[53]
Generation 4 polyamidoamine dendrimer	*S*-Nitrosothiol	2.1 ± 0.2	7–106[c]	0.5–12[c]	[65]
Porous materials					
Ni-CPO-27 (MOF)	Pure NO gas	7.0	NA	14[d]	[66]
Ni-vitamin B_3 (MOF)	Pure NO gas	2.6	NA	3.3[e]	[55]
Zeolite-A	Pure NO gas	1.3	NA	10[d]	[67]
ETS-4 (titanosilicate)	Pure NO gas	3.7	NA	1[e]	[57]

[a] *NO release half-life.*

[b] *Duration of NO release.*

[c] *Will depend on the NO release trigger (e.g., light or cooper).*

[d] *NO release triggered by exposure to humid nitrogen (11% RH) at 37°C.*

[e] *NO release in oxyhemoglobin solution at 25°C.*

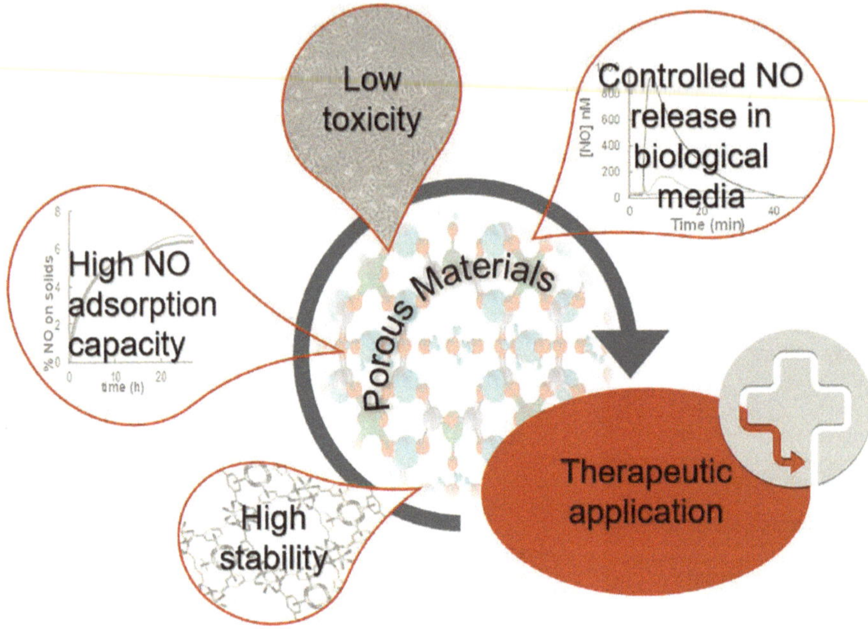

FIG. 3
Essential requirements for the use of nanoporous materials as NO carriers for therapeutic applications.

materials is triggered by contact with water, which for biological applications may be not the best option if long release rates are required. This difficult challenge can be solved by choosing an appropriate structure with the adequate metal site and turning the pore size to control the hydration kinetics and diffusion of NO. The high adsorption capacity is also important since it allows more opportunities for tailoring the NO release conditions (i.e., flux and rate) in order to meet the therapeutic need.

Another requirement to determine the clinical success of these NO-releasing solids is their biocompatibilities. These porous materials contain metal ions in their framework that should be a concern since they can induce potential toxicity in biological environments. Therefore, low-toxicity metals (i.e., iron and zinc) should be preferred as well as design of stable structures under biological conditions, avoiding the leaching of their components and potential side reactions in the surrounding tissues. The material's stability is also critical to ensure the storage of the gas and its controlled release. Thus, it is fundamental to understand the binding and decomposition pathways of the structures for a rational design of the storage materials.

Several types of porous materials have recently been studied for NO storage/release, including zeolites, clays, MOFs, and titanosilicates [55, 57, 67, 73].

In the following subchapters, a detailed description of all the models explored so far for each category is provided, and the mechanism of its NO loading is described.

NO Loading Mechanism Into the Porous Materials

Fig. 4 schematically represents the typical necessary steps to load the material with NO. This methodology, optimized at our laboratories, comprises a high-vacuum system associated with a NO bottle and a glass cell with a PTFE high-vacuum valve. In the first step, the adsorbent material is introduced in the cell and degassed under high-vacuum conditions (better than 10^{-2} Pa) to activate the pores (i.e., removal of residual solvents, water, and gas molecules from air). Heating temperatures and time for degassing depend on the thermal stability of the material, but usually do not exceed 300°C and 3 h, respectively. This is an important step since the small molecules (i.e., water or other solvents) previously coordinated in the metal sites will be removed, allowing NO to coordinate in the following step. After the outgassing and with the material already at room temperature, NO is admitted to the vacuum cell until reaching a pressure of 80 kPa. The valve is closed, and the sample remains in contact with NO for 3 days (this time can be reduced if the material exhibits fast NO adsorption, which can be previously evaluated by gravimetric adsorption experiments). After this period, the cell connected to the vacuum line is opened, and the remaining nonadsorbed gas is evacuated. Finally, the cell with the loaded material is filled immediately with helium up to atmospheric pressure. The material may be stored inside the cell closed with helium until be used.

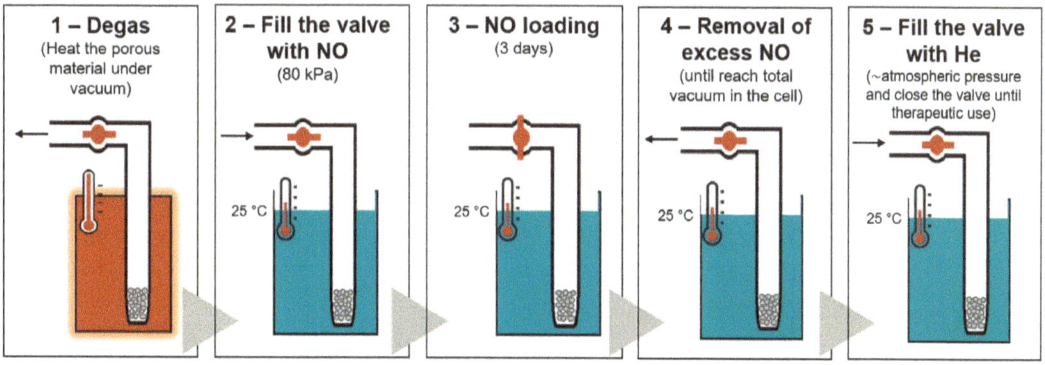

FIG. 4
Methodology used for loading NO inside the nanoporous materials.

NO-Releasing Nanoporous Materials

Zeolites

Zeolites are highly crystalline aluminosilicate microporous insoluble materials with a rigid three-dimensional open framework that may be of natural or synthetic origin. Their high stability facilitates considerable freedom of ion exchange and many other applications such as in catalysis and in gas adsorption/storage, including NO [11, 33, 67]. Exposure of zeolites to NO gas may result in both reversible and irreversible NO adsorptions. Reversibly adsorbed NO is considered to be physically absorbed, being easily released, while irreversibly adsorbed NO is not spontaneously released, since it interacts strongly with the extraframework cations in these porous materials [66, 68]. The pioneering research using zeolites to the field of NO release scaffolds has been led by the group of Morris [19, 67]. In an initial study, Wheatley et al. [67] used zeolite-A with exchanged cobalt cations to adsorb NO. Experiments showed that at 1.0 bar loading pressure, these materials were capable to store 1.2–1.3 μmol NO per milligram of zeolite. The NO releasing was triggered by exposing these zeolites to humidified atmosphere (11%), showing a release of 1.02 μmol NO per milligram of zeolite over 40 min with a NO half-life of ~5 min. However, changing the humidity of the gas in the release experiments displayed a great impact in the NO kinetics. It is important to note that this method of measuring NO would underestimate the true amount of NO released due to the poor solubility and diffusion of NO into the solution where it is measured [19]. Also, the water presence in biological media is immensely higher than in the release experiments with a controlled humidified atmosphere, which will give a significant faster NO release in biological media.

Additionally, in a subsequent study, the same group observed that changing the chemical composition within the same framework topology (type and quantity of the extraframework cations) gives a degree of control on the amount of NO that can be absorbed [19, 56]. In this specific study, typical zeolites were used, i.e., Linde type A and Faujasite, and various metal ions as extraframework cations (Cu^{2+}, Co^{2+}, Ni^{2+}, Mn^{2+}, Zn^{2+}, etc.). Results showed that NO is adsorbed and released based on the affinity of each metal, with Co storing the greatest amount [19].

The best result regarding the NO storage capacity in zeolites was achieved by copper-exchanged zeolite-X that led an adsorption capacity of 3.6 mmol of NO per gram of zeolite [74]. Moreover, it has been reported that NO-loaded zeolites are stable in the anhydrous state for several months and even for more than 1 year [68].

Metal Organic Frameworks

MOFs are porous hybrid inorganic-organic crystalline materials, developed in the last decades, built from metal ion oxoclusters (often just named metal clusters) connected by organic ligands, in a quasi-infinite array [75–77]. These structures offer countless combinations between a large number of possible organic linkers and metal clusters, which have drawn much attention for a wide range of applications. Two of the major advantages of this class of materials are their highly crystalline structure, which allows detailed structure characterization from diffraction techniques and very high porosities (with internal surface areas exceeding $5000\,m^2\,g^{-1}$ in some cases), being ideal for capture, storage, and delivery applications [68]. However, certain structures have low thermal and moisture stabilities compared with inorganic materials (e.g., zeolites), which may condition their long-term application [68]. The common methodologies to synthesize these materials include hydrothermal, solvothermal, electrochemical, mechanochemical, and sonochemical techniques [78]. By varying the synthesis conditions, it is possible to manipulate the amount of unsaturated metal sites where NO can bind the size and the shape of the porosity that may be very useful to achieve the ideal design for a specific application. For medical applications, a panoply of biocompatible MOFs already exists with great potential as nanoencapsulators and controlled-release agents [79–81]. Moreover, due to the gas storage capabilities, these materials were easily recognized as potential carriers for therapeutic gases, namely, NO.

The first demonstration of NO storage and release using MOFs was also performed by the Morris group, which already explored a variety of MOF structures, that allows the understanding of the requirements needed to design a suitable structure for this application. Xiao et al. [82] evaluated the ability of copper benzene tricarboxylate (HKUST-1) to store and release NO, showing a large adsorption capacity of $\sim 9\,mmol\,g^{-1}$ at 196 K (1 bar) and $\sim 3\,mmol\,g^{-1}$ at room temperature (1 bar), which was significantly greater than any other adsorption capacity previously reported on a porous solid. Desorption from the material revealed that $2.21\,mmol\,g^{-1}$ remained inside, which suggests an irreversible NO adsorption attributed to the stronger coordination bonds of NO with the copper. Release studies performed in the humidified atmosphere showed only a release of a small portion of NO adsorbed (<0.05%) [82]. In an effort to improve HKUST-1 NO release capabilities, Peikert et al. [83] developed four new secondary amino-functionalized trimesic acid ligands to be used in combination with the original trimesic acid, originating four new MOFs (UHM-36, UHM-37, UHM-38, and UHM-39), all isostructural to HKUST-1. Despite their NO release kinetics, they showed very slow release NO moieties comparing with other NO-releasing nanoporous materials; the highest NO released amount achieved from UHM-37 ($0.064\,mmol\,g^{-1}$) was 30 times lower than for HKUST-1. Moreover, the time of release was extended with these structural modifications, reaching 7 h for UHM-39.

In order to explore other alternatives to accomplish higher release capacities, McKinlay et al. [66] assessed two isostructural MOFs (Ni-CPO-27 and Co-CPO-27) that differ only in the nature of the metal cation, exhibiting high storage of up to (~7 mmol g^{-1}) under room temperature, which correspond to seven times more than zeolites. Considering the metal fraction (~7–8 mmol g^{-1}) calculated for these MOFs, it seems that all the unsaturated metal sites (Co or Ni) are accessible for NO coordination. NO release assessment showed a complete release of the absorbed amount over 15 h under exposure to humid nitrogen (11% RH). Slight differences were observed between the two cases regarding NO adsorption and release kinetics. The Co-CPO-27 stores less NO but presents a slower release rate than for Ni-CPO-27. For this type of structure, the nature of the metal determines these observed differences.

Although these materials present an exceptional behavior over the whole NO adsorption, storage, and delivery, biocompatibility concerns are eminent since cobalt and nickel ions are not toxicologically the best choices. Taking that in consideration, in a subsequent study, different structures with nontoxic ions, i.e., iron and calcium ions, were considered [84, 85]. The iron-containing structures were dicarboxylate MOFs with two different frameworks derived from fumaric (MIL-88A) and terephthalic (MIL-88B) acids that gave NO storage capacities of 2.5 and 1.6 mmol g^{-1}, respectively [85]. Release studies showed that, for MIL-88A, only 4.8% of NO adsorbed was released; however, that NO was released at an extremely slow rate (up to 24 h), in the presence of humid nitrogen (11%). For the case of MIL-88B, no release was observed under the same conditions. Another interesting structure based on calcium 3,3′,5,5′-azobenzenetetracarboxylate (BioMIL-3) with accessible metal-acid sites was also evaluated, showing a total amount of adsorbed NO (0.8 mmol g^{-1}) much lower than the theoretical number of available unsaturated Ca sites (4.6 mmol g^{-1}) suggesting an inefficient NO accessibility to the pores [84]. The authors justify that this inaccessibility is caused by the flexibility of the frameworks, which results in relatively narrow pore sizes upon solvent molecule removal [86]. Nevertheless, these solids exhibit an interesting NO release at the biological levels for an extended period (>16 h), which makes this biocompatible MOF of interest for medical applications.

Eubank et al. [87] also investigated another type of iron MOFs. However, unlike those previously presented, these show rigid structures and enhanced pore sizes. These materials are composed of inorganic trinuclear metal clusters (Fe and Cr) and tricarboxylate ((Fe and Cr) MIL-100) or tetracarboxylate ligands (Fe-MIL-127). These materials have higher NO adsorption capacities (2.7–3.28 mmol g^{-1}) than MIL-88A and MIL-88B, which corresponds to 75%–88% of the theoretical values of available unsaturated metal sites. However, regarding NO release profiles, these structures release less than 23% of the adsorbed amounts but exhibit a prolonged delivery at the biological level.

Another recent examples studied by us are the Co- and Ni-MOFs built from the ligand vitamin B_3 (nicotinic acid) [55]. They showed the capability to store and

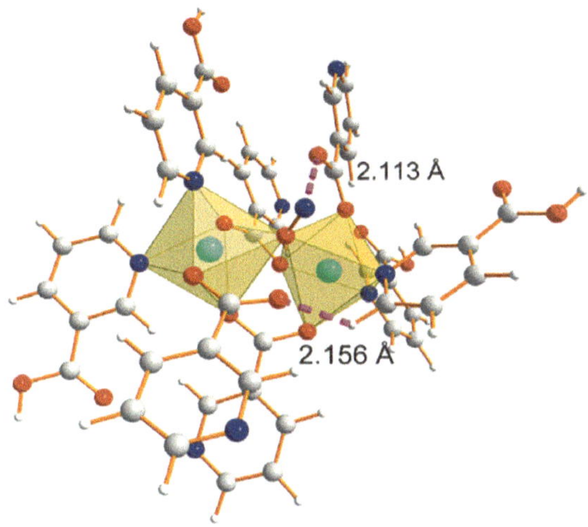

FIG. 5
Cluster model of the metal centers of Ni-MOF with a NO molecule bonded in the position that was previously occupied by H_2O. Hydrogen bond and O=N–O are represented in dashed purple lines. Color code for the atoms is oxygen in red, nitrogen in blue, carbon in gray, nickel in light blue inside the yellow octahedrons, and hydrogen in white [55].

release NO in a slow and reversible manner, releasing almost all the NO amount absorbed (i.e., 2.6 and 2.0 μmol NO mg^{-1} solid for the Ni and Co compound, respectively) [55]. Through computational methods, employing the density functional theory formalism to study these systems, it was possible to understand the NO interactions with the unsaturated metal sites inside the pores. The result showed that, apart from the predictable NO coordination with the metal site, the neighbor uncoordinated oxygen atom from the organic ligand also interacts with the N atom of the NO (Fig. 5). Such an interaction stabilizes the NO molecule and has inevitable influence in the NO release.

Although these various MOF systems present promising features, the engineering of an ideal structure for the delivery of NO in real biomedical applications is a challenging task. For example, Co- and Ni-CPO-27 remain to date the most interesting structures for the NO storage; however, their toxicity is certainly a limitation. Table 2 summarizes the NO adsorption and release properties of the MOFs explored so far.

Titanosilicates

Microporous titanosilicates are a type of materials with considerable similarities with traditional zeolites, since they also have a negatively charged

Table 2 Comparison of key parameters of NO adsorption/release between reported MOFs

MOF	NO Adsorption Capacity ($mmol\,g^{-1}$)	NO Released % From the Total Adsorbed Amount	Total Released Amount ($mmol\,g^{-1}$)	Time of Release (h)	Ref.
HKUST-1	3	0.02	0.002	1[a]	[82]
Co-CPO-27	6.5	100	6.5	14[a]	[66]
Ni-CPO-27	7.0	100	7.9	14[a]	[66]
Fe-MIL-88-A	2.5	4.8	0.12	24[a]	[85]
Fe-MIL-88-B	1.6	0	0	NA	[85]
BioMIL-3	0.8	15	0.12	10[a]	[84]
Cr-MIL-100	3.3	19.7	0.65	40[a]	[87]
Fe-MIL-100	4.5	12	0.55	14[a]	[87]
Fe-MIL-127	2.2	23	0.5	25[a]	[87]
UHM-36	NA	NA	0.016	2[a]	[83]
UHM-37	NA	NA	0.064	6[a]	[83]
UHM-38	NA	NA	0.026	3[a]	[83]
UHM-39	NA	NA	0.022	7[a]	[83]
Ni-vitamin B_3	2.6	100	2.6	20[b]	[55]
Co-vitamin B_3	2.4	82	2.0	20[b]	[55]

[a] *NO release triggered by exposure to humid nitrogen (11% RH) at 37°C.*

[b] *NO release under vacuum conditions at 25°C.*

framework compensated by exchangeable cations inside their pores. However, some of these structures have distinct characteristics that make them attractive for NO delivery and that were explored by our group [57, 88, 89]. In the case of titanosilicate ETS-4, the framework contains Ti^{4+} coordinated to six oxygens as depicted by the green spheres inside the green octahedrons in Fig. 6A. Some of these octahedrons are isolated with two opposing terminal oxygens. One of this oxygen atoms belongs to a water molecule that is hydrating the framework (the bright-red sphere in Fig. 6A indicates a possible position for the water molecule) and can be removed by heating and/or vacuum, producing an unsaturated (pentacoordinated) Ti^{4+} center. With this structural characteristic, ETS-4 is able to adsorb large amounts of NO (3.7 $mmol\,g^{-1}$) and to slowly release it over a period of almost 1 day [57]. The NO becomes coordinated to the Ti metal center inside the eight-member ring pores (Fig. 6B and C). The very small pore diameter (about 0.4 nm) and poor pore accessibility probably also help to explain the observed slow release.

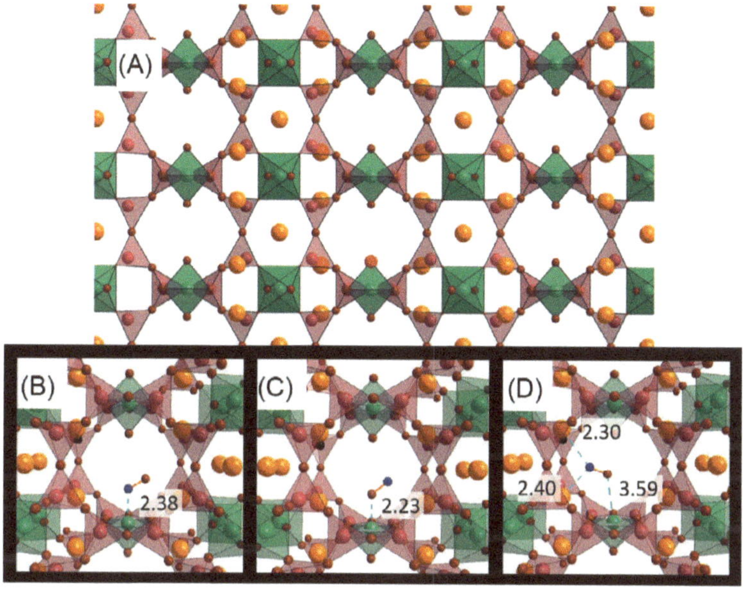

FIG. 6

Views of the ETS-4 structure along the eight-member ring pores where the water molecules that are coordinated to Ti can be replaced by NO. (A) Initial hydrated material with one water molecule coordinated to each Ti positioned between two neighboring eight-member ring pores. The water molecule is randomly coordinated at each axial position, for example, that represented by the bright-red sphere. (B) Optimized geometry of coordinated NO (N-down) interacting with the five-coordinate Ti atom. (C) Optimized geometry of coordinated NO (O-down) interacting with the five-coordinate Ti atom. (D) Optimized configuration of NO interacting simultaneously with Ti and Na atoms, which occurs when the Na is also dehydrated. Interatomic distances (*dashed blue lines*) are in angstroms and were calculated by full optimization of the structures with a periodic density functional theory method [57]. Colored spheres represent atoms of sodium (*yellow*), oxygen (*red*), nitrogen (*blue*), silicon (*plum inside the plum tetrahedrons*), and titanium (*green inside the green octahedrons*). Hydrogen atoms were not represented for simplicity.

In the following works, the storage capacity of the ETS-4 was improved by exchanging the extraframework cations (Na^+ or K^+) by Cu^{2+} or Co^{2+} to achieve exceptional NO storage/release properties, although the cytotoxicity also increased [89].

Another microporous titanosilicate bearing hexacoordinated Ti^{4+}, ETS-10, was explored for NO delivery, namely, by understanding the effect of the isomorphic substitution of Si by Al or Ga (ETAS-10 and ETGS-10, respectively) on the material's performance. Both ETAS-10 and ETGS-10 exhibited an increased NO loading capacity but a faster release kinetics relatively to ETS-4 [88]. The most notable characteristic of these materials is their exceptional very low toxicity (cell viability above 87%, after 72 h) at high concentrations of material (450 μg/mL). Thus, the ETS-10-type materials may be certainly interesting for biomedical applications where faster NO delivery is needed.

Clays

Porous clay materials have aroused interest for NO delivery. Clay materials are already used in the therapeutic field, achieving a status of pharmaceutical suitable materials either by being used as active substances or as excipients to other drugs [90]. These solids have shown significant advantages, such as high mechanical, thermal, and chemical stability and good adsorption and penetrability due to their regular structures and appreciable surface areas [91]. Considering all these advantages, new NO-releasing materials were designed at our labs by exploring and modifying different types of clays, namely, mineral clays (sepiolite and montmorillonite) [73, 92], synthetic clays (smectite clays with cobalt ions) [93], and organoclays (natural clays modified with L-histidine), following different strategies (Fig. 7) [94].

The study started with montmorillonite (MMT), which belongs to the group of smectites, being widely used in pharmacology due to their high cation-exchange capacity comparing with the common silicates [95]. MMT presents a low surface area, with a characteristic layered structure, which justifies the residual NO adsorption capacity [92]. This limitation was overcome by interlaying the structure with inorganic/organic pillars (Fig. 7). Pillared interlayered clays (PILCs) are obtained by exchanging the interlayered cations of layered

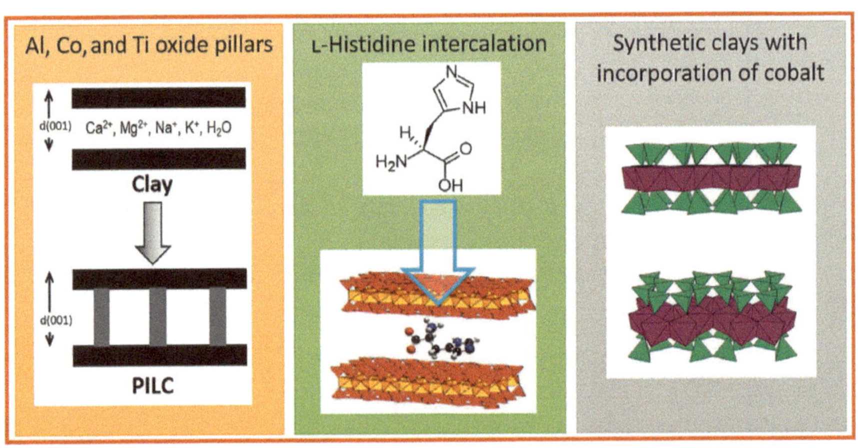

FIG. 7
Explored strategies to improve the NO storage/release properties of clays.

clays with bulky inorganic cationic species that are then calcinated, forming oxide pillars bound between the layers [96]. A set of materials were prepared by pillaring the natural MMT with various oligocations (aluminum, cobalt, and titanium), which are of special interest for NO storage, showing a regular microporosity (<2 nm). Data from NO adsorption and release revealed that the material with the aluminum oxide pillars exhibit the highest adsorbed amount (3.5% in mass) and a slower release profile under physiological medium [92]. Sepiolite, which is fine-grained silicates with either lamellar or fibrous morphology, was also explored. A sepiolite PILC was designed with aluminum, titanium, cobalt, and zinc pillars using the same method used for MMT. Some of the developed materials showed enhanced NO storage capacities, reaching 5% in mass of stored NO in the case of aluminum pillared sepiolite, which correspond to the same capacity found for type A zeolites [67].

Mineral clays may also be substantially modified by substituting the interlamellar cations by charged organic molecules. These hybrid materials, often named organoclays, have been studied in several scientific domains due to their easy availability and the possibility to incorporate a wide variety of functional molecules [97]. With this in mind, MMT and sepiolite were subject to modification with L-histidine, which is a natural amino acid with three amine groups that can bind NO. There is only another amino acid with a similar structure-arginine, which is the precursor used by nitric oxide synthetases for the formation of NO inside the body. Two concentrations of L-histidine were tested, giving thus four different organoclays. Sepiolite modified with high concentrations of L-histidine adsorbed the largest amount (approximately 4% NO). The remaining samples showed similar capacities among them, with adsorbed NO amounts about 3% in mass.

Since natural clays do not provide a wide variety of transition metals in their structure that can bind NO, synthetic clays were prepared with cobalt ions because of their proved high affinity for NO adsorption [98]. The storage capacity for these materials ranged 3.5%–5.1% in mass of adsorbed NO but with an interesting very slow release profile under oxyhemoglobin solution.

The toxicity of all the explored clay-based materials was assessed with HeLa cells, revealing high survival rates even when using high concentrations of material (450 μg/mL). Although storing less NO than the other reported NO-releasing porous solids and displaying in some cases faster NO release, these materials can still deliver NO amounts that may trigger positive biological responses.

Therapeutic Studies on NO-Delivering Porous Materials

Nitric oxide has multiple therapeutic effects, and the development of this new generation of nanoporous NO donors will provide an excellent alternative for several conventional therapies. So far, studies using these new donors for the control of biological processes are still poorly explored, being essential to validate these new approaches as viable alternatives. It should be clearly demonstrated that a relation between the NO release and the effect is observed in biological systems. The intensity and duration of the response must be controlled by the amount and type of porous material, which should be related to its NO release properties.

Mowbray et al. [99] have been conducting studies to demonstrate the potentialities of topical application of NO-releasing zeolites (combined with an aqueous cream BP, 0.02 mL 33%, wt/wt) on the human skin by comparison with acidified nitrite (NO_2^-) creams (an inorganic NO donor). Biologically, equivalent concentrations of NO_2^- and NO-loaded manganese-exchanged zeolite were used by adjusting the volumes and concentrations of each product. The zeolite proposed for this study was the manganese ion-exchanged zeolite-A, which presents a high adsorption NO capacity (up to 1 mmol of NO per gram of zeolite) [67]. NO delivered to the superficial dermis by 33% NO-zeolite was quantified by cutaneous microdialysis with further measurement of the NO_2^- present in the dialysate, reaching 1000 μM after 60 min. Topical application of NO-releasing zeolites (0.02 mL 33%, wt/wt) induced local vasodilatation and no significant inflammatory response. Parallel studies compared the biological effect of NO released by acidified nitrite creams (0.04 mL aqueous cream BP with ascorbic acid (2%, wt/wt) + NO_2^- (5%, wt/wt)), which caused an intense cutaneous inflammatory reaction. Of note, this acidified nitrite concentration releases approximately the same NO concentration as 33% NO-zeolite. This contrast of results can be explained by the fact that acidified nitrite creams release toxic coproducts in addition to NO, highlighting the great capacity of zeolites to deliver pure NO.

Further research developed by Neidrauer et al. [100] evaluated the healing potential and antimicrobial efficacy of topical ointment containing NO-loaded zinc-exchanged zeolites in Zucker obese rats. Zn-zeolites were synthetized and loaded with NO according to the procedure described by Wheatley et al. [67]. The approach for the in vivo studies consisted in applying the solution (aqueous cream BP + 33%, wt/wt of NO-loaded zeolite) three times per week in the wounded area. NO was released from the solution when in contact with water, showing an initial burst in the first minute reaching a total release after 3 h, with a cumulative NO release of 104 ± 14 nmol (mg zeolite)$^{-1}$. An enhanced wound healing rate of 3.4% day^{-1} in the treated wounds with NO-loaded zeolite solution was observed in comparison with the control (wounds treated with aqueous cream BP). Moreover, in vitro microbial studies were performed proving the potential of these materials in the killing of five common wound pathogens (*Escherichia coli*, *Acinetobacter baumannii*, *Staphylococcus epidermidis*, methicillin-resistant *S. aureus* (MRSA), and *Candida albicans fungus*). Five to eight log reductions in bacterial viability and three log reductions in fungal viability were observed after 8 h in the presence of the NO-releasing zeolite ointment.

The anti-infective properties of these new donors were also explored by Fox et al. [101] using the Zn^{2+}-exchanged zeolite material at 50 wt% composition in a polytetrafluoroethylene polymer. This study demonstrated that this zeolite is an efficient bactericidal over different strains of bacteria, namely, Gram-negative *Pseudomonas aeruginosa* and Gram-positive methicillin-sensitive and methicillin-resistant *S. aureus* and *Clostridium difficile,* killing all the bacteria after 45 min of exposure.

NO has antithrombotic properties, and the NO released from Coexchanged zeolite-A and HKUST-1 (MOF) demonstrated great potential in the inhibition of platelet aggregation and adhesion [67, 82]. Briefly, a mixture of 75% zeolite or MOF and 25% PTFE was pressed into disks (20 mg) and loaded with NO. The loaded disks were incubated with platelet plasma, and the platelet aggregation was initiated using the thromboxane A2 analogue, U46619 (8 μM). Results were compared with the unloaded NO disk and without any material. Fig. 8 illustrates the results obtained for the Coexchanged zeolite-A, showing that only the NO-loaded zeolite disk completely inhibits platelet aggregation in platelet-rich plasma. These preliminary results make these new donors extremely attractive for antithrombotic therapies.

McKinlay et al. [66] evaluated the potentiality of the CPO-27-Ni (MOF with the best NO storage capacity developed up to now) to relax smooth muscle from blood vessels. To this end, compressed pellets of NO-loaded Ni-MOF (5 mg)

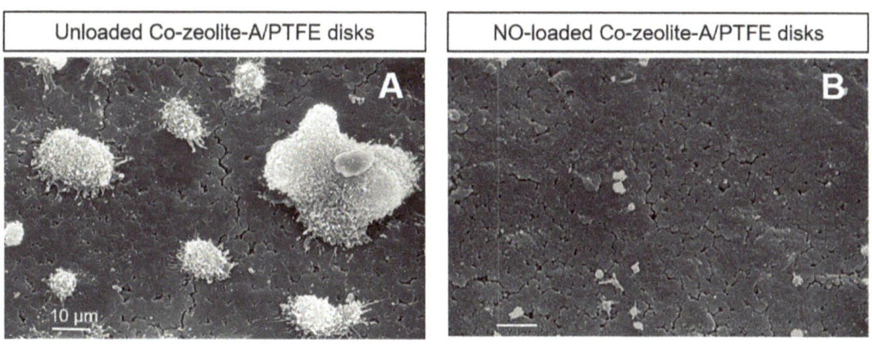

FIG. 8

Effect of NO released from Co-zeolite-A on platelet aggregation. SEM images represent a comparison between (A) the surface treated with unloaded Co-zeolite-A/PTFE disks and (B) the surface treated with NO-loaded Co-zeolite-A/PTFE disks, showing high platelet aggregation in image A and only few isolated platelets in image B. *Adapted from Wheatley PS, Butler AR, Crane MS, Fox S, Xiao B, Rossi AG, Megson IL, Morris RE, NO-releasing zeolites and their antithrombotic properties. J Am Chem Soc 2006;128:502–509. doi:https://doi.org/10.1021/ja0503579 (web archive link). with ACS permission.*

were added to the organ bath at a distance of 2 mm from precontracted pig coronary arteries, showing a fast relaxation (100%) of the vessel. After 10 min, the pellet was removed from the organ bath resulting in a subsequent recovery of the muscle tension, clearly indicating the potential use of these NO-releasing MOFs in vasodilatation.

Recently, our group has been demonstrating the potentialities in the use of ETS-4 loaded with NO in the promotion of cell migration. Material synthesis and NO loading were described in previous works [57]. This NO-loaded material presented a sustained NO release under biological medium and in vitro cell migration studies using HeLa cells displayed an enhanced fast migration after 6 h of about 8% in cells treated with NO-releasing ETS-4 comparing with the untreated controls (unpublished results).

CONCLUSION AND FUTURE DEVELOPMENTS

Given to the multiple and essential activities that NO provides during several therapeutic processes, a wide range of exogenous NO delivery therapies have been proposed as an alternative for the current therapeutic approaches. Hybrid NO donors combining NO with drug delivery systems have been emerged, presenting malleable features, longer NO release profiles, and reduced side effects. Several materials have been proposed as systems for NO delivery, including polymers, nanoparticles, liposomes, dendrimers, and more recently nanoporous materials. This last strategy has gradually demonstrated its superiority, being unique that releases pure NO directly to the target site avoiding potential toxic by-products, and given to its exceptional NO storage capacity, it allows more opportunities for tailoring NO release conditions that may cover a variety of therapeutic needs. Additionally, a number of structural architectures and chemical compositions with distinct NO release kinetics have been considered, namely, zeolites, MOFs, titanosilicates, and clay-based materials. Among them, MOFs present the highest NO storage capacity, but their instability and the possible presence of toxic metals in their structure may affect therapies that require a prolonged exposure to the solids. For these cases, zeolites and titanosilicates are preferred because, although they do not release so large NO amounts, they allow prolonged release with structural integrity. Clay-based materials reveal low storage and NO release capacity, but for therapeutic applications that require low NO amounts, these platforms may be considered.

Most of the cases here described for this new generation of nitric oxide donors present, however, some knowledge gaps regarding NO release parameters (NO payload, maximum NO flux, NO release half-life, time required to reach maximum flux, and total duration of NO release) under biological conditions. These parameters are fundamental to compare with other strategies disclosed in the literature. Often, the reported methods of measuring the released NO from these porous solids are not comparable with the real biological environments. Thus, exploring the behavior of these materials is needed to get a more comprehensive evaluation of their profiles, in order to be able to combine the desirable characteristics for achieving the best application performance.

Another concern about these strategies is related to their storage conditions since NO gas is notoriously difficult to handle. Morris's group already demonstrates that some of these new donors are stable inside vials with controlled atmospheric conditions (i.e., dry argon) over several months. However, it would be fundamental to evaluate more practical and costly alternatives without compromising their effectiveness.

Based on preliminary biological tests effectuated so far for the described NO-releasing porous materials, it is notorious for their numerous benefits for clinical therapy. However, long-term assessments using these materials for controlling biological effects are still poorly explored, being essential to validate these new approaches as viable alternatives. Overall, the results for this new generation of NO donors present an important contribution for a new therapy approach; however, significant research still needs to be performed before considering these materials for medical commercialization. We hope that this chapter stimulates other researchers working in this field to consider nanoporous materials as promising vehicles for NO delivery.

Acknowledgment

RVP acknowledges for the grant

- PhD grant 16/BAD/2017 provided by Colégio de Química, University of Lisbon
- Project grants IF/00993/2012/CP0172/CT0013, PTDC/MED-QUI/28721/2017 and UID/ECI/04028/2013 provided by Fundaçdão para a Ciência e a Tecnologia.

References

[1] Lee RH, Efron D, Tantry U, Barbul A. Nitric oxide in the healing wound: a time-course study. J Surg Res 2001;101:104–8. https://doi.org/10.1006/jsre.2001.6261.

[2] Ignarro L, Freeman B. Nitric oxide: biology and pathobiology. 3rd ed. Elsevier Science; 2017.

[3] Napoli C, de Nigris F, Williams-Ignarro S, Pignalosa O, Sica V, Ignarro LJ. Nitric oxide and atherosclerosis: an update. Nitric Oxide Biol Chem 2006;15:265–79. https://doi.org/10.1016/j.niox.2006.03.011.

[4] Jaffrey SR, Snyder SH. Nitric oxide: a neural messenger. Annu Rev Cell Dev Biol 1995;11:417–40. https://doi.org/10.1146/annurev.cb.11.110195.002221.

[5] Witte MB, Barbul A. Role of nitric oxide in wound repair. Am J Surg 2002;183:406–12. https://doi.org/10.1016/S0002-9610(02)00815-2.

[6] Fang FC. Perspectives series: host/pathogen interactions. Mechanisms of nitric oxide-related antimicrobial activity, J Clin Invest 1997;99:2818–25. http://www.ncbi.nlm.nih.gov/pmc/articles/PMC508130/.

[7] Brüne B. Nitric oxide: NO apoptosis or turning it ON? Cell Death Differ 2003;10:864. https://doi.org/10.1038/sj.cdd.4401261.

[8] Morbidelli L, Donnini S, Ziche M. Role of nitric oxide in the modulation of angiogenesis, Curr Pharm Des 9 (2003) 521–530. https://doi.org/10.2174/1381612033391405.

[9] Mason RP, Cockcroft JR. Targeting nitric oxide with drug therapy. J Clin Hypertens 2006;8:40–52. https://doi.org/10.1111/j.1524-6175.2006.06041.x.

[10] Al-sa'doni H, Ferro A. S-Nitrosothiols: a class of nitric oxide-donor drugs, Clin Sci 2000;98: 507LP–520LP. content/98/5/507.abstract.

[11] Miller MR, Megson IL. Recent developments in nitric oxide donor drugs. Br J Pharmacol 2007;151:305–21. https://doi.org/10.1038/sj.bjp.0707224.

[12] Kim J, Saravanakumar G, Choi HW, Park D, Kim WJ. A platform for nitric oxide delivery. J Mater Chem B 2014;2:341–56. https://doi.org/10.1039/C3TB21259A.

[13] Seabra A. Nitric oxide donors: novel biomedical applications and perspectives. Elsevier Science; 2017 https://books.google.pt/books?id=iY3fDQAAQBAJ.

[14] Gregg ST, Yuan Q, Morris RE, Xiao B. Functionalised solids delivering bioactive nitric oxide gas for therapeutic applications. Mater Today Commun 2017;12:95–105. https://doi.org/10.1016/j.mtcomm.2017.07.007.

[15] Ghaffari A, Jalili R, Ghaffari M, Miller C, Ghahary A. Efficacy of gaseous nitric oxide in the treatment of skin and soft tissue infections. Wound Repair Regen 2007;15:368–77. https://doi.org/10.1111/j.1524-475X.2007.00239.x.

[16] Clark P, Ekekezie I, Kaftan H, Castor C, Truog W. Safety and efficacy of nitric oxide in chronic lung disease. Arch Dis Child Fetal Neonatal Ed 2002;86:F41–5. https://doi.org/10.1136/fn.86.1.F41.

[17] Brisbois EJ, Bayliss J, Wu J, Major TC, Xi C, Wang SC, Bartlett RH, Handa H, Meyerhoff ME. Optimized polymeric film-based nitric oxide delivery inhibits bacterial growth in a mouse burn wound model. Acta Biomater 2014;10:4136–42. https://doi.org/10.1016/j.actbio.2014.06.032.

[18] Macherla C, Sanchez DA, Ahmadi MS, Vellozzi EM, Friedman AJ, Nosanchuk JD, Martinez LR. Nitric oxide releasing nanoparticles for treatment of Candida Albicans burn infections. Front Microbiol 2012;3:193https://doi.org/10.3389/fmicb.2012.00193.

[19] Wheatley PS, Butler AR, Crane MS, Rossi AG, Megson IL, Morris RE. Zeolites for storage and delivery of nitric oxide in human physiology. In: Čejka J, Žilková N, Nachtigall P, editors. Mol. Sieves from basic res. To Ind. Appl. Elsevier; 2005. p. 2033–40. https://doi.org/10.1016/S0167-2991(05)80570-1.

[20] Hinks NJ, McKinlay AC, Xiao B, Wheatley PS, Morris RE. Metal organic frameworks as {NO} delivery materials for biological applications. Micropor Mesopor Mater 2010;129:330–4. https://doi.org/10.1016/j.micromeso.2009.04.031.

[21] Smith DJ, Chakravarthy D, Pulfer S, Simmons ML, Hrabie JA, Citro ML, Saavedra JE, Davies KM, Hutsell TC, Mooradian DL, Hanson SR, Keefer LK. Nitric oxide-releasing polymers containing the [N(O)NO]-group. J Med Chem 1996;39:1148–56. https://doi.org/10.1021/jm950652b.

[22] Champeau M, Seabra AB, de Oliveira MG. Chapter 13—hydrogels for topical nitric oxide delivery. In: Seabra AB, editor. Nitric oxide donors. Academic Press; 2017. p. 313–30. https://doi.org/10.1016/B978-0-12-809275-0.00013-2.

[23] Zhao Q, Zhang J, Song L, Ji Q, Yao Y, Cui Y, Shen J, Wang PG, Kong D. Polysaccharide-based biomaterials with on-demand nitric oxide releasing property regulated by enzyme catalysis. Biomaterials 2013;34:8450–8. https://doi.org/10.1016/j.biomaterials.2013.07.045.

[24] Lowe A, Bills J, Verma R, Lavery L, Davis K, Balkus KJ. Electrospun nitric oxide releasing bandage with enhanced wound healing. Acta Biomater 2015;13:121–30. https://doi.org/10.1016/j.actbio.2014.11.032.

[25] Lijinsky W. Chemical structure of nitrosamines related to carcinogenesis. In: Nitrosamines Relat. N-Nitroso Compd. ACS Symposium Series. American Cancer Society; 1994. p. 250–66. https://doi.org/10.1021/bk-1994-0553.ch020.

[26] Reynolds MM, Frost MC, Meyerhoff ME. Nitric oxide-releasing hydrophobic polymers: preparation, characterization, and potential biomedical applications. Free Radic Biol Med 2004;37:926–36. https://doi.org/10.1016/j.freeradbiomed.2004.06.019.

[27] Wo Y, Xu L-C, Li Z, Matzger AJ, Meyerhoff ME, Siedlecki CA. Antimicrobial nitric oxide releasing surfaces based on S-nitroso-N-acetylpenicillamine impregnated polymers combined with submicron-textured surface topography. Biomater Sci 2017;5:1265–78. https://doi.org/10.1039/C7BM00108H.

[28] Goudie MJ, Brisbois EJ, Pant J, Thompson A, Potkay JA, Handa H. Characterization of an S-nitroso-N-acetylpenicillamine-based nitric oxide releasing polymer from a translational perspective. Int J Polym Mater 2016;65:769–78. https://doi.org/10.1080/00914037.2016.1163570.

[29] Zhang H, Annich GM, Miskulin J, Stankiewicz K, Osterholzer K, Merz SI, Bartlett RH, Meyerhoff ME. Nitric oxide-releasing fumed silica particles: synthesis, characterization, and biomedical application. J Am Chem Soc 2003;125:5015–24. https://doi.org/10.1021/ja0291538.

[30] Shin JH, Schoenfisch MH. Inorganic/organic hybrid silica nanoparticles as a nitric oxide delivery scaffold. Chem Mater 2008;20:239–49. https://doi.org/10.1021/cm702526q.

[31] Shin JH, Metzger SK, Schoenfisch MH. Synthesis of nitric oxide-releasing silica nanoparticles. J Am Chem Soc 2007;129:4612–9. https://doi.org/10.1021/ja0674338.

[32] Wang Y, Zhao Q, Han N, Bai L, Li J, Liu J, Che E, Hu L, Zhang Q, Jiang T, Wang S. Mesoporous silica nanoparticles in drug delivery and biomedical applications. Nanomed Nanotechnol Biol Med 2015;11:313–27. https://doi.org/10.1016/j.nano.2014.09.014.

[33] Riccio DA, Schoenfisch MH. Nitric oxide release: Part I. Macromolecular scaffolds. Chem Soc Rev 2012;41:3731–41. https://doi.org/10.1039/C2CS15272J.

[34] Nam J, Won N, Bang J, Jin H, Park J, Jung S, Jung S, Park Y, Kim S. Surface engineering of inorganic nanoparticles for imaging and therapy. Adv Drug Deliv Rev 2013;65:622–48. https://doi.org/10.1016/j.addr.2012.08.015.

[35] Rothrock AR, Donkers RL, Schoenfisch MH. Synthesis of nitric oxide-releasing gold nanoparticles. J Am Chem Soc 2005;127:9362–3. https://doi.org/10.1021/ja052027u.

[36] Polizzi MA, Stasko NA, Schoenfisch MH. Water-soluble nitric oxide-releasing gold nanoparticles. Langmuir 2007;23:4938–43. https://doi.org/10.1021/la0633841.

[37] Amedea ND, Seabra B, Manosalva N, de Araujo Lima B, Pelegrino MT, Brocchi M, Rubilar O. Antibacterial activity of nitric oxide releasing silver nanoparticles. J Phys Conf Ser 2017;838: https://doi.org/10.1088/1742-6596/838/1/012031.

[38] Neuman D, Ostrowski AD, Mikhailovsky AA, Absalonson RO, Strouse GF, Ford PC. Quantum dot fluorescence quenching pathways with Cr(III) complexes. Photosensitized NO production from trans-Cr(cyclam)(ONO)(2)(+). J Am Chem Soc 2008;130:168–75. https://doi.org/10.1021/ja074164s.

[39] Tan L, Wan A, Li H. Ag2S quantum dots conjugated chitosan nanospheres toward light-triggered nitric oxide release and near-infrared fluorescence imaging. Langmuir 2013;29:15032–42. https://doi.org/10.1021/la403028j.

[40] Zhang XF, Mansouri S, Mbeh DA, Yahia L, Sacher E, Veres T. Nitric oxide delivery by core/shell superparamagnetic nanoparticle vehicles with enhanced biocompatibility. Langmuir 2012;28:12879–85. https://doi.org/10.1021/la302357h.

[41] Molina MM, Seabra AB, de Oliveira MG, Itri R, Haddad PS. Nitric oxide donor superparamagnetic iron oxide nanoparticles. Mater Sci Eng C 2013;33:746–51. https://doi.org/10.1016/j.msec.2012.10.027.

[42] Derfus AM, Chan WCW, Bhatia SN. Probing the cytotoxicity of semiconductor quantum dots. Nano Lett 2004;4:11–8. https://doi.org/10.1021/nl0347334.

[43] Chang E, Thekkek N, Yu WW, Colvin VL, Drezek R. Evaluation of quantum dot cytotoxicity based on intracellular uptake. Small 2006;2:1412–7. https://doi.org/10.1002/smll.200600218.

[44] Huang S-L, Kee PH, Kim H, Moody MR, Chrzanowski SM, MacDonald RC, McPherson DD. Nitric oxide loaded echogenic liposomes for nitric oxide delivery and inhibition of intimal hyperplasia. J Am Coll Cardiol 2009;54:652–9. https://doi.org/10.1016/j.jacc.2009.04.039.

[45] Tai L-A, Wang Y-C, Yang C-S. Heat-activated sustaining nitric oxide release from zwitterionic diazeniumdiolate loaded in thermo-sensitive liposomes. Nitric Oxide 2010;23:60–4. https://doi.org/10.1016/j.niox.2010.04.003.

[46] Suchyta DJ, Schoenfisch MH. Controlled release of nitric oxide from liposomes. ACS Biomater Sci Eng 2017;3:2136–43. https://doi.org/10.1021/acsbiomaterials.7b00255.

[47] Koehler JJ, Zhao J, Jedlicka SS, Porterfield DM, Rickus JL. Compartmentalized nanocomposite for dynamic nitric oxide release. J Phys Chem B 2008;112:15086–93. https://doi.org/10.1021/jp803276u.

[48] Mazur F, Bally M, Städler B, Chandrawati R. Liposomes and lipid bilayers in biosensors. Adv Colloid Interface Sci 2017;249:88–99. https://doi.org/10.1016/j.cis.2017.05.020.

[49] Pattni BS, Chupin VV, Torchilin VP. New developments in liposomal drug delivery. Chem Rev 2015;115:10938–66. https://doi.org/10.1021/acs.chemrev.5b00046.

[50] Suchyta DJ, Schoenfisch MH. Encapsulation of N-diazeniumdiolates within liposomes for enhanced nitric oxide donor stability and delivery. Mol Pharm 2015;12:3569–74. https://doi.org/10.1021/acs.molpharmaceut.5b00248.

[51] Seabra AB, Duran N. Nitric oxide-releasing vehicles for biomedical applications. J Mater Chem 2010;20:1624–37. https://doi.org/10.1039/B912493B.

[52] Yang T, Zelikin AN, Chandrawati R. Progress and promise of nitric oxide-releasing platforms. Adv Sci 2018;5:1701043https://doi.org/10.1002/advs.201701043.

[53] Stasko NA, Schoenfisch MH. Dendrimers as a scaffold for nitric oxide release. J Am Chem Soc 2006;128:8265–71. https://doi.org/10.1021/ja060875z.

[54] Worley BV, Schilly KM, Schoenfisch MH. Anti-biofilm efficacy of dual-action nitric oxide-releasing alkyl chain modified poly(amidoamine) dendrimers. Mol Pharm 2015;12:1573–83. https://doi.org/10.1021/acs.molpharmaceut.5b00006.

[55] Pinto RV, Antunes F, Pires J, Graça V, Brandão P, Pinto ML. Vitamin {B3} metal-organic frameworks as potential delivery vehicles for therapeutic nitric oxide. Acta Biomater 2017;51:66–74. https://doi.org/10.1016/j.actbio.2017.01.039.

[56] Wheatley PS, McKinlay AC, Morris RE. A comparison of zeolites and metal organic frameworks as storage and delivery vehicles for biologically active nitric oxide. In: Antoine Gédéon PM, Babonneau F, editors. Zeolites Relat. Mater. Trends. Targets challenges proceedings 4th Int FEZA ConfElsevier; 2008. p. 441–6. https://doi.org/10.1016/S0167-2991(08)80236-4.

[57] Pinto ML, Rocha J, Gomes JRB, Pires J. Slow release of NO by microporous titanosilicate ETS-4. J Am Chem Soc 2011;133:6396–402. https://doi.org/10.1021/ja200663e.

[58] Lutzke A, Tapia JB, Neufeld MJ, Reynolds MM. Sustained nitric oxide release from a tertiary S-nitrosothiol-based polyphosphazene coating. ACS Appl Mater Interfaces 2017;9:2104–13. https://doi.org/10.1021/acsami.6b12888.

[59] Lutzke A, Neufeld BH, Neufeld MJ, Reynolds MM. Nitric oxide release from a biodegradable cysteine-based polyphosphazene. J Mater Chem B 2016;4:1987–98. https://doi.org/10.1039/C6TB00037A.

[60] Parzuchowski PG, Frost MC, Meyerhoff ME. Synthesis and characterization of polymethacrylate-based nitric oxide donors. J Am Chem Soc 2002;124:12182–91. https://doi.org/10.1021/ja020268l.

[61] Coneski PN, Nash JA, Schoenfisch MH. Nitric oxide-releasing electrospun polymer microfibers. ACS Appl Mater Interfaces 2011;3:426–32. https://doi.org/10.1021/am101010e.

[62] Lautner G, Meyerhoff ME, Schwendeman SP. Biodegradable poly(lactic-co-glycolic acid) microspheres loaded with S-nitroso-N-acetyl-D-penicillamine for controlled nitric oxide delivery. J Control Release 2016;225:133–9. https://doi.org/10.1016/j.jconrel. 2015.12.056.

[63] Soto RJ, Yang L, Schoenfisch MH. Functionalized mesoporous silica via an aminosilane surfactant ion exchange reaction: controlled scaffold design and nitric oxide release. ACS Appl Mater Interfaces 2016;8:2220–31. https://doi.org/10.1021/acsami.5b10942.

[64] Backlund CJ, Worley BV, Schoenfisch MH. Anti-biofilm action of nitric oxide-releasing alkyl-modified poly(amidoamine) dendrimers against Streptococcus mutans. Acta Biomater 2016;29:198–205. https://doi.org/10.1016/j.actbio.2015.10.021.

[65] Stasko NA, Fischer TH, Schoenfisch MH. S-Nitrosothiol-modified dendrimers as nitric oxide delivery vehicles. Biomacromolecules 2008;9:834–41. https://doi.org/10.1021/bm7011746.

[66] McKinlay AC, Xiao B, Wragg DS, Wheatley PS, Megson IL, Morris RE. Exceptional behavior over the whole adsorption-storage-delivery cycle for NO in porous metal organic frameworks. J Am Chem Soc 2008;130:10440–4. https://doi.org/10.1021/ja801997r.

[67] Wheatley PS, Butler AR, Crane MS, Fox S, Xiao B, Rossi AG, Megson IL, Morris RE. NO-releasing zeolites and their antithrombotic properties. J Am Chem Soc 2006;128:502–9. https://doi.org/10.1021/ja0503579.

[68] Morris RE, Wheatley PS. Gas storage in nanoporous materials. Angew Chem Int Ed 2008;47:4966–81. https://doi.org/10.1002/anie.200703934.

[69] Zhao XS. Novel porous materials for emerging applications. J Mater Chem 2006;16:623–5. https://doi.org/10.1039/B600327N.

[70] Leffler CW, Parfenova H, Jaggar JH, Wang R. Carbon monoxide and hydrogen sulfide: gaseous messengers in cerebrovascular circulation. J Appl Physiol 2006;100:1065–76. https://doi.org/10.1152/japplphysiol.00793.2005.

[71] Vandiver MS, Snyder SH. Hydrogen sulfide: a gasotransmitter of clinical relevance. J Mol Med (Berl) 2012;90:255–63. https://doi.org/10.1007/s00109-012-0873-4.

[72] Mustafa AK, Gadalla MM, Snyder SH. Signaling by gasotransmitters. Sci Signal 2009;2:re2. https://doi.org/10.1126/scisignal.268re2.

[73] Fernandes AC, Antunes F, Pires J. Sepiolite based materials for storage and slow release of nitric oxide. New J Chem 2013;37:4052–60. https://doi.org/10.1039/C3NJ00452J.

[74] Boes A-K, Wheatley PS, Xiao B, Megson IL, Morris RE. Simultaneous and cooperative gas storage and gas production using bifunctional zeolites. Chem Commun 2008;6146–8. https://doi.org/10.1039/B813976H.

[75] "Joe" Zhou H-C, Kitagawa S. Metal–organic frameworks (MOFs). Chem Soc Rev 2014;43:5415–8. https://doi.org/10.1039/C4CS90059F.

[76] Maurin G, Serre C, Cooper A, Ferey G. The new age of MOFs and of their porous-related solids. Chem Soc Rev 2017;46:3104–7. https://doi.org/10.1039/C7CS90049J.

[77] Zhou H-C, Long JR, Yaghi OM. Introduction to metal–organic frameworks. Chem Rev 2012;112:673–4. https://doi.org/10.1021/cr300014x.

[78] Stock N, Biswas S. Synthesis of metal-organic frameworks (MOFs): routes to various MOF topologies, morphologies, and composites. Chem Rev 2012;112:933–69. https://doi.org/10.1021/cr200304e.

[79] Li B, Wen H-M, Zhou W, Chen B. Porous metal–organic frameworks for gas storage and separation: what, how, and why? J Phys Chem Lett 2014;5:3468–79. https://doi.org/10.1021/jz501586e.

[80] Keskin S, Kızılel S. Biomedical applications of metal organic frameworks. Ind Eng Chem Res 2011;50:1799–812. https://doi.org/10.1021/ie101312k.

[81] McKinlay AC, Morris RE, Horcajada P, Férey G, Gref R, Couvreur P, Serre C. BioMOFs: metal–organic frameworks for biological and medical applications. Angew Chem Int Ed 2010;49:6260–6. https://doi.org/10.1002/anie.201000048.

[82] Xiao B, Wheatley PS, Zhao X, Fletcher AJ, Fox S, Rossi AG, Megson IL, Bordiga S, Regli L, Thomas KM, Morris RE. High-capacity hydrogen and nitric oxide adsorption and storage in a metal-organic framework. J Am Chem Soc 2007;129:1203–9. https://doi.org/10.1021/ja066098k.

[83] Peikert K, McCormick LJ, Cattaneo D, Duncan MJ, Hoffmann F, Khan AH, Bertmer M, Morris RE, Fröba M. Tuning the nitric oxide release behavior of amino functionalized HKUST-1. Micropor Mesopor Mater 2015;216:118–26. https://doi.org/10.1016/j.micromeso.2015.06.020.

[84] Miller SR, Alvarez E, Fradcourt L, Devic T, Wuttke S, Wheatley PS, Steunou N, Bonhomme C, Gervais C, Laurencin D, Morris RE, Vimont A, Daturi M, Horcajada P, Serre C. A rare example of a porous Ca-MOF for the controlled release of biologically active NO. Chem Commun 2013;49:7773–5. https://doi.org/10.1039/C3CC41987H.

[85] McKinlay AC, Eubank JF, Wuttke S, Xiao B, Wheatley PS, Bazin P, Lavalley J-C, Daturi M, Vimont A, De Weireld G, Horcajada P, Serre C, Morris RE. Nitric oxide adsorption and delivery in flexible MIL-88(Fe) metal–organic frameworks. Chem Mater 2013;25:1592–9. https://doi.org/10.1021/cm304037x.

[86] Serre C, Mellot-Draznieks C, Surblé S, Audebrand N, Filinchuk Y, Férey G. Role of solvent-host interactions that lead to very large swelling of hybrid frameworks. Science 2007;315:1828–31. https://doi.org/10.1126/science.1137975.

[87] Eubank JF, Wheatley PS, Lebars G, McKinlay AC, Leclerc H, Horcajada P, Daturi M, Vimont A, Morris RE, Serre C. Porous, rigid metal(III)-carboxylate metal-organic frameworks for the delivery of nitric oxide. APL Mater 2014;2:124112https://doi.org/10.1063/1.4904069.

[88] Pinto ML, Fernandes AC, Antunes F, Pires J, Rocha J. Storage and delivery of nitric oxide by microporous titanosilicate ETS-10 and Al and Ga substituted analogues. Micropor Mesopor Mater 2016;229:83–9. https://doi.org/10.1016/j.micromeso.2016.04.021.

[89] Pinto ML, Fernandes AC, Rocha J, Ferreira A, Antunes F, Pires J. Microporous titanosilicates Cu^{2+}– And Co^{2+}–ETS-4 for storage and slow release of therapeutic nitric oxide. J Mater Chem B 2014;2:224–30. https://doi.org/10.1039/C3TB20929F.

[90] Viseras C, Cerezo P, Sanchez R, Salcedo I, Aguzzi C. Current challenges in clay minerals for drug delivery. Appl Clay Sci 2010;48:291–5. https://doi.org/10.1016/j.clay.2010.01.007.

[91] Zhang X, Ju H, Wang J. Electrochemical sensors, biosensors and their biomedical applications, Elsevier Science; 2011. https://books.google.pt/books?id=Evx8Ee6I7fIC.

[92] Fernandes A, Antunes F, Pinto ML, Pires J. Clay based materials for storage and therapeutic release of nitric oxide. J Mater Chem B 2013;1:3287–94. https://doi.org/10.1039/c3tb20535e.

[93] Fernandes AC, Pinto ML, Antunes F, Pires J. Synthetic cobalt clays for the storage and slow release of therapeutic nitric oxide. RSC Adv 2016;6:41195–203. https://doi.org/10.1039/C6RA05794B.

[94] Fernandes AC, Pinto ML, Antunes F, Pires J. L-Histidine-based organoclays for the storage and release of therapeutic nitric oxide. J Mater Chem B 2015;3:3556–63. https://doi.org/10.1039/C4TB01913J.

[95] Anand V, Kandarapu R, Garg S. Ion-exchange resins: carrying drug delivery forward. Drug Discov Today 2001;6:905–14. https://doi.org/10.1016/S1359-6446(01)01922-5.

[96] Gil A, Gandía LM, Vicente MA. Recent advances in the synthesis and catalytic applications of pillared clays. Catal Rev 2000;42:145–212. https://doi.org/10.1081/CR-100100261.

[97] de Paiva LB, Morales AR, Díaz FRV. Organoclays: properties, preparation and applications. Appl Clay Sci 2008;42:8–24. https://doi.org/10.1016/j.clay.2008.02.006.

[98] Chien M, Pearson I, Nobe K. Reduction and absorption kinetics of nitric oxide on colbalt perovskite catalysts. Prod R&D 1975;14:131–4. https://doi.org/10.1021/i360054a601.

[99] Mowbray M, Tan X, Wheatley PS, Morris RE, Weller RB. Topically applied nitric oxide induces T-lymphocyte infiltration in human skin, but minimal inflammation. J Invest Dermatol 2008;128:352–60. https://doi.org/10.1038/sj.jid.5701096.

[100] Neidrauer M, Ercan UK, Bhattacharyya A, Samuels J, Sedlak J, Trikha R, Barbee KA, Weingarten MS, Joshi SG. Antimicrobial efficacy and wound-healing property of a topical ointment containing nitric-oxide-loaded zeolites. J Med Microbiol 2014;63:203–9. https://doi.org/10.1099/jmm.0.067322-0.

[101] S. Fox, T.S. Wilkinson, P.S. Wheatley, B. Xiao, R.E. Morris, A. Sutherland, a J. Simpson, P.G. Barlow, A.R. Butler, I.L. Megson, A.G. Rossi, NO-loaded Zn^{2+}-exchanged zeolite materials: a potential bifunctional anti-bacterial strategy., Acta Biomater 6 (2010) 1515–1521. doi:10.1016/j.actbio.2009.10.038.

3

PART

Abstracts (Presentations at Conference in Siena, October 2018)

Serum Nitric Oxide Metabolites and the Incidence of Metabolic Syndrome and Its Phenotypes: A Population-Based Prospective Study

Zahra Bahadoran*, Parvin Mirmiran*, Fereidoun Azizi†, Asghar Ghasemi‡

*Nutrition and Endocrine Research Center, Research Institute for Endocrine Sciences, Shahid Beheshti University of Medical Sciences, Tehran, Iran

†Endocrine Research Center, Research Institute for Endocrine Sciences, Shahid Beheshti University of Medical Sciences, Tehran, Iran

‡Endocrine Physiology Research Center, Research Institute for Endocrine Sciences, Shahid Beheshti University of Medical Sciences, Tehran, Iran

Abstract

Background and Aim: Although nitric oxide (NO) is emerging as a central regulator of metabolism, its endogenous overproduction, especially originated by inducible NO synthase, is now considered as a marker and potential mediator of metabolic diseases. Current knowledge regarding the prognostic relevance of circulating NO is limited and inconsistent. This study was conducted to investigate whether serum NO metabolites (NOx) could predict the occurrence of a metabolic syndrome (MetS) and its phenotypes, in a general population.

Methods: We measured serum NOx concentrations in the Tehran Lipid and Glucose Study participants (aged $\geq$19 years) at baseline examination (2006–08) and followed them for a median 7.7 years for MetS and its phenotypes. The receiver operator characteristic (ROC) curve analysis was used to determine the optimal cutoff points of serum NOx for predicting MetS. Multivariate Cox proportional hazard models were used to estimate the hazard ratios (HRs) with 95% confidence intervals (95% CIs) of MetS in response to serum NOx values. The incidence (%) of MetS phenotypes across optimal cutoff points of serum NOx was compared using the chi-square test.

Results: The median (interquartile range, IQR) follow-up of MetS was 7.7 (7.3–8.0) years, and the incidence rate (95% CI) was 5.18 (4.77–5.62) per 1000 person-years. The optimal cutoff points of serum NOx levels for predicting MetS were 26.5 μmol/L. The area under the curve (AUC) of serum NOx for MetS was 0.56 ($P = 0.001$). A significant positive association was observed between

Therapeutic Application of Nitric Oxide in Cancer and Inflammatory Disorders. https://doi.org/10.1016/B978-0-12-816545-4.00015-3

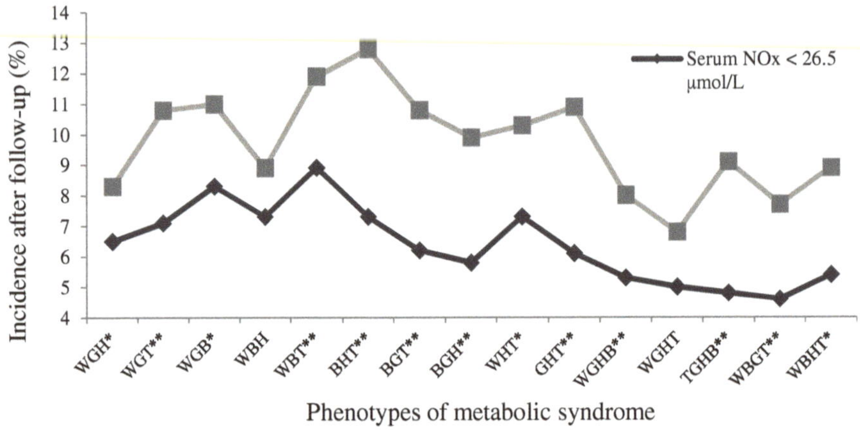

FIG. 1

The incidence of metabolic syndrome phenotypes below and above the defined optimal cutoff points for serum NOx (≥ 26.5 μmol/L). W, elevated waist circumference; G, elevated blood glucose; T, elevated triglyceride levels; B, elevated blood pressure; H, low HDL-C.*$P<0.01$ and **$P<0.001$.

Ln-transformed serum NOx levels and the incidence of MetS (HR = 1.20, 95% CI = 1.03–1.39). After the adjustment of potential confounding variables, participants with serum NOx levels ≥26.5 μmol/L had a significant increased risk of MetS (HR = 1.19, 95% CI = 1.01–1.40). The incidence of MetS phenotypes above and below the cutoff point of serum NOx is illustrated in Fig. 1.

Conclusion: Our findings imply that increased serum NOx, most probably due to endogenous overproduction of NO, may be considered as an independent factor contributing to the development of MetS and its phenotypes. With respect to these findings, serum NOx may be suggested as a novel predictor of future risk of metabolic disorders.

Nitric Oxide (NO): A Multifaceted Target for Reversal of Cancer Cell Pleiotropic Properties by NO-Modulating Therapies

Stavroula Baritaki*, Benjamin Bonavida†
*Division of Surgical Oncology, School of Medicine, University of Crete, Crete, Greece
†Department of Microbiology, Immunology, and Molecular Genetics, David Geffen School of Medicine, University of California, Los Angeles (UCLA), Los Angeles, CA, United States

Abstract

During the last two decades, nitric oxide (NO) has emerged as a molecule of interest in cancer cell biology. There are several lines of experimental evidence showing that NO, produced by cancer cells or released by surrounding cells in the tumor milieu, contributes critically to various functions in cancer cells and the tumor microenvironment. These functions, among others, include carcinogenesis, tumor cell viability, growth and invasiveness, maintenance of cancer stem cell properties, angiogenesis, immunoediting of immune signals, and tumor resistance to conventional chemo-/immunotherapeutics. NO may affect, either positively or negatively, the above processes in various cancer types, thus creating a considerable controversy and confusion in the understanding of the role of NO in cancer biology. In the context of the contrasting roles of NO in cancer, recent findings clarified this controversy, namely, that NO exerts pro- or antineoplastic properties in the cancer milieu depending on the NO concentration and the location and timing of its release.

Several studies have shown that NO, at high doses, acts as an antineoplastic agent, when used either as a single agent or in combination with other antineoplastic compounds. These findings demonstrate the importance of NO modulation in the response to cancer therapeutics. We and others have reported that compounds able to release high concentrations of NO, such as NO donors, are able, among others, to overcome tumor cell resistance to immune-mediated cytotoxicity and conventional chemotherapy and to inhibit the initial metastatic steps of tumor cells, known as the "oncogenic epithelial-to-mesenchymal transition (EMT)." NO mediates the above antineoplastic properties via modulation of signaling pathways and feedback loops composed of multiple interactions among gene products involved in the regulation of cellular apoptosis and tumor invasive and migratory properties. We will present findings on

Therapeutic Application of Nitric Oxide in Cancer and Inflammatory Disorders. https://doi.org/10.1016/B978-0-12-816545-4.00016-5

(1) the range of antineoplastic activities of NO in cancer cells and the tumor microenvironment, (2) the various molecular mechanisms by which NO mediates its antineoplastic activities, and (3) NO's novel potential applications in cancer treatment and prevention.

Pharmacology of Hepatoselective Nitric Oxide: Donors in NAFLDD in Mice

Stefan Chlopicki
Jagiellonian Centre for Experimental Therapeutics (JCET), Jagiellonian University, Krakow, Poland

Abstract

The impairment of the function of liver sinusoidal endothelial cells (LSECs) associated with the impairment of NO production by LSECs plays an important role in the pathogenesis of NAFLD [1]. We provided evidence that the treatment of mice fed a high-fat diet (HFD) with a liver-selective NO donor such as V-PYRRO/NO was protected against the development of liver steatosis and attenuated insulin resistance [2]. The effects of V-PYRRO/NO were stronger as compared with metformin, and the protection was mediated by a different mechanism as that afforded by metformin [2]. V-PYRRO/NO inhibited the synthesis of de novo fatty acid and increased insulin resistance, while metformin increased mitochondrial biogenesis. Distinct mechanisms of action of V-PYRRO/NO and metformin suggest a possible additive value of the combined treatment based on the hepatoselective NO delivery and metformin. Interestingly, V-PROLI/NO, despite its similar structure to V-PYRRO/NO, displayed a completely different pharmacokinetic profile, including weak NO release by a cytochrome P450-mediated metabolism in the major part of the kidney metabolism and subsequently was not effective in the prevention of liver steatosis and insulin resistance in mice fed an HFD [3, 4]. In summary, our studies [1–4] were complemented recently by the demonstration of the beneficial effects of V-PYRRO/NO-based treatment on the liver microcirculation perfusion in NAFLD [5]. These suggested that liver-targeted delivery of NO represents an effective approach to normalize metabolic disturbances and impairment of liver microcirculation of the blood flow occurring in NAFLD. Our ongoing studies are aimed to better understand the regulatory role of LSEC-derived NO in the progression of NAFLD.

Therapeutic Application of Nitric Oxide in Cancer and Inflammatory Disorders. https://doi.org/10.1016/B978-0-12-816545-4.00017-7

Acknowledgments

This work was supported by the grant from the resources of the European Regional Development Fund under the Innovative Economy Program (grant coordinated by JCET-UJ, no. POIG.01.01.02-00-069/09) and by National Science Centre Symfonia grant no. DEC-2015/16/W/NZ4/00070.

References

[1] Maslak E, Gregorius A, Chlopicki S. Liver sinusoidal endothelial cells (LSECs) function and NAFLD; NO-based therapy targeted to the liver. Pharmacol Rep 2015;67:689–94.

[2] Maslak E, Zabielski P, Kochan K, Kus K, Jasztal A, Sitek B, Proniewski B, Wojcik T, Gula K, Kij A, Walczak M, Baranska M, Chabowski A, Holland RJ, Saavedra JE, Keefer LK, Chlopicki S. The liver-selective NO donor, V-PYRRO/NO, protects against liver steatosis and improves postprandial glucose tolerance in mice fed high fat diet. Biochem Pharmacol 2015;93:389–400.

[3] Kus K, Walczak M, Maslak E, Zakrzewska A, Gonciarz-Dytman A, Zabielski P, Sitek B, Wandzel K, Kij A, Chabowski A, Holland RJ, Saavedra JE, Keefer LK, Chlopicki S. Hepatoselective nitric oxide (NO) donors, V-PYRRO/NO and V-PROLI/NO, in nonalcoholic fatty liver disease: a comparison of antisteatotic effects with the biotransformation and pharmacokinetics. Drug Metab Dispos 2015;43:1028–36.

[4] Kus K, Kus E, Zakrzewska A, Jawien W, Sitek B, Walczak M, Chlopicki S. Differential effects of liver steatosis on pharmacokinetic profile of two closely related hepatoselective NO-donors; V-PYRRO/NO and V-PROLI/NO. Pharmacol Rep 2017;69:560–5.

[5] Kus E, Jasinski K, Skorka T, Czyzynska-Cichon I, Chlopicki S. Short-term treatment with hepatoselective NO donor V-PYRRO/NO improves blood flow in hepatic microcirculation in liver steatosis in mice. Pharmacol Rep 2018;70:463–9.

Anticancer Effect of a Novel H_2S-Hybrid Molecule on Human Breast Adenocarcinoma (MFC-7) and Human Breast Epithelial (MCF-10A) Cell Lines

Valentina Citi*, Alma Martelli*,†, Eugenia Piragine*, Elisabetta Barresi*, Sabrina Taliani*, Lara Testai*,†, Federico Da Settimo*, Vincenzo Calderone*,†

**Department of Pharmacy, University of Pisa, Pisa, Italy*

†Interdepartmental Research Centre "Nutraceuticals and Food for Health (NUTRAFOOD)," University of Pisa, Pisa, Italy

Abstract

Hydrogen sulfide (H_2S) is a gasotransmitter produced endogenously in the mammalian system. Many in vitro studies described the role of H_2S in regulating cell growth and survival, resulting in apoptosis of cancer cells mediated by the activation of the caspase pathway. Recently, our group described some synthetic and natural isothiocyanates (ITCs) as H_2S-donor agents. Noteworthy, there is an overlap between the anticancer effects widely attributed to many ITCs and those exhibited by the treatment with some known H_2S donors, such as the inhibition of human cancer cell viability and the arrest of the cell cycle in the G2/M phase [1, 2]. The development of H_2S-hybrid drugs, obtained through the conjugation of H_2S-releasing moieties to conventional drugs, is aimed to reduce possible adverse effects of the old drugs and/or improve their therapeutic impact. In this work, an innovative molecule (named "hybrid") has been obtained from the combination of a well-known drug (the "native" drug) exerting anticancer effect and the H_2S-releasing moiety ITC. Firstly, the characterization of the H_2S-generating properties of the hybrid and the native drug has been carried out in a cell-free model by an amperometric approach. Then, the antiproliferative effect of the two molecules at different concentrations (5 mM, 1 mM, 500 μM, 100 μM, 50 μM, 10 μM, 5 μM, 1 μM, 500 nM, and 100 nM) after 72 h treatment has been assessed in the MCF7 cell line (human breast adenocarcinoma cells) and the MCF10A cell line (human breast epithelial cells). Furthermore, the intracellular H_2S release of the two compounds (1 mM, 100 μM, and 10 μM) has

Therapeutic Application of Nitric Oxide in Cancer and Inflammatory Disorders. https://doi.org/10.1016/B978-0-12-816545-4.00019-0

been recorded, using a cell-based spectrofluorometric method. Finally, in order to investigate the anticancer mechanism of action, the flow cytometry analyses of the cell cycle and annexin V (early apoptosis) and caspase-3/caspase-7 (mild apoptosis) activities have been performed on the MCF7 cells after 72 h treatment at the concentration of 20 μM.

As concerns of the amperometric results, the incubation of the hybrid 1 mM in the presence of L-cysteine 4 mM led to a slow increase of the H_2S concentration (about 8 μM); in contrast, after the incubation of the hybrid 1 mM in the absence of L-cysteine, no H_2S release has been recorded. The incubation of the native drug did not lead to the formation of H_2S either in the presence or in the absence of L-cysteine. The cell treatment with the hybrid for 72 h showed a concentration-dependent antiproliferative effect on MCF7 cells, with a maximum efficacy value (100% inhibition of cell viability) at the concentration of 500 μM and a parameter of potency of 20 μM (IC50, concentration evoking the inhibition of the 50% of cell viability). In contrast, the active drug showed a dramatic decrease in the antiproliferative effect on MCF7, showing an inhibition of cell viability of about 50% at the highest concentration (5 mM). Furthermore, the hybrid was able to release H_2S into MCF7 cells: in fact, at the concentration of 1 mM, the amount of H_2S released by the hybrid was similar to that evoked by the reference H_2S-donor diallyl disulfide (DADS) at 100 μM. The antiproliferative effect of the two molecules has been also assessed in the nontumorigenic MCF10A: while the native drug showed an antiproliferative behavior similar to that exhibited in MCF7, the hybrid showed a lower parameter of potency (about 170 μM) and a maximum efficacy value only at the concentration of 1 mM of about the 90% and, interestingly, did not promote any detectable H_2S release when incubated with MCF10A. On MCF7 cell cycle, the hybrid induced a significant reduction of the G0/G1 phase cells and an increase of the G2/M phase; moreover, an increased number of cells in early and mild apoptosis, respectively, of about the 10% and 20% were recorded in cells treated with the hybrid compared with vehicle-treated cell. In conclusion, this hybrid molecule showed a significant improvement of the native drug anticancer properties, indicating that ITC can be viewed as a versatile chemical moiety for developing novel hybrid compounds in this field.

References

[1] Martelli A, et al. Vascul Pharmacol 2014;60(1):32–41.

[2] Prelowska M, et al. Pharmacol Rep 2017;69(5):1059–66.

Heme Oxygenase-1 in Duchenne Muscular Dystrophy: Interaction With Nitric Oxide Synthase Pathway in Satellite Cells Differentiation

Józef Dulak

Department of Medical Biotechnology, Faculty of Biochemistry, Biophysics and Biotechnology, Jagiellonian University, Krakow, Poland

Abstract

Muscle damage in Duchenne muscular dystrophy (DMD), caused by the lack of dystrophin, is strongly linked to inflammation. Accordingly, modulation of inflammation can ameliorate the disease progression. Heme oxygenase-1 (HO-1, *Hmox1*) is the crucial anti-inflammatory protein affecting also muscle stem cell differentiation. Surprisingly, the role of HO-1 has not been so far deeply addressed in DMD.

The results of our recent studies [1, 2] showed that HO-1 is strongly upregulated in response to cardiotoxin-induced muscle injury and in DMD muscles. The lack of HO-1 augments the injury, and consequently, multiple rounds of degeneration/regeneration in conditions of HO-1 deficiency may lead to exhaustion of satellite cell (SC) pool. Accordingly, the number of SCs is decreased in old mice lacking HO-1.

The lack of HO-1 affects the severity of DMD in *mdx* mice deficient of dystrophin. Accordingly, double-knockout animals (*Hmox1*$^{-/-}$ *mdx)* demonstrate strongly impaired exercise capacity. Moreover, *mdx* SCs show disturbed and enhanced differentiation, which are further aggravated by *Hmox1* deficiency. Interestingly, in contrast to the effect observed in *mdx* striated muscles, in which the *Hmox1* is strongly induced in comparison with healthy animals, the expression of *Hmox1* is decreased in dystrophin-deficient SCs. Importantly, SCs of *mdx* mice demonstrate disturbed and enhanced differentiation, which are further intensified by *Hmox1* deficiency. RNA sequencing revealed downregulation of Atf3, MafK, Foxo1, and Klf2 transcription factors, known to activate *Hmox1* expression, as well as attenuation of nitric oxide-mediated cGMP-dependent signaling in *mdx* SCs. Accordingly, treatment with NO donor induces *Hmox1* expression

Therapeutic Application of Nitric Oxide in Cancer and Inflammatory Disorders. https://doi.org/10.1016/B978-0-12-816545-4.00020-7

and inhibits differentiation. Finally, differentiation of *mdx* SCs was normalized by CO, a product of HO-1 activity.

In summary, HO-1 can be considered as a therapeutic target to alleviate muscle injury and DMD severity. Our study indicates the interaction between nitric oxide and heme oxygenase-1 in SC differentiation.

Acknowledgment

This work was supported by the NCN grant MAESTRO-2012/06/A/NZ1/00004.

References

[1] Kozakowska M, Pietraszek-Gremplewicz K, Ciesla M, Seczynska M, Bronisz-Budzynska I, Podkalicka P, Bukowska-Strakova K, Loboda A, Jozkowicz A, Dulak J. Lack of heme oxygenase-1 induces inflammatory reaction and proliferation of muscle satellite cells after cardiotoxin-induced skeletal muscle injury. Am J Pathol 2018;188(2):491–506. https://doi.org/10.1016/j.ajpath.2017.10.017.

[2] Pietraszek-Gremplewicz K, Kozakowska M, Bronisz-Budzynska I, Ciesla M, Mucha O, Podkalicka P, Madej M, Glowniak U, Szade K, Stepniewski J, Jez M, Andrysiak K, Bukowska-Strakova K, Kaminska A, Kostera-Pruszczyk A, Jozkowicz A, Loboda A, Dulak J. Heme Oxygenase-1 influences satellite cells and progression of Duchenne muscular dystrophy in mice. Antioxid Redox Signal 2018;29(2):128–48. https://doi.org/10.1089/ars.2017.7435.

In Vitro Assessment of NitDox Toxicity Toward Vasculature

Miriam Durante*, Maria Frosini*, Fabio Fusi*, Barbara Rolando†, Konstantin Chegaev†, Simona Saponara*

**Department of Life Sciences, University of Siena, Siena, Italy*
†Department of Science and Drug Technology, University of Turin, Turin, Italy

Abstract

Doxorubicin (Dox) is one of the most powerful anticancer drugs, but its therapeutic efficacy is hampered by the development of drug resistance, mainly due to the overexpression of the ATP-binding cassette (ABC) efflux transporters [1].

The conjugation of Dox with nitric oxide (NO)-releasing groups gives rise to novel multitarget anthracyclines [2–4], such as nitrooxy-Dox (NitDox), capable to overcome drug resistance by decreasing the activity of ABC transporters via nitration of critical tyrosine residues on the pumps [2]. Moreover, NitDox preferentially accumulates in mitochondria and affects its function, thus representing a prototype of novel anthracyclines, which have cellular targets different from and of greater efficacy against drug-resistant tumor cells than the parent compound [5–7].

The widely described anthracycline cardiovascular toxicity, however, might limit their clinical use. The aim of this study was to investigate the NitDox-induced vascular effects, as potential hazard, on the vascular smooth muscle A7r5 and the endothelial EA.hy926 cell viability, on the mechanical activity of freshly and cultured rat aorta rings, and on L-type Ca^{2+} current [$I_{Ca(L)}$] of A7r5 cells [8]. Dox was used as a reference compound.

Both A7r5 and EAhy.926 cells proved to be more sensitive to Dox than to NitDox at concentrations comparable with those accumulating and exerting antitumor activity in drug-resistant cells after 24 h exposure. Both compounds, however, promoted similar apoptotic effects in A7r5 cells. Dox was more active than its derivative in augmenting the subdiploid-, dapi-, and annexin V-positive cell number in endothelial cells. The proapoptotic activity of both Dox and NitDox was due to the generation of intracellular ROS and changes in the

Therapeutic Application of Nitric Oxide in Cancer and Inflammatory Disorders. https://doi.org/10.1016/B978-0-12-816545-4.00021-9

mitochondrial potential. Finally, NitDox doubled basal NO cell content after 24 h of treatment in both A7r5 and EAhy.926 cells.

Dox exhibited a modest contracturing effect in endothelium intact rings, while NitDox induced a significant vasodilation in endothelium-denuded rings that disappeared in the presence of ODQ. NitDox and Dox did not significantly affect KCl-induced contraction. In arteries cultured with both drugs for 7 days, NitDox blocked both phenylephrine- and KCl-induced contractions at a concentration 10-fold higher than that of Dox. NitDox exhibited weak Ca^{2+} antagonist properties in single A7r5 cells.

In conclusion, NitDox is an NO-releasing Dox with a more favorable vascular toxicity profile and a greater efficacy against drug-resistant cells than the parent compound, worthy of further investigations in preclinical and clinical settings.

Acknowledgment

This work was supported by Italian Ministry for Instruction, University and Research (Futuro in Ricerca 2012, RBFR12SOQ).

References

[1] Szakács et al., Chem Rev 2014;114:5753–74.

[2] Chegaev et al., ACS Med Chem Lett 2011;2:494–7.

[3] Chegaev et al., ACS Med Chem Lett 2017;8:361–5.

[4] Gazzano et al., Bioorg Med Chem 2016;24:967–75.

[5] Riganti et al., Mol Pharm 2013;10:161–74.

[6] Pedrini et al., Mol Pharm 2014;11:3068–79.

[7] Gazzano et al., J Control Release 2018;270:37–52.

[8] Durante et al., Toxicol In Vitro 2017;44:241–7.

The Regulation of GSNOR Expression Discloses the Role of S-Nitrosylation in Aging and Disease

Giuseppe Filomeni[*,†]

*Redox Signaling and Oxidative Stress Research Group, Cell Stress and Survival Unit, Center for Autophagy, Recycling and Disease (CARD), Danish Cancer Society Research Center, Copenhagen, Denmark

†Department of Biology, University of Rome Tor Vergata, Rome, Italy

Abstract

S-Nitrosylation, the reversible modification of cysteine residues by nitric oxide (NO), represents a prototypic redox-based signal that is frequently dysregulated in disease.

By examining the literature, we were struck by a constellation of changes in mice deficient in the denitrosylase *S*-nitrosoglutathione reductase (GSNOR) that typifies aging.

Here, we go into depth on the molecular mechanisms underlying this phenotype. In particular, we show that GSNOR deficiency—and excessive *S*-nitrosylation resulting from it—drives cell senescence and aging in mammals by affecting mitochondrial dynamics and homeostasis. We report that GSNOR null cells show an accumulation of fragmented and dysfunctional mitochondria that are not properly recognized by the autophagic machinery. As a result, they are not removed and degraded by mitophagy, with this alteration being a marker of cell aging.

In order to provide the biological relevance to our results, we demonstrate that GSNOR expression physiologically decreases in senescent cells in mice and humans during their life span owing to an epigenetic regulation of the methylation status of the *GSNOR* promoter. In stark contrast, but in line with the hypothesis that GSNOR represents a new longevity protein, exceptionally long-lived individuals maintain GSNOR levels.

Therapeutic Application of Nitric Oxide in Cancer and Inflammatory Disorders. https://doi.org/10.1016/B978-0-12-816545-4.00022-0

Finally, we also provide preliminary evidence about GSNOR expression being regulated at translational levels by the ataxia-telangiectasia mutated (ATM)-induced signaling cascade in response to nitro-oxidative insults. We speculate that this event is crucial to rapidly modulate intracellular glutathione levels and, in turn, face up oxidative stress, providing new cues for the comprehension of the ataxia-telangiectasia etiopathogenesis.

Effects of Long-Term Nitrite Supplementation on Gene Expressions of GLUT2, GLUT4, and Glucokinase in Male Obese Type 2 Diabetic Rats

Asghar Ghasemi, Sevda Gheibi, Sajad Jeddi, Hanieh Gholami
Endocrine Physiology Research Center, Research Institute for Endocrine Sciences, Shahid Beheshti University of Medical Sciences, Tehran, Iran

Abstract

Purpose: Reduced bioavailability of nitric oxide (NO) is associated with the pathogenesis of type 2 diabetes. Nitrite can act as a substrate for boosting the nitrate-nitrite-NO pathway in diabetes. The aim of this study was to determine the effects of nitrite on gene expressions of GLUT2, GLUT4, and glucokinase in male obese type 2 diabetic rats.

Methods: Male Wistar rats were divided into four groups: control, control plus nitrite, diabetes, and diabetes plus nitrite. Diabetes was induced using a high-fat diet combined with a low dose of streptozotocin. Sodium nitrite (50 mg/L in drinking water) was administered for 2 months. At the end of the study, mRNA expressions of GLUT2 and glucokinase were measured in the pancreas, and mRNA expression of GLUT4 was measured in the soleus muscle and epididymal adipose tissue.

Results: Compared with the controls, in diabetic rats, the mRNA expressions of glucokinase (70% and $p < 0.001$) and GLUT4 (52% and 40%, in the soleus muscle and epididymal adipose tissue, respectively) were significantly lower, while there was no significant change in GLUT2 mRNA expression. Nitrite administration increased GLUT4 mRNA expression in both the soleus muscle (25% and 54%, in control and diabetic rats, respectively) and the epididymal adipose tissue (120% and 70%, respectively) and increased glucokinase (231% and 153% in control and diabetic rats, respectively) mRNA expression in the pancreas. Nitrite administration increased mRNA expression of GLUT2 only in control rats (123% and $p < 0.001$) but not in diabetic rats.

Therapeutic Application of Nitric Oxide in Cancer and Inflammatory Disorders. https://doi.org/10.1016/B978-0-12-816545-4.00023-2

Conclusion: Chronic supplementation with nitrite in a rat model of obesity and type 2 diabetes increased mRNA expression of GLUT4 in insulin-sensitive tissues and glucokinase in the pancreas, while it had no effect on GLUT2 mRNA expression.

Nitric Oxide Reinstates Doxorubicin Cytotoxic and Proimmunogenic Effects in Refractory Breast Cancer

Joanna Kopecka*, Elena Gazzano*, Iris C. Salaroglio*, Massimo Donadelli[†], Menachem Rubinstein[‡], Chiara Riganti*

**Department of Oncology, University of Torino, Torino, Italy*

[†]Department of Neurosciences, Biomedicine and Movement Sciences, University of Verona, Verona, Italy

[‡]Department of Molecular Genetics, The Weizmann Institute of Science, Rehovot, Israel

Abstract

Doxorubicin fails in triple-negative breast cancer because of chemoresistance, mainly due to the presence of membrane drug transporters, such as P-glycoprotein (Pgp), which limit intracellular doxorubicin accumulation and prevent cytotoxic and immunogenic cell death (ICD) mediated by this drug. ICD occurs in the condition of endoplasmic reticulum (ER) stress when calreticulin (CRT) is translocated from ER to plasma membrane. Here, CRT works as a bait for dendritic cells (DC), which phagocytose tumor cells and activate antitumor CD8 T cells. This type of tumor death promotes long-term protection of the host immune system against tumor [1].

We have shown previously that nitric oxide (NO) is a very effective inhibitor of Pgp activity because it nitrates tyrosine in the active center of this transporter. NO in the cells is produced by three NO synthases (NOS) that can be stimulated by doxorubicin in sensitive cells, but not in resistant ones [2]. Moreover, NO causes ER stress and induces ICD-promoting CRT translocation [3].

Here, we show that the proteasome inhibitor bortezomib and the lysosome inhibitor chloroquine activate NOS I and NOS III, respectively, in human and murine triple-negative breast cancers, expressing Pgp and resistant to doxorubicin. The increased NO levels caused ER stress, and the activation of the C/EBP-β LIP transcription factor, that triggers apoptosis via a CHOP/TRB3/caspase-3 pathway, downregulates Pgp expression and upregulates CRT transcription, which mediate ICD. In parallel, NO nitrates Pgp and inhibits its catalytic efficacy, increasing the intracellular accumulation of doxorubicin and the drug-mediated cytotoxicity via ER stress-dependent cell death

Therapeutic Application of Nitric Oxide in Cancer and Inflammatory Disorders. https://doi.org/10.1016/B978-0-12-816545-4.00024-4

and ICD in resistant cells. The combination of bortezomib, chloroquine, and doxorubicin was effective in immunocompetent models of triple-negative breast cancers refractory to doxorubicin, where it reduced tumor growth, induced intratumoral ER stress-dependent apoptosis, and increased the amount and activity of intratumoral DC and $CD8^+$ T cells.

In conclusion, the increase of intratumoral NO levels by the FDA-approved drugs bortezomib and chloroquine was an excellent adjuvant to doxorubicin chemotherapy in refractory triple-negative breast cancer, representing an effective approach to overcome doxorubicin resistance.

References

[1] Galluzzi L, Buqué A, Kepp O, Zitvogel L, Kroemer G. Immunogenic cell death in cancer and infectious disease. Nat Rev Immunol 2017;17(2):97–111. https://doi.org/10.1038/nri.2016.107. Epub 2016 Oct 17.

[2] De Boo S, Kopecka J, Brusa D, Gazzano E, Matera L, Ghigo D, Bosia A, Riganti C. iNOS activity is necessary for the cytotoxic and immunogenic effects of doxorubicin in human colon cancer cells. Mol Cancer 2009;8:108. https://doi.org/10.1186/1476-4598-8-108.

[3] Kopecka J, Campia I, Brusa D, Doublier S, Matera L, Ghigo D, Bosia A, Riganti C. Nitric oxide and P-glycoprotein modulate the phagocytosis of colon cancer cells. J Cell Mol Med 2011;15(7):1492–504. https://doi.org/10.1111/j.1582-4934.2010.01137.x.

Anticancer Activities of Erucin a H_2S-Donor Isothiocyanate From *Eruca Sativa* Mill.: Is H_2S the Real Player?

Alma Martelli*,†, Valentina Citi*, Eugenia Piragine*, Lara Testai*,†, Eleonora Pagnotta‡, Luisa Ugolini‡, Luca Lazzeri‡, Vincenzo Calderone*,†

**Department of Pharmacy, University of Pisa, Pisa, Italy*

†*Interdepartmental Research Centre "Nutraceuticals and Food for Health (NUTRAFOOD)," University of Pisa, Pisa, Italy*

‡*Council for Agricultural Research and Economics, Research Centre for Cereal and Industrial Crops (CREA CI), Bologna, Italy*

Abstract

Erucin (4-methylthiobutyl isothiocyanate) is produced by the enzymatic hydrolysis of the glucosinolate glucoerucin present in *Eruca sativa* Mill. seeds (Brassicaceae *or* Cruciferae). Erucin has structural analogies with sulforaphane (SFN), an isothiocyanate derived from glucoraphanin, a glucosinolate present in some edible *crucifers*, and known in the literature for its chemopreventive properties. Indeed, more in general, different isothiocyanates exert anticancer properties on many tumor types (liver, breast, bladder, lung, and pancreatic) [1], and in particular, SFN is actually employed in a pilot randomized controlled clinical trial in advanced pancreatic cancer [2]. In this work, we evaluated the anticancer properties of erucin on human pancreatic adenocarcinoma cells (AsPc-1) with particular focus on the possible role of its hydrogen sulfide (H_2S)-releasing properties [3]. Indeed, also the gasotransmitter H_2S is recognized as antiproliferative agent in melanoma and intestinal cancers [4].

Methods: Erucin (purified from *E. sativa* seeds and available in *Brassica Collection* at CREA CI of Bologna, Italy) antiproliferative effects at 72 h on AsPc-1 viability were evaluated; then, its H_2S-donor properties were also verified both in cell-based and cell-free assays. Moreover, the erucin effects on cell cycle, mitochondrial potential, and caspase-3/7 were investigated. Finally, an evaluation of a possible involvement of the mitogen-activated protein kinase (MAPK) pathway and in particular of ERK1/2 was carried out.

Results: Erucin inhibited AsPc-1 viability with a potency index (IC_{50}) of about 30 μM, and its ability to release H_2S was evident both in an *in vitro* assay buffer

Therapeutic Application of Nitric Oxide in Cancer and Inflammatory Disorders. https://doi.org/10.1016/B978-0-12-816545-4.00025-6

and on the AsPc-1 cells. Moreover, erucin evoked a reduction of the cell cycle G0/G1 phase, due to an enlargement of both the S and G2/M phases, and increased the number of cells exhibiting depolarization of the mitochondrial potential as a hallmark of early apoptosis. These effects were coupled with an increase of the number of apoptotic cells, recorded by the evaluation of caspase-3/7, as markers of a more advanced stage of apoptosis. Finally, the investigation of a possible mechanism of action highlighted a significant reduction of p-ERK1/2 activation suggesting an involvement of the MAPK pathway in erucin an antiproliferative effect on AsPc-1.

Conclusion: Our data demonstrated that erucin induced antiproliferative effect on AsPc-1 cell altering cell cycle phases, inducing mitochondrial depolarization, and increasing caspase-3/7 levels as markers of apoptosis. A significant decrease of p-ERK1/2 activation accounted for an involvement of the MAPK pathway in erucin anticancer effect. Moreover, the H_2S-donor property of erucin may represent a reliable explanation of its anticancer activity that is worthy to be further investigated.

References

[1] Fofaria et al., Enzyme 2015;37:111–37.
[2] Lozanovski et al., Trials 2014;15:204.
[3] Citi et al., Planta Med 2014;80(8–9):610–3.
[4] Ianaro et al., Pharmacol Res 2016;111:652–8.

Hydrogen Sulfide as a Signaling Molecule in the Cardiovascular System: Actions and Interactions With NO

Andreas Papapetropoulos

National and Kapodistrian University of Athens, Laboratory of Pharmacology, Faculty of Pharmacy, Athens, Greece

Abstract

Hydrogen sulfide (H_2S) has emerged as an important endogenous gasotransmitter, which regulates homeostasis and affects the function of most organs in the body. Endogenously produced H_2S is generated through cystathionine γ-lyase (CSE) and cystathionine β-synthase (CBS), which use cysteine as a substrate, and 3-mercaptopyruvate sulfurtransferase (3-MST), which utilizes 3-mercaptopyruvate as a substrate. CBS expression is higher in neuronal tissues, while CSE predominates in the heart and blood vessels. H_2S production and metabolism have been shown to be deregulated under pathophysiological conditions, contributing to the development and progression of disease.

H_2S exhibits a variety of effects in the cardiovascular system that impact on the endothelial, smooth muscle, and cardiomyocyte functions. H_2S promotes endothelial cell growth, migration and organization into capillary networks, and enhanced angiogenesis. In addition, H_2S acts on smooth muscle cells to reduce the vascular tone leading to hypotension. Both endogenously produced and exogenously administrated H_2S exhibit cardioprotective actions by exerting antiapoptotic, antiinflammatory, and antioxidant effects. H_2S ameliorates the damage that occurs after ischemia-reperfusion in many organs, in addition to the heart. H_2S also protects against the development of heart failure and atherosclerosis. The actions of H_2S in cells and tissues are mediated by modulation of ion channel activity, inhibition of phosphodiesterases, activation of kinases, and nitric oxide synthase activation. H_2S has the ability to trigger post-translational modifications of proteins by sulfhydrating cysteine residues to modify protein functions.

A better understanding of the regulation of H_2S production and signaling in physiological and pathophysiological conditions will help harness the therapeutic potential of this gasotransmitter.

Therapeutic Application of Nitric Oxide in Cancer and Inflammatory Disorders. https://doi.org/10.1016/B978-0-12-816545-4.00026-8

Nitric Oxide in Cancer Resistance to Cisplatin: Tumor-Associated Macrophages as Key Players

Cristiana Perrotta, Emilio Clementi

Department of Biomedical and Clinical Sciences, "L. Sacco," Università degli Studi di Milano, Milan, Italy

Abstract

Tumor-associated macrophages (TAMs) represent the largest population of infiltrating inflammatory cells in malignant tumors, promoting their growth, invasion, immune evasion, and chemoresistance. Using different tumor models in vitro and in vivo, we found that nitric oxide (NO) generated at low levels by inducible NO synthase of M2-polarized TAMs is able to protect tumor cells from apoptosis induced by cisplatin (CDDP). The NO effect depends on the inhibition of acid sphingomyelinase (A-SMase), which is activated by CDDP. Mechanistic insights indicate that NO actions occur via the generation of cGMP and activation of protein kinase G (PKG), inducing phosphorylation of syntaxin-4 (synt4), a SNARE protein responsible for A-SMase trafficking and activation. Noteworthy, phosphorylation of synt4 at serine 78 by PKG is responsible for the proteasome-dependent degradation of synt4, leading to inhibition of A-SMase-triggered apoptosis.

This is the first demonstration that the NO system is a key mechanism through which M2-polarized TAMs protect cancer cells from CDDP-induced apoptosis leading to tumor chemoresistance. Moreover, this is the proof that NO produced at low levels has proneoplastic properties and has to be taken into account when designing a therapeutic strategy based on NO donors.

Therapeutic Application of Nitric Oxide in Cancer and Inflammatory Disorders. https://doi.org/10.1016/B978-0-12-816545-4.00027-X

Therapeutic Potential of Nitric Oxide Modulation in Ocular Diseases: A Focus on Novel NO-Releasing Molecules

Chiara Bianca Maria Platania*, Valeria Pittalà†, Gian Marco Leggio*,‡, Salvatore Salomone*,‡, Filippo Drago*,‡, Claudio Bucolo*,‡

**Department of Biomedical and Biotechnological Sciences, Section of Pharmacology, School of Medicine, University of Catania, Catania, Italy*

†Department of Drug Sciences, University of Catania, Catania, Italy

‡Center for Research in Ocular Pharmacology—CERFO, University of Catania, Catania, Italy

Abstract

Nitric oxide is a gaseous small molecule, considered for a long time as an environment pollutant. Since the discovery of the physiological role of NO, NO-donor molecules have been developed as therapeutic agents for the treatment of ischemic cardiac disease [1] and recently for treatment of glaucoma, an ocular neurodegenerative disease [2]. In fact, latanoprostene bunod (NO-donor prostaglandin analogue—Vyzulta) was approved for the treatment of glaucoma, working as an ocular hypotensive agent, with greater efficacy than latanoprost [3]. NO has been proved to regulate the vascular tone of ophthalmic artery through a cross talk with other gases CO and H_2S [4]. Additionally, several studies suggested the role of NO as a neuroprotectant in preclinical models of glaucoma [5–7], although clinical evidence of the NO-donor molecule neuroprotection in glaucoma has not already been proved [3, 8]. Additionally, NO can regulate the expression of heme oxygenase 1 (HO-1) [9]. HO-1 is also known as heat-shock protein 32 (HSP32) and is one of the components of cellular defense mechanisms against oxidative stress-mediated injury. Recently, we explored the effect of novel NO-releasing molecules in an in vitro model of age-related macular degeneration, i.e., caffeic acid phenethyl ester NO-releasing derivative (CAPE-NO) and the curcumin NO-releasing derivative (curcumin-NO) [10]. The CAPE-NO was found to be efficacious in decreasing oxidative stress on the retinal pigmented epithelial cells and to induce heme oxygenase 1 (HO-1), through the promotion of Nrf2 translocation. In conclusion, the drug design and development of bitopic ligands (e.g., Nrf2 inducers or prostaglandin analogues) with NO-releasing moieties could be a good strategy for the treatment of ocular disease.

Therapeutic Application of Nitric Oxide in Cancer and Inflammatory Disorders. https://doi.org/10.1016/B978-0-12-816545-4.00028-1

Acknowledgment

This work was supported by the National Grant PRIN 2015JXE7E8 from the Ministry of Education, University and Research (MIUR).

References

[1] Nossaman VE, Nossaman BD, Kadowitz PJ. Nitrates and nitrites in the treatment of ischemic cardiac disease. Cardiol Rev 2010;18:190–7.

[2] Kaufman MB. Pharmaceutical approval update. P T 2018;43:22–60.

[3] Bucolo C, Platania CBM, Drago F, Reibaldi M, Bonfiglio V, Avitabile T, et al. Novel therapeutics in glaucoma management. Curr Neuropharmacol 2018;16:978–92.

[4] Salomone S, Foresti R, Villari A, Giurdanella G, Drago F, Bucolo C. Regulation of vascular tone in rabbit ophthalmic artery: cross talk of endogenous and exogenous gas mediators. Biochem Pharmacol 2014;92:661–8.

[5] Cavet ME, Vittitow JL, Impagnatiello F, Ongini E, Bastia E. Nitric oxide (NO): an emerging target for the treatment of glaucoma. Invest Ophthalmol Vis Sci 2014;55:5005–15.

[6] Karim MZ, Sawada A, Mizuno K, Kawakami H, Ishida K, Yamamoto T. Neuroprotective effect of nipradilol [3,4-dihydro-8-(2-hydroxy-3-isopropylamino)-propoxy-3-nitroxy-2H-1-benzopyran] in a rat model of optic nerve degeneration. J Glaucoma 2009;18:26–31.

[7] Miyamoto N, Izumi H, Miyamoto R, Kubota T, Tawara A, Sasaguri Y, et al. Nipradilol and timolol induce Foxo3a and peroxiredoxin 2 expression and protect trabecular meshwork cells from oxidative stress. Invest Ophthalmol Vis Sci 2009;50:2777–84.

[8] Araie M, Shirato S, Yamazaki Y, Kitazawa Y, Ohashi Y. Clinical efficacy of topical nipradilol and timolol on visual field performance in normal-tension glaucoma: a multicenter, randomized, double-masked comparative study. Jpn J Ophthalmol 2008;52:255–64.

[9] Hartsfield CL, Alam J, Cook JL, Choi AM. Regulation of heme oxygenase-1 gene expression in vascular smooth muscle cells by nitric oxide. Am J Physiol 1997;273:L980–8.

[10] Pittala V, Fidilio A, Lazzara F, Platania CBM, Salerno L, Foresti R, et al. Effects of novel nitric oxide-releasing molecules against oxidative stress on retinal pigmented epithelial cells. Oxid Med Cell Longev 2017;2017:1420892.

Proapoptotic Effects of *S*-Nitrosothiols *O*-Chloro and *M*-Chloro *S*-Nitroso-Aryl-Amides in Human Breast Cancer Cell Lines

Arnold Stern*, Adriana K.C.A. Reis†, Rafaella C. da Silva‡, Hugo P. Monteiro‡

**New York University School of Medicine, New York, NY, United States*

†Department of Chemistry, UNIFESP, Diadema, Brazil

‡Department of Biochemistry, UNIFESP, São Paulo, Brazil

Abstract

Breast cancer affects women worldwide. Breast cancer tumors are classified as estrogen receptor-positive (ER+) or receptor-negative (ER−). ER− tumors are generally more aggressive, and patients with ER− tumors have a poor prognosis due to unresponsiveness of their tumors to drugs that interfere with estrogen production. Nitric oxide (NO) plays an important role in regulating tumor signaling processes associated with aggressiveness and progression. Increased expression of inducible NO synthase is a poor prognostic marker for breast cancer. Control of intracellular NO production and exposure of breast cancer cells to NO donors are potential chemotherapeutic strategies. *S*-Nitrosothiols, a specific class of NO donors, react with GSH/cysteine and thiol proteins making them potential chemotherapeutic agents in the treatment of various cancer types. A series of *S*-nitroso-aryl-amide derivatives of penicillamine that are fairly water-soluble were synthesized and used for biological testing (Scheme 1—derivatives **1–4**) [1]. Conformational analysis of the compounds indicated that the *o*-chloro (**2**) and *m*-chloro (**3**) derivatives were the most effective as NO-donating compounds. Release of NO by these compounds in PBS, pH 7.4, confirmed this.

O-Chloro and *m*-chloro derivatives were evaluated for them in *in vitro* cytotoxicity against human breast cancer cells (MCF7 (ER+) and MDA-MB-231 (ER−)), nontumor human mammary epithelial cells (MCF10), and human umbilical vein endothelial cells (HUVEC). The *m*-chloro derivative displayed a potent cytotoxic effect against both breast cancer cell lines. However, the compound did not affect the viability of MCF10 nontumor epithelial cells. After 2 h of treatment with the *o*-chloro and *m*-chloro derivatives, the cancer cells lines were positive for Annexin V and propidium iodide labeling. HUVEC were

Therapeutic Application of Nitric Oxide in Cancer and Inflammatory Disorders. https://doi.org/10.1016/B978-0-12-816545-4.00029-3

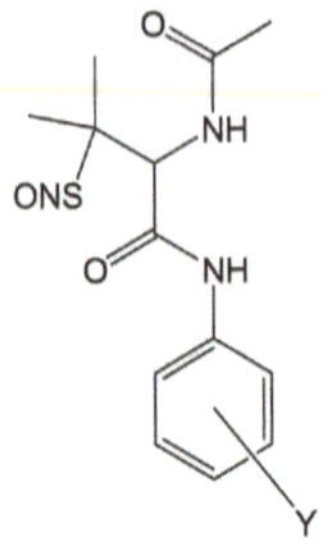

Compound	Y	$C_{Nitrosothiol}$ (g/ml)	C_{NO} (g/ml)	% NO
1	H	4×10^{-4}	8×10^{-6}	19.0
2	*o*-Cl	$5 \text{ X } 10^{-4}$	13.1×10^{-6}	27.0
3	*m*-Cl	$8 \text{ X } 10^{-4}$	14.4×10^{-6}	30.5
4	MeO	17×10^{-4}	12.8×10^{-6}	7.9

SCHEME 1

negative for Annexin V/propidium iodide. Both compounds produced higher intracellular levels of NO in breast cancer cells when compared with normal cells. Apparently, these high concentrations of NO are major contributors to the differential cytotoxicity displayed by these compounds. These findings suggest that the chloro derivatives of the nitrosothiols *S*-nitroso-aryl-amides may aid in the design of chemotherapeutic agents used for the treatment of human breast cancer.

Acknowledgment

Funding provided by FAPESP, Brazil, to AKCAR 2010/51784-3 and to HPM 2012/10470-1.

Reference

[1] Santana RG, Paiva DR, da Silva Gomes R, Reis AK. 1H and 13C NMR analysis of 2-acetamido-3-mercapto-3-methyl-N-aryl-butanamides and 2-acetamido-3-methyl-3-nitrososulfanyl-N-aryl-butanamide derivatives. MagnResonChem 2013;51:316–9.

Light-Guided Production of Nitric Oxide and Singlet Oxygen for the Multimodal Treatment of Cancer

Greta Varchi
Institute of Organic Synthesis and Photoreactivity, Italian National Research Council, Bologna, Italy

Abstract

Combined cancer therapies exploit either synergistic or additive effects arising from appropriate drug combinations [1]. These treatment modalities have long been adopted as the standard of care for improving the therapeutic efficacy against many cancer types [2]. However, there are significant drawbacks that still limit the use of combination regimens, such as the inadequate control of the delivery process; the insurgence of multidrug resistance (MDR); and, most importantly, the different pharmacokinetic properties of the coadministered drugs, which strongly limit the therapeutic effectiveness. Indeed, these problems require the design of novel multimodal treatments that possibly overcome the paradigm of targeted antioncogenic treatment, moving toward a more selective but generalized damaging effect of tumor tissue.

In this respect, the light-controlled generation of reactive oxygen and nitrogen species, such as singlet oxygen ($^1O_2$) and nitric oxide (NO), represents a fascinating alternative.

In recent years, we have explored the possibility of delivering both singlet oxygen and nitric oxide through different chemical and (nano) technological approaches, [3, 4] both comprising the use of light as a tool for the production and control of the cytotoxic effect (Fig. 1). In this conference, we will provide an overview of our most recent results in this field highlighting the potential impact of our approach on cancer therapy.

Therapeutic Application of Nitric Oxide in Cancer and Inflammatory Disorders. https://doi.org/10.1016/B978-0-12-816545-4.00030-X

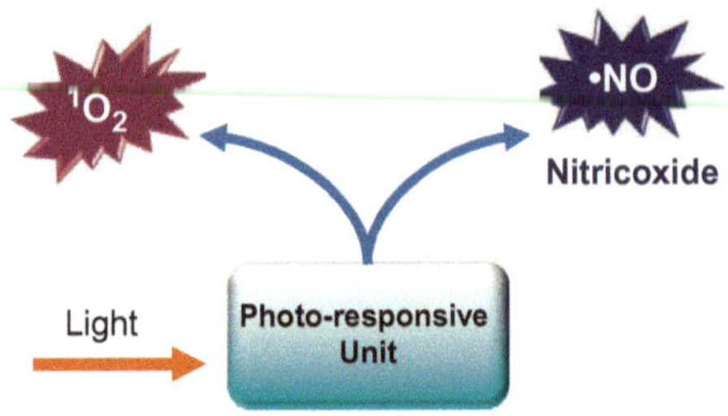

FIG. 1

Schematic representation of a molecular or macromolecular entity able to produce both nitric oxide and single oxygen upon light irradiation.

References

[1] Zhang X-Y, Zhang P. Combinations in multimodality treatments and clinical outcomes during cancer. Oncol Lett 2016;12:4301–4.

[2] McGuire J, Utset-Ward TJ, Reed DR, Lynch CC. Re-calculating! Navigating through the osteosarcoma treatment roadblock. Pharmacol Res 2017;117:54–64.

[3] Fraix A, Manet I, Ballestri M, et al. Polymer nanoparticles with electrostatically loaded multi-cargo for combined cancer phototherapy. J Mater Chem B 2015;3(15):3001–10.

[4] Rapozzi V, Varchi G, Della Pietra E, Ferroni C, Xodo LE. A photodynamic bifunctional conjugate for prostate cancer: an in vitro mechanistic study. Invest New Drugs 2017;35(1):115–23.

H_2S Metabolism in Colon Cancer Cells Exposed to Hypoxia

Karim Zuhra[*,†], Francesca Malagrinò[*,†], Ludovica Mascolo[*], Daniela Mastronicola[*,†], João B. Vicente[‡], Elena Forte[*], Alessandro Giuffrè[†]

[*]Department of Biochemical Sciences, Sapienza University of Rome, Rome, Italy
[†]CNR Institute of Molecular Biology and Pathology, Rome, Italy
[‡]Instituto de Tecnologia Química e Biológica António Xavier, Universidade Nova de Lisboa, Oeiras, Portugal

Abstract

Hydrogen sulfide (H_2S), after nitric oxide (NO) and carbon monoxide (CO), is known as the third gasotransmitter [1]. Involved in antioxidant protection and regulation of different physiological functions, H_2S is synthesized by cystathionine β-synthase (CBS) and few other enzymes and is oxidatively catabolized in the mitochondrion, where H_2S-derived electrons are transferred to coenzyme Q by sulfide-quinone oxidoreductase (SQR), stimulating respiration and ATP production. Metabolic and oncological diseases are related to dysregulation of the H_2S metabolism. Hence, it is important to understand how H_2S bioavailability is regulated under (patho)physiological conditions. We previously reported that both CO and NO negatively modulate CBS with high affinity [2] especially in the presence of its allosteric activator *S*-adenosyl-L-methionine [3] and that a CBS variant responsible for classical homocystinuria shows an unusual high propensity to CO inhibition, suggesting a novel pathogenic mechanism [4]. Here, by investigating the effect of hypoxia on H_2S catabolism in colon cancer, we report that hypoxia-treated SW480 cells have overall reduced ability to detoxify H_2S and less mitochondria but enriched in SQR. These data suggest that under hypoxic conditions, while the lower sulfide-detoxifying activity contributes to ensure higher protective H_2S levels against O_2 deprivation, the mitochondria get enriched in SQR to afford protection from H_2S poisoning.

Therapeutic Application of Nitric Oxide in Cancer and Inflammatory Disorders. https://doi.org/10.1016/B978-0-12-816545-4.00032-3

References

[1] Giuffrè and Vicente. Oxid Med Cell Longev 2018;2018(629):093.

[2] Vicente et al., J Biol Chem 2014;289:8579–87.

[3] Vicente et al., J Biol Chem 2016;291:572–81.

[4] Vicente et al., Oxid Med Cell Longev 2017;8(940):321.

Index

Note: Page numbers followed by *f* indicate figures, and *t* indicate tables.

Kolibri 360 GmbH
Ferdinand-Jühlke-Straße 7
99095 Erfurt, Deutschland
produktsicherheit@kolibri360.de